SOLUTIONS MANUAL

for the

ENGINEER-IN-TRAINING REFERENCE MANUAL

8th Edition

ENGLISH UNITS

Michael R. Lindeburg, P.E.

PROFESSIONAL PUBLICATIONS, INC.
Belmont, CA 94002

In the ENGINEERING REFERENCE MANUAL SERIES

Engineer-In-Training Reference Manual
    Engineering Fundamentals Quick Reference Cards
    Engineer-In-Training Sample Examinations
    Mini-Exams for the E-I-T Exam
    1001 Solved Engineering Fundamentals Problems
    E-I-T Review: A Study Guide
Civil Engineering Reference Manual
    Civil Engineering Quick Reference Cards
    Civil Engineering Sample Examination
    Civil Engineering Review Course on Cassettes
    Seismic Design of Building Structures
    Seismic Design Fast
    Timber Design for the Civil P.E. Examination
    Fundamentals of Reinforced Masonry Design
    246 Solved Structural Engineering Problems
Mechanical Engineering Reference Manual
    Mechanical Engineering Quick Reference Cards
    Mechanical Engineering Sample Examination
    101 Solved Mechanical Engineering Problems
    Mechanical Engineering Review Course on Cassettes
    Consolidated Gas Dynamics Tables
Electrical Engineering Reference Manual
    Electrical Engineering Quick Reference Cards
    Electrical Engineering Sample Examination
Chemical Engineering Reference Manual
    Chemical Engineering Quick Reference Cards
    Chemical Engineering Practice Exam Set
Land Surveyor Reference Manual
Petroleum Engineering Practice Problem Manual
Expanded Interest Tables
Engineering Law, Design Liability, and Professional Ethics
Engineering Unit Conversions

In the ENGINEERING CAREER ADVANCEMENT SERIES

How to Become a Professional Engineer
The Expert Witness Handbook—A Guide for Engineers
Getting Started as a Consulting Engineer
Intellectual Property Protection—A Guide for Engineers
E-I-T/P.E. Course Coordinator's Handbook
Becoming a Professional Engineer

**SOLUTIONS MANUAL FOR THE
ENGINEER-IN-TRAINING REFERENCE MANUAL
Eighth Edition
ENGLISH UNITS**

Printed in the United States of America

ISBN: 0-912045-39-6

Professional Publications, Inc.
1250 Fifth Avenue, Belmont, CA 94002
(415) 593-9119

Current printing of this edition (last number): 6 5 4 3 2 1

PREFACE
TO THE EIGHTH EDITION

For many years I considered the solutions manual a necessary evil, something that every reference manual had to have and which always represented a tremendous amount of work on my part. Over the years my expectations and the standards of Professional Publications have changed considerably, and the solutions manuals to my books have taken on lives of their own. My current belief is that a solutions manual should *teach* the material as well as document the solutions to practice problems.

This solutions manual now has a structure and features that even the *Engineer-In-Training Reference Manual* didn't have two editions ago. Two main features differentiate this solutions manual from its predecessor. First, 165 solutions have been added to parallel the addition of as many problems to the *Engineer-In-Training Reference Manual*. The second, and certainly more obvious, difference is that this eighth edition of the solutions manual is now available as two separate texts. In keeping with the dual-dimensioned problems now included in the reference manual, one manual provides solutions worked out in traditional English units and the other manual has the same problems solved in SI units.

The decision to publish two smaller solutions manuals rather than a single large one was based on a phone survey of existing customers. Most indicated that they would prefer to work in either English units or in SI units, but not in both. Approximately 20 percent of the engineers surveyed wanted both solutions, while the other 80 percent would have used only 50 percent of the book. Therefore, to keep the bulk and cost of this book down, two different solutions manuals were produced.

My attention is continually directed toward improving this book and its companion, the *Engineer-In-Training Reference Manual*. I hope that the improvements made to this edition will benefit you. If not, I ask that you let me know by sending in one of the reply cards from the back of the book.

Michael R. Lindeburg, P.E.
Belmont, CA
January 1992

PROFESSIONAL PUBLICATIONS, INC. ● Belmont, CA

PREFACE
TO THE SEVENTH EDITION

This book contains solutions to all of the practice problems in the seventh edition of the *Engineer-In-Training Reference Manual*. The solutions draw upon the same theories and use the same symbols presented in the corresponding chapters.

More so than in any of my previous solutions manuals, I have tried to present all intermediate solution steps, even those steps that may be obvious to some engineers. Although this has increased the size of the book, it should help you study subjects outside of your areas of specialty.

To help you understand how I solved the problems, I have made many other departures from the format of the previous *Solutions Manual*. The equations used are given prior to the problem values being inserted. Except in rare cases, units are carried along in each calculation. And as you have probably already discovered, the solutions are typeset and professionally illustrated for easier reading.

I have also tried to make it easy to find particular solutions. In addition to the subject names (which match the chapter titles) appearing at the top of every other page, this book is the first solutions manual in the *Reference Manual Series* to have page tabs.

To alleviate some of the inherent confusion (in the English unit system) between mass, force, and weight, I routinely distinguish between lbm and lbf, as well as between g and g_c.

You may also be pleased to note that I have returned to standard scientific notation (e.g., 10^5) instead of the awkward EE notation which never "caught on" in the engineering community.

When I wrote my first engineering book, a seasoned author told me, "It is tough not to make mistakes in these books." He was referring both to the unforgiving nature of engineering material (it is either right or wrong), and to the difficulty in proofreading technical and numerical information. I have kept this warning in mind while writing this manual, but occasional typographical errors seem to be inevitable. If you find what you think is an error, I would be grateful if you would pass it on to me so that your correction can benefit other readers. (A card is included in the back of this book for this purpose.) Of course, feedback of all types is always appreciated.

I am sure you understand that to get the most out of this manual you should not refer to the solutions until you have attempted a problem. You will not retain important data, formulas, and solution techniques as well if you skim through the solutions without actually trying the problems.

Whatever the reasons you are reviewing engineering fundamentals, I wish you great success in your career, and I hope that this book will contribute to that success.

Michael R. Lindeburg, P.E.
Belmont, CA
June 1990

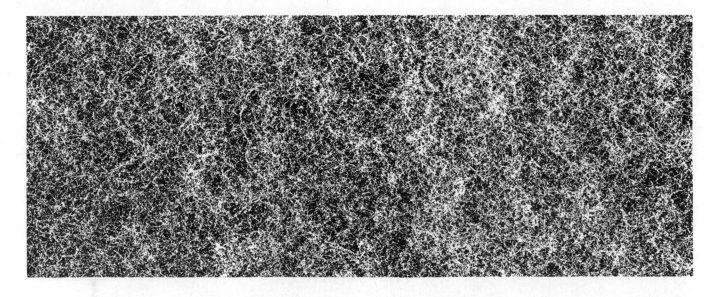

PROFESSIONAL PUBLICATIONS, INC. ● Belmont, CA

Make the Most of Your Exam Study Time!

These handy supplements let you customize your study program to meet your needs. Call (415) 593-9119 for current prices; use the order card at the back of this book.

Solutions Manual for the Engineer-In-Training Reference Manual, SI Units

$8\frac{1}{2} \times 11$, softcover, 256 pages

Given enough time, you can probably solve almost any engineering problem put in front of you. Unfortunately, the E-I-T exam is a fast exam, and you probably won't have all the time you would like. This solutions manual contains the full solutions in SI units to the practice problems in the *Engineer-In-Training Reference Manual*, and is indispensable in training you to solve problems fast. Each solution is typeset and professionally illustrated.

Engineer-In-Training Sample Examinations

$8\frac{1}{2} \times 11$, softcover, 144 pages

This book contains two complete E-I-T exams, with detailed solutions to all the problems provided in the back of the book. Just like the E-I-T exam, each of the simulated eight-hour tests has 140 morning questions and 70 afternoon questions, all in the NCEES format. There is no better preparation for the rigorous E-I-T exam than working these two realistic sample examinations.

Engineering Fundamentals Quick Reference Cards

$8\frac{1}{2} \times 11$, spiral bound, 44 pages

If you are like the majority of examinees, you will diligently study the *Engineer-In-Training Reference Manual*, but you will want something more concise to use during the examination. To speed your recall of important formulas and data, each exam subject is summarized on one or two pages of card stock. For your added convenience, the quick reference cards use the same symbols, variable names, and nomenclature as the reference manual.

1001 Solved Engineering Fundamentals Problems

6×9, softcover, 760 pages

This is the most complete collection of typical E-I-T problems ever published. There is a chapter for each different exam subject, and similar problems are grouped together for maximum reinforcement of key concepts. Each question is an accurate simulation of exam-type problems, and typeset solutions are provided to improve your skills.

Professional Publications, Inc.
1250 Fifth Avenue
Department 77
Belmont, CA 94002
(415) 593-9119

TABLE OF CONTENTS

PROFESSIONAL PUBLICATIONS, INC. ● Belmont, CA

Notice to Examinees

SYSTEMS OF UNITS

1. $a = \dfrac{F}{m} = \left[\dfrac{(4\text{ lbf})\left(32.2\,\frac{\text{lbm-ft}}{\text{lbf-sec}^2}\right)}{10\text{ lbm}}\right]\left(\dfrac{60\text{ sec}}{\text{min}}\right)^2$

$= \boxed{4.637 \times 10^4 \text{ ft/min}^2}$

2. Assume 100 lbm mud, consisting of 70 lbm sand and 30 lbm water. The volumes of the mud and water are

$V_{\text{mud}} = \dfrac{m}{\rho} = \dfrac{70\text{ lbm}}{140\,\frac{\text{lbm}}{\text{ft}^3}} = 0.5\text{ ft}^3$

$V_{\text{water}} = \dfrac{m}{\rho} = \dfrac{30\text{ lbm}}{62.4\,\frac{\text{lbm}}{\text{ft}^3}} = 0.48\text{ ft}^3$

$V_{\text{total}} = V_{\text{mud}} + V_{\text{water}} = 0.5\text{ ft}^3 + 0.48\text{ ft}^3 = 0.98\text{ ft}^3$

The density is

$\rho_{\text{mud}} = \dfrac{m}{V} = \dfrac{100\text{ lbm}}{(0.98\text{ ft}^3)\left(32.2\,\frac{\text{lbm}}{\text{slug}}\right)}$

$= \boxed{3.167\text{ slug/ft}^3}$

3. $W = mg = (10\text{ lbm})\left(\dfrac{\text{slug}}{32.2\text{ lbm}}\right)\left(5.47\,\dfrac{\text{ft}}{\text{sec}^2}\right)$

$= 1.699\,\dfrac{\text{slug-ft}}{\text{sec}^2} = \boxed{1.699\text{ lbf}}$

4. (a) $W_{\text{moon}} = mg_{\text{moon}} = (10\text{ slug})\left(5.47\,\dfrac{\text{ft}}{\text{sec}^2}\right)$

$= 54.7\,\dfrac{\text{slug-ft}}{\text{sec}^2} = \boxed{54.7\text{ lbf}}$

(b) $W_{\text{earth}} = mg_{\text{earth}} = (10\text{ slug})\left(32.2\,\dfrac{\text{ft}}{\text{sec}^2}\right)$

$= 322\,\dfrac{\text{slug-ft}}{\text{sec}^2} = \boxed{322\text{ lbf}}$

5. $F = ma = (40\text{ slug})\left(8\,\dfrac{\text{ft}}{\text{sec}^2}\right)$

$= 320\,\dfrac{\text{slug-ft}}{\text{sec}^2} = \boxed{320\text{ lbf}}$

6. $F = mg = (7\text{ kg})\left(9.81\,\dfrac{\text{m}}{\text{s}^2}\right)$

$= 68.67\,\dfrac{\text{kg}\cdot\text{m}}{\text{s}^2} = \boxed{68.67\text{ N}}$

7. $a = \dfrac{\text{v} - \text{v}_0}{t} = \dfrac{40\,\frac{\text{m}}{\text{s}} - 20\,\frac{\text{m}}{\text{s}}}{3\text{ s}}$

$= \boxed{6.67\text{ m/s}^2}$

8. $F = ma = (3\text{ kg})\left(4\,\dfrac{\text{m}}{\text{s}^2}\right)$

$= 12\,\dfrac{\text{kg}\cdot\text{m}}{\text{s}^2} = \boxed{12\text{ N}}$

9. $h = \dfrac{\Delta E}{mg} = \dfrac{100\text{ J} - 20\text{ J}}{(4\text{ kg})\left(9.81\,\frac{\text{m}}{\text{s}^2}\right)}$

$= 2.039\,\dfrac{\text{J}}{\frac{\text{kg}\cdot\text{m}}{\text{s}^2}} = \boxed{2.039\text{ m}}$

10. $\Delta E = mgh = (10\text{ kg})\left(9.81\,\dfrac{\text{m}}{\text{s}^2}\right)(10\text{ m})$

$= 981\,\dfrac{\text{kg}\cdot\text{m}^2}{\text{s}^2} = \boxed{981\text{ J}}$

11. $F_f = |ma - F| = |(7\text{ kg})\left(2\,\dfrac{\text{m}}{\text{s}^2}\right) - 25\text{ N}|$

$= \left|14\,\dfrac{\text{kg}\cdot\text{m}}{\text{s}^2} - 25\text{ N}\right| = |14\text{ N} - 25\text{ N}|$

$= \boxed{11\text{ N}}$

12. $x = x_0 + \dfrac{1}{2}at^2 = 0 + \left(\dfrac{1}{2}\right)\left(9.81\,\dfrac{\text{m}}{\text{s}^2}\right)(6\text{ s})^2$

$= \boxed{176\text{ m}}$

13. (a) $\Delta E = mgl\sin(\tan^{-1}\tfrac{1}{5})$

$= (10\text{ kg})\left(9.81\,\dfrac{\text{m}}{\text{s}^2}\right)(10\text{ m})\left(\dfrac{1}{\sqrt{(1)^2 + (5)^2}}\right)$

$= 192.4\,\dfrac{\text{kg}\cdot\text{m}^2}{\text{s}^2} = \boxed{192.4\text{ J}}$

(b) $\Delta E = \dfrac{1}{2}m\text{v}^2 = mgh$

$= \boxed{192.4\text{ J}}$

(c) $\qquad v = \sqrt{\dfrac{2\Delta E}{m}} = \sqrt{\dfrac{(2)(192\,\text{J})}{10\,\text{kg}}}$

$\qquad\qquad = 6.203\sqrt{\dfrac{\text{J}}{\text{kg}}} = \boxed{6.203\ \text{m/s}}$

14. (a) $\quad p = \dfrac{F}{A} = \dfrac{150\,\text{N}}{\pi\left(\dfrac{0.035\,\text{m}}{2}\right)^2}$

$\qquad\qquad = 1.56\ \times\ 10^5\,\dfrac{\text{N}}{\text{m}^2} = \boxed{0.156\ \text{MPa}}$

(b) $\quad m = \dfrac{F}{a} = \dfrac{\left(1.56\ \times\ 10^5\,\dfrac{\text{N}}{\text{m}^2}\right)\left[\pi\left(\dfrac{0.35\,\text{m}}{2}\right)^2\right]}{9.81\,\dfrac{\text{m}}{\text{s}^2}}$

$\qquad\qquad = \boxed{1530\ \text{kg}}$

15. $\quad d_2 = d_1(1 + \alpha\Delta T)$

$\qquad = (0.48\,\text{m})\left[1 + \left(1.6 \times 10^{-5}\,\dfrac{1}{°\text{C}}\right)(80°\text{C} - 20°\text{C})\right]$

$\qquad = \boxed{0.48046\ \text{m}}$

16. $\quad p_2 = \dfrac{p_1 T_2}{T_1} = \dfrac{(1500\,\text{kPa})(276\text{K})}{293\text{K}}$

$\qquad = \boxed{1413\ \text{kPa}}$

17. $q_{\text{loss},\ 80°\ \text{water}} = q_{\text{gain},\ 15°\ \text{container + water}}$

$(mc_p\Delta T)_{80°\ \text{water}} = [(mc_p)_{\text{container}} + (mc_p)_{15°\ \text{water}}]$
$\qquad\qquad\qquad\qquad\qquad \times \Delta T_{\text{container}}$

$(2l)\left(1\,\dfrac{\text{kg}}{l}\right)\left(4186.8\,\dfrac{\text{J}}{\text{kg·°C}}\right)(80° - T_f)$

$\qquad = \left[(1\,\text{kg})\left(500\,\dfrac{\text{J}}{\text{kg·°C}}\right) + (5\,\text{kg})\left(4186.8\,\dfrac{\text{J}}{\text{kg·°C}}\right)\right](T_f - 15°)$

$669{,}888 - 8373.6 T_f = 21{,}434 T_f - 321{,}510$

$\qquad\qquad 991{,}398 = 29{,}807.6 T_f$

$\qquad\qquad\qquad T_f = \boxed{33.3°\text{C}}$

18. $\delta = \dfrac{LF}{AE}$

$\qquad = \dfrac{(10\,\text{m})(1000\,\text{N})}{\pi\left(\dfrac{0.05\,\text{m}}{2}\right)^2\left(30\times10^6\,\dfrac{\text{lbf}}{\text{in}^2}\right)\left(6.895\times10^3\,\dfrac{\text{Pa}}{\dfrac{\text{lbf}}{\text{in}^2}}\right)}$

$\qquad = 2.46\ \times\ 10^{-5}\,\dfrac{\text{N}}{\text{m·Pa}} = \boxed{24.6\ \mu\text{m}}$

19. $\dfrac{q}{A} = \dfrac{\Delta T}{\Sigma\,\dfrac{L_i}{k_i} + \Sigma\,\dfrac{1}{h_j}} =$

$\qquad \dfrac{(25°\text{C} - 15°\text{C})\left(\dfrac{1\text{K}}{°\text{C}}\right)}{\dfrac{0.1\,\text{m}}{0.69\,\dfrac{\text{W}}{\text{m·K}}} + \dfrac{0.03\,\text{m}}{0.043\,\dfrac{\text{W}}{\text{m·K}}} + \dfrac{0.06\,\text{m}}{0.21\,\dfrac{\text{W}}{\text{m·K}}} + \dfrac{1}{34\,\dfrac{\text{W}}{\text{m}^2\text{·K}}} + \dfrac{1}{9.37\,\dfrac{\text{W}}{\text{m}^2\text{·K}}}}$

$\qquad = \boxed{7.91\ \text{W/m}^2}$

20. (a) $\quad x = \dfrac{s - s_f}{s_{fg}} = \dfrac{7.8992\,\dfrac{\text{kJ}}{\text{kg·K}} - 0.4367\,\dfrac{\text{kJ}}{\text{kg·K}}}{8.0174\,\dfrac{\text{kJ}}{\text{kg·K}}}$

$\qquad\qquad = \boxed{0.931}$

(b) $P_0 = |\eta\,\dot{m}(h_o - h_i)|$

$\qquad = \left|(0.85)\left(1000\,\dfrac{\text{kg}}{\text{s}}\right)\left[(1 - 0.931)\left(125.8\,\dfrac{\text{kJ}}{\text{kg}}\right)\right.\right.$

$\qquad\qquad \left.\left. + (0.931)\left(2430.4\,\dfrac{\text{kJ}}{\text{kg}}\right) - 3273.4\,\dfrac{\text{kJ}}{\text{kg}}\right]\right|$

$\qquad = 8.52\ \times\ 10^5\,\dfrac{\text{kJ}}{\text{s}} = \boxed{852\ \text{MW}}$

21. $\qquad F = V\rho g + p_{\text{atm}}A$

$\qquad = (20\,\text{m})(15\,\text{m})(4\,\text{m})\left(10^3\,\dfrac{\text{kg}}{\text{m}^3}\right)\left(9.81\,\dfrac{\text{m}}{\text{s}^2}\right)$

$\qquad\quad + \left(1.0\ \times\ 10^5\,\dfrac{\text{N}}{\text{m}^2}\right)(20\,\text{m})(15\,\text{m})$

$\qquad = 1.18\ \times\ 10^7\,\dfrac{\text{kg·m}}{\text{s}^2} + 3.0\ \times\ 10^7\,\text{N}$

$\qquad = 1.18\ \times\ 10^7\,\text{N} + 3.0\ \times\ 10^7\,\text{N}$

$\qquad = \boxed{41.8\ \text{MN}}$

22. $v = v_f + x\, v_{fg}$

$$= \left[0.01613\, \frac{\text{ft}^3}{\text{lbm}} + (0.95)\left(350.3\, \frac{\text{ft}^3}{\text{lbm}} \right) \right]$$

$$\times \left(\frac{0.3048\, \text{m}}{\text{ft}} \right)^3 \left(\frac{\text{lbm}}{0.4536\, \text{kg}} \right)$$

$$= \boxed{20.78\ \text{m}^3/\text{kg}}$$

23. (a) $a_0 = \sqrt{kRT}$

$$= \sqrt{(1.4)\left(287.0304\, \frac{\text{J}}{\text{kg} \cdot \text{K}} \right)(273.15\text{K})}$$

$$= 331.3 \sqrt{\frac{\text{J}}{\text{kg}}} = \boxed{331.3\ \text{m/s}}$$

(b) $a_{25} = \sqrt{kRT}$

$$= \sqrt{(1.4)\left(287.0304\, \frac{\text{J}}{\text{kg} \cdot \text{K}} \right)(273.15 + 25)\text{K}}$$

$$= 346.1 \sqrt{\frac{\text{J}}{\text{kg}}} = 346.1\, \frac{\text{m}}{\text{s}}$$

$$\frac{a_{25}}{a_0} - 1 = \frac{346\, \frac{\text{m}}{\text{s}}}{331\, \frac{\text{m}}{\text{s}}} - 1 = 0.045$$

$$= \boxed{4.5\ \%}$$

24. $Q = \rho V c (T_2 - T_1)$

$$= \left(625\, \frac{\text{kg}}{\text{m}^3} \right)(2\,\text{in})(4\,\text{in})(60\,\text{in})$$

$$\times \left(\frac{25.4 \times 10^{-3}\, \text{m}}{\text{in}} \right)^3 \left(2.5\, \frac{\text{kJ}}{\text{kg} \cdot {}^\circ\text{C}} \right)$$

$$\times (205^\circ\text{C} - 25^\circ\text{C})$$

$$= \boxed{2212\ \text{kJ}}$$

25. $M = Fd = (1\ \text{N})\left(\frac{0.06\ \text{m}}{2} \right)$

$$= \boxed{0.03\ \text{N} \cdot \text{m}}$$

26. $v_p = \sqrt{\dfrac{L_p}{L_m}}\, v_m$

$$= (\sqrt{50})\left(30\, \frac{\text{cm}}{\text{s}} \right)\left(\frac{10^{-2}\, \text{m}}{\text{cm}} \right)$$

$$\times \left(\frac{1\, \text{km}}{10^3\, \text{m}} \right)\left(\frac{3600\, \text{s}}{\text{hr}} \right)$$

$$= \boxed{7.64\ \text{km/hr}}$$

27. $I_x = \dfrac{1}{3}\text{bh}^3 = \left(\dfrac{1}{3} \right)(5\,\text{cm})(9\,\text{cm})^3$

$$= \boxed{1215\ \text{cm}^4}$$

28. $W = fmgd = (0.3)(100\,\text{kg})\left(9.81\, \frac{\text{m}}{\text{s}^2} \right)(70\,\text{m})$

$$= 2.06 \times 10^4\, \frac{\text{kg} \cdot \text{m}^2}{\text{s}^2} = \boxed{20.6\ \text{kJ}}$$

29. (a) $E_k = \dfrac{1}{2}\,\text{mv}^2$

$$= \left(\frac{1}{2} \right)(8 \times 10^3\,\text{kg})$$

$$\times \left[\left(100\, \frac{\text{km}}{\text{hr}} \right)\left(\frac{10^3\, \text{m}}{\text{km}} \right)\left(\frac{\text{hr}}{3600\, \text{s}} \right) \right]^2$$

$$= 3.086 \times 10^6\, \frac{\text{kg} \cdot \text{m}}{\text{s}^2} = \boxed{3.086\ \text{MJ}}$$

(b) $h = \dfrac{E_k}{mg} = \dfrac{3.1 \times 10^6\, \text{J}}{(8 \times 10^3\,\text{kg})\left(9.81\, \frac{\text{m}}{\text{s}^2} \right)}$

$$= 39.5\, \frac{\text{J}}{\frac{\text{kg} \cdot \text{m}}{\text{s}^2}} = \boxed{39.5\ \text{m}}$$

30. $M = mgd = (73\,\text{kg})\left(9.81\, \frac{\text{m}}{\text{s}^2} \right)(0.20\,\text{m})$

$$= 143\, \frac{\text{kg} \cdot \text{m}^2}{\text{s}^2} = \boxed{143\ \text{N} \cdot \text{m}}$$

ENGINEERING DRAWING PRACTICE

DRAWING

1.

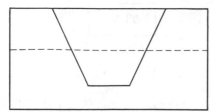

2.

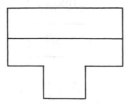

3.

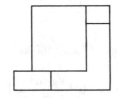

4. (a)

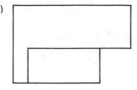

(b) 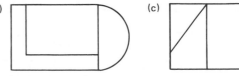 (c)

5.

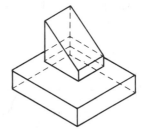

PROFESSIONAL PUBLICATIONS, INC. ● Belmont, CA

ALGEBRA

1. (a) $\boxed{\text{5: All digits are significant.}}$

(b) $\boxed{\text{7: If the decimal zeros are not significant, they should not be there.}}$

(c) $\boxed{\text{4: Leading zeros are not significant.}}$

(d) $\boxed{\text{3: Only the 7.93 is significant.}}$

2. (a) 22.52; $\boxed{\text{22.5 with 3 significant digits.}}$

(b) 0.11346; $\boxed{\text{0.1 with 1 significant digit.}}$

(c) 3.4913; $\boxed{\text{3 with 1 significant digit.}}$

(d) -0.00229; $\boxed{-0.002 \text{ with 1 significant digit.}}$

3. (a) $x^2 + 6x + 8 = x^2 + 6x + \left(\dfrac{6}{2}\right)^2 + 8 - \left(\dfrac{6}{2}\right)^2$

$\quad = (x+3)^2 - (1)^2$

$\quad = (x+3+1)(x+3-1)$

$\quad = \boxed{(x+4)(x+2)}$

(b) $3x^3 - 3x^2 - 18x$

$\quad = 3x(x^2 - x - 6)$

$\quad = 3x\left[x^2 - x + \left(\dfrac{1}{2}\right)^2 - 6 - \left(\dfrac{1}{2}\right)^2\right]$

$\quad = 3x\left[\left(x - \dfrac{1}{2}\right)^2 - \left(\dfrac{5}{2}\right)^2\right]$

$\quad = 3x\left(x - \dfrac{1}{2} + \dfrac{5}{2}\right)\left(x - \dfrac{1}{2} - \dfrac{5}{2}\right)$

$\quad = \boxed{3x(x+2)(x-3)}$

(c) $x^4 + 7x^2 + 12$

$\quad = x^4 + 7x^2 + \left(\dfrac{7}{2}\right)^2 + 12 - \left(\dfrac{7}{2}\right)^2$

$\quad = \left(x^2 + \dfrac{7}{2}\right)^2 - \left(\dfrac{1}{2}\right)^2$

$\quad = \left(x^2 + \dfrac{7}{2} + \dfrac{1}{2}\right)\left(x^2 + \dfrac{7}{2} - \dfrac{1}{2}\right)$

$\quad = \boxed{(x^2 + 4)(x^2 + 3)}$

(d) $16 - 10x + x^2$

$\quad = x^2 - 10x + \left(\dfrac{10}{2}\right)^2 + 16 - \left(\dfrac{10}{2}\right)^2$

$\quad = (x-5)^2 - (3)^2$

$\quad = (x - 5 + 3)(x - 5 - 3)$

$\quad = \boxed{(x-2)(x-8)}$

4. (a) $x^2 - 9 = \boxed{(x+3)(x-3)}$

(b) $3x^2 - 12 = (3)(x^2 - 4)$

$\quad = \boxed{(3)(x+2)(x-2)}$

(c) $1 - x^8 = (1 + x^4)(1 - x^4)$

$\quad = (1 + x^4)(1 + x^2)(1 - x^2)$

$\quad = \boxed{(1 + x^4)(1 + x^2)(1 + x)(1 - x)}$

(d) $x^4 - y^4 = (x^2 + y^2)(x^2 - y^2)$

$\quad = \boxed{(x^2 + y^2)(x + y)(x - y)}$

5. (a) $x^2 + 8x + 16 = \boxed{(x+4)^2}$

(b) $x^2 - 4x + 4 = \boxed{(x-2)^2}$

(c) $1 + 4y + 4y^2 = \boxed{(1 + 2y)^2}$

(d) $25x^2 + 60xy + 36y^2 = \boxed{(5x + 6y)^2}$

PROFESSIONAL PUBLICATIONS, INC. ● Belmont, CA

6. (a) $x^3 + 8 = x^3 + (2)^3$

$$= \boxed{(x+2)(x^2 - 2x + 4)}$$

(b) $x^6 - y^6 = (x^3 + y^3)(x^3 - y^3)$

$$= (x+y)(x^2 - xy + y^2)(x - y)$$
$$\times \ (x^2 + xy + y^2)$$

$$= \boxed{\begin{array}{l} (x+y)(x-y)(x^2 + xy + y^2) \\ \times \ (x^2 - xy + y^2) \end{array}}$$

(c) $(x-2)^3 - 8y^3 = (x-2)^3 - (2y)^3$

$$= \boxed{\begin{array}{l} (x-2-2y)[(x-2)^2 \\ + (x-2)(2y) + 4y^2] \end{array}}$$

(d) $6x^2 - 4ax - 9bx + 6ab = 6x^2 - (4a + 9b)x$
$$+ 6ab$$

$$= \boxed{(2x - 3b)(3x - 2a)}$$

(e) $64 + y^3 = y^3 + (4)^3$

$$= \boxed{(y+4)(y^2 - 4y + 16)}$$

(f) $z^5 + 32 = z^5 + 2^5$

$$= \boxed{(z+2)(z^4 - 2z^3 + 4z^2 - 8z + 16)}$$

7. (a) $$4x^2 = 12x - 7$$
$$4x^2 - 12x = -7$$
$$4\left[x^2 - 3x + \left(\frac{3}{2}\right)^2\right] = -7 + (4)\left(\frac{3}{2}\right)^2$$
$$4\left(x - \frac{3}{2}\right)^2 = 2$$
$$\left(x - \frac{3}{2}\right)^2 = \frac{1}{2}$$
$$x - \frac{3}{2} = \pm\sqrt{\frac{1}{2}} = \pm 0.707$$
$$x = \frac{3}{2} \pm 0.707$$
$$= \boxed{2.207 \text{ or } 0.793}$$

(b) $$2x^2 - 400 = 0$$
$$x^2 - 200 = 0$$
$$x^2 = 200$$
$$x = \boxed{\pm 10\sqrt{2}}$$

(c) $$x^2 + 36 = 9 - 2x^2$$
$$3x^2 = 9 - 36 = -27$$
$$x^2 = -9$$
$$x = \boxed{\pm 3i}$$

(d) $$6x^2 - 7x - 5 = 0$$
$$x = \frac{-b \pm \sqrt{b^2 - 4ac}}{2a}$$
$$= \frac{7 \pm \sqrt{49 + 120}}{12} = \frac{7 \pm 13}{12}$$
$$= \boxed{\frac{5}{3} \text{ or } -\frac{1}{2}}$$

8.

	$\log(x)$	characteristic	mantissa
(a)	1.60853	1	0.60853
(b)	0.96755	0	0.96755
(c)	−2.14874	−3	0.85126
(d)	3.60969	3	0.60969

9. (a) $$\log(38.5^x) = \log(6.5^{x-2})$$
$$x\log(38.5) = (x - 2)\log(6.5)$$
$$1.5855x = (0.8129)(x - 2)$$
$$x = \boxed{-2.10}$$

(b) $$\log(1.4) = \log\left(\frac{0.0613}{x}\right)^{1.32}$$
$$0.1461 = (1.32)[\log(0.0613) - \log(x)]$$
$$\log x = -1.3232$$
$$x = \boxed{0.0475}$$

(c) $$\ln 7.4 \times 10^{-4} = \ln e^{-9.7x}$$
$$-7.2089 = -9.7x$$
$$x = \boxed{0.7432}$$

10. (a) $\dfrac{x+2}{x^2-7x+12} = \dfrac{x+2}{(x-3)(x-4)}$

$= \dfrac{A_1}{x-4} + \dfrac{A_2}{x-3}$

$= \dfrac{A_1(x-3) + A_2(x-4)}{(x-4)(x-3)}$

$x+2 = A_1(x-3) + A_2(x-4)$

$= (A_1 + A_2)x - (3A_1 + 4A_2)$

$A_1 + A_2 = 1$
$3A_1 + 4A_2 = -2$

$A_1 = 6$

$A_2 = -5$

$\dfrac{x+2}{x^2-7x+12} = \boxed{\dfrac{6}{x-4} - \dfrac{5}{x-3}}$

(b) $\dfrac{5x+4}{x^2+2x} = \dfrac{5x+4}{x(x+2)}$

$= \dfrac{A_1}{x} + \dfrac{A_2}{x+2}$

$= \dfrac{A_1(x+2) + A_2x}{x(x+2)}$

$5x+4 = A_1(x+2) + A_2x$

$= (A_1 + A_2)x + 2A_1$

$A_1 + A_2 = 5$
$2A_1 = 4$

$A_1 = 2$

$A_2 = 3$

$\dfrac{5x+4}{x^2+2x} = \boxed{\dfrac{2}{x} + \dfrac{3}{x+2}}$

(c) $\dfrac{3x^2-8x+9}{(x-2)^3} = \dfrac{A_1}{(x-2)} + \dfrac{A_2}{(x-2)^2}$

$+ \dfrac{A_3}{(x-2)^3}$

$= \dfrac{A_1(x-2)^2 + A_2(x-2) + A_3}{(x-2)^3}$

$3x^2-8x+9 = A_1(x^2-4x+4)$

$+ A_2(x-2) + A_3$

From the coefficients of the x^2 terms,

$A_1 = 3$

$-4A_1 + A_2 = -8$
$4A_1 - 2A_2 + A_3 = 9$

$A_1 = 3$

$A_2 = 4$

$A_3 = 5$

$\dfrac{3x^2-8x+9}{(x-2)^3} = \boxed{\dfrac{3}{x-2} + \dfrac{4}{(x-2)^2} + \dfrac{5}{(x-2)^3}}$

(d) $\dfrac{3x}{x^3-1}$

$= \dfrac{3x}{(x-1)(x^2+x+1)}$

$= \dfrac{A_1}{x-1} + \dfrac{A_2 + A_3x}{x^2+x+1}$

$= \dfrac{A_1(x^2+x+1) + A_2(x-1) + A_3x(x-1)}{(x-1)(x^2+x+1)}$

$3x = A_1(x^2+x+1) + A_2(x-1) + A_3x(x-1)$

$A_1 + A_3 = 0$
$A_1 + A_2 - A_3 = 3$
$A_1 - A_2 = 0$

$A_1 = 1$

$A_2 = 1$

$A_3 = -1$

$\dfrac{3x}{x^3-1} = \boxed{\dfrac{1}{x-1} + \dfrac{1-x}{x^2+x+1}}$

11. (a) 5 times Eq. 1: $20x + 10y = 25$ (Eq. 1′)

4 times Eq. 2: $20x - 12y = -8$ (Eq. 2′)

Subtract Eq. 2′ from Eq. 1′:

$22y = 33$

$y = \boxed{\dfrac{3}{2}}$

Substitute $y = \dfrac{3}{2}$ into Eq. 1′:

$20x + 15 = 25$

$x = \boxed{\dfrac{1}{2}}$

(b) Subtract Eq. 1 from Eq. 2:

$$6x = 6$$

$$x = \boxed{1}$$

Substitute $x = 1$ into Eq. 1:

$$3 - y = 6$$

$$y = \boxed{-3}$$

(c) 2 times Eq. 2:

$$2x - 6y - 4z = -18 \text{ (Eq. 2')}$$

Subtract Eq. 2' from Eq. 1:

$$4y + 7z = 19 \text{ (Eq. 4)}$$

Subtract Eq. 2 from Eq. 3:

$$4y + 3z = 15 \text{ (Eq. 5)}$$

Subtract Eq. 5 from Eq. 4:

$$4z = 4$$

$$z = 1$$

Substitute $z = 1$ into Eq. 4:

$$4y + 3 = 15$$

$$y = \boxed{3}$$

Substitute $y = 3$ and $z = 1$ into Eq. 3:

$$x + 3 + 1 = 6$$

$$x = \boxed{2}$$

(d) Add Eq. 1 to Eq. 3:

$$4x + 2z = 0 \text{ (Eq. 4)}$$

2 times Eq. 2: $2x + 2z - 2 = 0$ (Eq. 2')

Subtract Eq. 2' from Eq. 4:

$$2x + 2 = 0$$

$$x = \boxed{-1}$$

Substitute $x = -1$ into Eq. 4:

$$-4 + 2z = 0$$

$$z = \boxed{2}$$

Substitute $x = -1$ and $z = 2$ into Eq. 3:

$$4 - y - 3 = 4$$

$$y = \boxed{-3}$$

12. (a) Substitute $y = x - 1$ into Eq. 1:

$$2x^2 - (x - 1)^2 - 14 = 0$$
$$2x^2 - x^2 + 2x - 1 - 14 = 0$$
$$x^2 + 2x - 15 = 0$$
$$(x + 5)(x - 3) = 0$$

$$x = \boxed{3 \text{ or } -5}$$

Since $y = x - 1$, $y = \boxed{2 \text{ or } -6}$

(b) Multiply Eq. 2 by y:

$$y^2 + x^2 = 24 \text{ (Eq. 2')}$$

Substitute Eq. 1 into Eq. 2':

$$(3x - 4)^2 + x^2 = 24$$
$$9x^2 - 24x + 16 + x^2 = 24$$
$$10x^2 - 24x - 8 = 0$$
$$5x^2 - 12x - 4 = 0$$

$$x = \frac{12 \pm \sqrt{144 + 80}}{10}$$

$$= \frac{12 \pm 4\sqrt{14}}{10}$$

$$x = \boxed{\frac{6 + 2\sqrt{14}}{5}} \text{ or}$$

$$= \boxed{\frac{6 - 2\sqrt{14}}{5}}$$

Since $y = 3x - 4$, $y = \boxed{\frac{-2 + 6\sqrt{14}}{5}} \text{ or}$

$$= \boxed{\frac{-2 - 6\sqrt{14}}{5}}$$

(c) Substitute Eq. 2 into Eq. 1:

$$(10 - 2y)^2 + y^2 = 25$$
$$100 - 40y + 4y^2 + y^2 = 25$$
$$5y^2 - 40y + 75 = 0$$
$$y^2 - 8y + 15 = 0$$
$$(y - 3)(y - 5) = 0$$
$$y = \boxed{3 \text{ or } 5}$$

Since $x = 10 - 2y$, $x = \boxed{4 \text{ or } 0}$

(d) 3 times Eq. 1 : $6x^2 - 9y^2 = 18$ (Eq. 1')

2 times Eq. 2 : $6x^2 + 4y^2 = 70$ (Eq. 2')

Subtract Eq. 1' from Eq. 2':

$$13y^2 = 52$$
$$y^2 = 4$$
$$y = \boxed{2 \text{ or } -2}$$

Substitute $y^2 = 4$ into Eq. 1':

$$6x^2 - (9)(4) = 18$$
$$x^2 = 9$$
$$x = \boxed{3 \text{ or } -3}$$

13. $\boxed{14e^{i30°}}$

14. $r = \sqrt{3^2 + 6^2} = \sqrt{45} = 6.7$

$$\theta = \arctan\left(\frac{6}{3}\right) = 63.43°$$

$3 + 6i = \boxed{6.7\angle 63.43°}$

15.   For $12e^{-i60°}$, $r = 12$, $\theta = -60°$

$$a_1 = r\cos\theta = 12\cos(-60°) = 6$$
$$b_1 = r\sin\theta = 12\sin(-60°) = -10.392$$

For $4e^{i60°}$, $r = 4$, $\theta = 60°$

$$a_2 = 4\cos(60°) = 2$$
$$b_2 = 4\sin(60°) = 3.464$$

$(a_1 + a_2) + (b_1 + b_2)i = \boxed{8 - 6.928i}$

16. $(16 \text{ cis } 20°)^{\frac{1}{4}} = \sqrt[4]{16} \text{ cis } \left(\dfrac{20 + k\,360°}{4}\right)$

$$k = 0, 1, 2, 3$$

$$= \boxed{\begin{array}{l} 2 \text{ cis } (5 + 90°\,k) \\ k = 0, 1, 2, 3 \end{array}}$$

17. (a) $10\angle 45° = 10\cos 45° + i10\sin 45°$

$$= 7.071 + 7.071i$$

$$6\angle 30° = 6\cos 30° + i6\sin 30°$$

$$= 5.196 + 3i$$

$$10\angle 45° + 6\angle 30° = 12.267 + 10.071i$$

$$r = \sqrt{(12.267)^2 + (10.071)^2} = 15.871$$

$$\theta = \arctan\left(\frac{10.071}{12.267}\right) = 39.4°$$

$10\angle 45 + 6\angle 30 = \boxed{15.87\angle 39.4°}$

(b) $(4 + 3i) + (7 + 2i) = (4 + 7) + (3 + 2)i$

$$= \boxed{11 + 5i}$$

(c) $(4\angle 120°)(1\angle 30°) = (4e^{i120°})(e^{i30°})$

$$= 4e^{i(120° + 30°)}$$

$$= 4e^{i150°}$$

$$= \boxed{4\angle 150°}$$

(d) $\dfrac{6 - 7i}{2 + 3i} = \dfrac{(6 - 7i)(2 - 3i)}{(2 + 3i)(2 - 3i)}$

$$= \dfrac{-9 - 32i}{4 + 9}$$

$$= \boxed{\dfrac{-9}{13} - \dfrac{32}{13}i}$$

(e) $(8e^{i30°})(2e^{i60°}) = (8 \times 2)e^{i(30° + 60°)}$

$$= 16e^{i90°}$$

$$= \boxed{16i}$$

PROFESSIONAL PUBLICATIONS, INC. ● Belmont, CA

(f) $5\angle 60° = 5\cos 60° + i5\sin 60°$

$= 2.5 + 4.33i$

$2\angle 35° = 2\cos 35° + i2\sin 35°$

$= 1.64 + 1.15i$

$5\angle 60° - 2\angle 35° = 0.86 + 3.18i$

$r = \sqrt{(0.86)^2 + (3.18)^2} = 3.3$

$\theta = \arctan\left(\dfrac{3.18}{0.86}\right) = 74.87°$

$5\angle 60° - 2\angle 35° = \boxed{3.3\angle 74.87°}$

(g) $(9 + 5i) - (6 + 6i) = (9 - 6) + (5 - 6)i$

$= \boxed{3 - i}$

(h) $(2 - i)(-1 - 4i) = -2 + i - 8i + 4i^2$

$= \boxed{-6 - 7i}$

(i) $\dfrac{15\angle 90°}{3\angle 30°} = \left(\dfrac{15}{3}\right)\angle(90° - 30°)$

$= \boxed{5\angle 60°}$

(j) $5e^{i15°} = 5\cos 15° + 5i\sin 15°$

$= 4.83 + 1.29i$

$6e^{i60°} = 6\cos 60° + 6i\sin 60°$

$= 3 + 5.20i$

$5e^{i15°} + 6e^{i60°} = 7.83 + 6.49i$

$r = \sqrt{(7.83)^2 + (6.49)^2} = 10.2$

$\theta = \arctan\left(\dfrac{6.49}{7.83}\right) = 39.7°$

$5e^{i15°} + 6e^{i60°} = \boxed{10.2\angle 39.7°}$

18. (a) Since both the numerators and denominators approach zero, L'Hôpital's rule should be used.

$$\lim_{x\to 0}\left(\dfrac{1 - \cos x}{x^2}\right) = \lim_{x\to 0}\left(\dfrac{\sin x}{2x}\right)$$

$$= \lim_{x\to 0}\left(\dfrac{\cos x}{2}\right) = \boxed{\dfrac{1}{2}}$$

(b) $\lim_{x\to 2}\left(\dfrac{x^2 - 4}{x - 2}\right) = \lim_{x\to 2}\left[\dfrac{(x - 2)(x + 2)}{x - 2}\right]$

$= \lim_{x\to 2}(x + 2) = \boxed{4}$

(c) $\lim_{x\to 3}\left(\dfrac{x^3 - 27}{x - 3}\right) = \lim_{x\to 3}\left[\dfrac{(x - 3)(x^2 + 3x + 9)}{x - 3}\right]$

$= \lim_{x\to 3}(x^2 + 3x + 9) = \boxed{27}$

(d) $\lim_{x\to\infty}\left(\dfrac{2x + 1}{5x - 2}\right) = \lim_{x\to\infty}\left(\dfrac{2 + \dfrac{1}{x}}{5 - \dfrac{2}{x}}\right) = \boxed{\dfrac{2}{5}}$

19. (a) The numerators are always 1. The denominators begin at 1 and increase by 2. Therefore, the general term is

$$a_n = \boxed{\dfrac{1}{2n - 1}}$$

(b) The numerators begin at 4 and increase by 1. The denominators are 1 less than perfect squares, beginning at 3. Therefore, the general term is

$$a_n = \dfrac{3 + n}{(n + 1)^2 - 1} = \dfrac{3 + n}{n^2 + 2n}$$

$$= \boxed{\dfrac{3 + n}{n(n + 2)}}$$

(c) The numerators begin at 1 and increase by 2. The denominators begin at 5 and increase by 2. Therefore, the general term is

$$a_n = \boxed{\dfrac{2n - 1}{2n + 3}}$$

(d) $a_n = \boxed{3n - 1}$

20. (a) Each term in the series is smaller than $1/2^n$.

$$\boxed{\text{Since } \sum_{n=0}^{\infty}\dfrac{1}{2^n} \text{ converges, the series converges.}}$$

(b) $\dfrac{1}{2} + \dfrac{1}{4} + \dfrac{1}{6} + \cdots + \dfrac{1}{2n} + \cdots$

$= \left(\dfrac{1}{2}\right)\left(1 + \dfrac{1}{2} + \dfrac{1}{3} + \cdots + \dfrac{1}{n} + \cdots\right)$

Since a harmonic series diverges, the series diverges.

(c) By ratio test,

$$\lim_{n \to \infty}\left|\frac{a_{n=1}}{a_n}\right| = \lim_{n \to \infty}\left|\frac{\dfrac{2^{n+1}}{(n+1)!}}{\dfrac{2^n}{n!}}\right|$$

$$= \lim_{n \to \infty}\left|\frac{2}{n+1}\right| = 0 < 1$$

The series converges.

(d) The general term is $(-1)^{n+1}/(3^n + 1)$. Its absolute value is smaller than $1/3^n$.

Since $\displaystyle\sum_{n=1}^{\infty}\dfrac{1}{3^n}$ converges, the series converges.

(e) The series is

$$\sum_{n=1}^{\infty}\frac{(2n-1)(-1)^{n+1}}{2(n+1)}$$

$$\lim_{n \to \infty}\frac{(2n-1)(-1)^{n+1}}{2(n+1)} = \lim_{n \to \infty}\frac{\left(2 - \dfrac{1}{n}\right)(-1)^{n+1}}{(2)\left(1 + \dfrac{1}{n}\right)}$$

$$= \lim_{n \to \infty}(-1)^{n+1}$$

This does not approach a specific value, so the series diverges.

MATHEMATICS
Algebra

LINEAR ALGEBRA

1. Using the notation A_{ij} for the cofactor of a_{ij},

$$A_{21} = -\begin{vmatrix} 1 & 2 \\ -1 & 0 \end{vmatrix} = \boxed{-2}$$

$$A_{33} = \begin{vmatrix} 4 & 1 \\ 3 & -1 \end{vmatrix} = \boxed{-7}$$

2.

$$A_{11} = \boxed{2}$$

$$A_{12} = \boxed{-3}$$

$$A_{21} = \boxed{1}$$

$$A_{22} = \boxed{1}$$

3. (a) $\quad (3)(-4) - (1)(2) = \boxed{-14}$

(b) $\quad (4)(1) - (-6)(1) = \boxed{10}$

(c) Expand by cofactors using the second row.

$$-(-1)\begin{vmatrix} 1 & 2 \\ 4 & -2 \end{vmatrix} = (1)(-2) - (2)(4)$$

$$= \boxed{-10}$$

(d) Expand by cofactors using the first column.

$$(2)\begin{vmatrix} 1 & 2 \\ -2 & 1 \end{vmatrix} + (3)\begin{vmatrix} -1 & 3 \\ 1 & 2 \end{vmatrix} = (2)[(1)(1) - (2)(-2)]$$

$$+ (3)[(-1)(2) - (3)(1)]$$

$$= \boxed{-5}$$

(e) Expand by cofactors using the last row.

$$(s-4)\begin{vmatrix} s-2 & 4 \\ 1 & s+1 \end{vmatrix} = (s-4)[(s-2)(s+1) - (1)(4)]$$

$$= (s-4)(s^2 - s - 6)$$

$$= \boxed{(s-4)(s-3)(s+2)}$$

(f) Expand by cofactors using the second row.

$$-\begin{vmatrix} 2 & 2 & 3 \\ -1 & 1 & -2 \\ -3 & 0 & 2 \end{vmatrix} - (-2)\begin{vmatrix} 1 & 2 & 3 \\ 3 & -1 & -2 \\ 4 & -3 & 2 \end{vmatrix}$$

Expand the first 3×3 matrix by the third row and expand the second 3×3 matrix by the first row.

$$- [(-3)(-4-3) + (2)(2+2)]$$

$$+ (2)[(-2-6) - (2)(6+8) + (3)(-9+4)]$$

$$= -(21+8) + (2)(-8-28-15)$$

$$= -29 - 102 = \boxed{-131}$$

4. (a) $\begin{bmatrix} 1 & 7 \\ 2 & 6 \end{bmatrix} + \begin{bmatrix} 3 & 5 \\ 1 & 1 \end{bmatrix} = \begin{bmatrix} 1+3 & 7+5 \\ 2+1 & 6+1 \end{bmatrix}$

$$= \boxed{\begin{bmatrix} 4 & 12 \\ 3 & 7 \end{bmatrix}}$$

(b) $\begin{bmatrix} 7 & 6 & 6 \\ 5 & 4 & 1 \\ 9 & 8 & 2 \end{bmatrix} - \begin{bmatrix} 1 & 0 & 0 \\ 0 & 1 & 0 \\ 0 & 0 & 1 \end{bmatrix}$

$$= \begin{bmatrix} 7-1 & 6 & 6 \\ 5 & 4-1 & 1 \\ 9 & 8 & 2-1 \end{bmatrix}$$

$$= \boxed{\begin{bmatrix} 6 & 6 & 6 \\ 5 & 3 & 1 \\ 9 & 8 & 1 \end{bmatrix}}$$

(c) $\begin{bmatrix} 2 & 1 \\ 0 & 0 \end{bmatrix} \times \begin{bmatrix} 3 & 2 \\ 1 & 1 \end{bmatrix}$

$$= \begin{bmatrix} (2)(3)+(1)(1) & (2)(2)+(1)(1) \\ (0)(3)+(0)(1) & (0)(2)+(0)(1) \end{bmatrix}$$

$$= \boxed{\begin{bmatrix} 7 & 5 \\ 0 & 0 \end{bmatrix}}$$

(d) $\begin{bmatrix} 1 & 9 \\ 7 & 2 \end{bmatrix} \times \begin{bmatrix} 2 & 1 & 3 \\ 4 & -1 & -7 \end{bmatrix} =$

$$\begin{bmatrix} (1)(2)+(9)(4) & (1)(1)+(9)(-1) & (1)(3)+(9)(-7) \\ (7)(2)+(2)(4) & (7)(1)+(2)(-1) & (7)(3)+(2)(-7) \end{bmatrix}$$

$$= \boxed{\begin{bmatrix} 38 & -8 & -60 \\ 22 & 5 & 7 \end{bmatrix}}$$

(e) $\begin{bmatrix} 1 & 2 & -3 & 4 \\ 0 & -5 & -1 & -2 \end{bmatrix} + \begin{bmatrix} 2 & -5 & 6 & 1 \\ 3 & 0 & -2 & -3 \end{bmatrix}$

$= \begin{bmatrix} 1+2 & 2+(-5) & -3+6 & 4+1 \\ 0+3 & -5+0 & -1+(-2) & -2+(-3) \end{bmatrix}$

$= \begin{bmatrix} 3 & -3 & 3 & 5 \\ 3 & -5 & -3 & -5 \end{bmatrix}$

(f) $\mathbf{A} - \mathbf{B} + \mathbf{C} = \begin{bmatrix} 2-1+0 & 3-0+1 \\ -5+2+1 & 0+1-1 \\ 1+3-2 & -4-5-1 \end{bmatrix}$

$= \begin{bmatrix} 1 & 4 \\ -2 & 0 \\ 2 & -10 \end{bmatrix}$

(g) $\begin{bmatrix} 1 & 6 \\ -3 & 4 \end{bmatrix} \begin{bmatrix} -1 & 0 \\ 3 & 2 \end{bmatrix} = \boxed{\begin{bmatrix} 17 & 12 \\ 15 & 8 \end{bmatrix}}$

(h) $[2, -1] \begin{bmatrix} -4 \\ -5 \end{bmatrix} = \boxed{-3}$

(i) $\begin{bmatrix} 3 \\ -1 \end{bmatrix} [2, 5] = \boxed{\begin{bmatrix} 6 & 15 \\ -2 & -5 \end{bmatrix}}$

5. Using the left distributive law,

$$(\mathbf{A}+\mathbf{B})(\mathbf{C}+\mathbf{D}) = [(\mathbf{A}+\mathbf{B})\mathbf{C}] + [(\mathbf{A}+\mathbf{B})\mathbf{D}]$$

Using the right distributive law,

$$= [\mathbf{AC}+\mathbf{BC}] + [\mathbf{AD}+\mathbf{BD}]$$

Using the associative law of addition,

$$= \mathbf{AC}+\mathbf{BC}+\mathbf{AD}+\mathbf{BD}$$

Using the commutative law of addition,

$$= \mathbf{AC}+\mathbf{AD}+\mathbf{BC}+\mathbf{BD}$$

6. (a) The number of columns of the left matrix $\mathbf{A}$, q, must equal the number of rows of the right matrix $\mathbf{B}$, r. Therefore,

$$\boxed{q = r}$$

(b) The product $\mathbf{AB}$ has shape

$$\boxed{p \times s}$$

7. (a) $\mathbf{C} = \mathbf{AB}$.

$\mathbf{C}$ is a 3×3 matrix.

$$c_{11} = (-3)(3) + (-1)(1) = -10$$
$$c_{12} = (-3)(4) + (-1)(-2) = -10$$
$$c_{13} = (-3)(-5) + (-1)(0) = 15$$
$$c_{21} = (1)(3) + (0)(1) = 3$$
$$c_{22} = (1)(4) + (0)(-2) = 4$$
$$c_{23} = (1)(-5) + (0)(0) = -5$$
$$c_{31} = (2)(3) + (4)(1) = 10$$
$$c_{32} = (2)(4) + (4)(-2) = 0$$
$$c_{33} = (2)(-5) + (4)(0) = -10$$

$$\mathbf{C} = \boxed{\begin{bmatrix} -10 & -10 & 15 \\ 3 & 4 & -5 \\ 10 & 0 & -10 \end{bmatrix}}$$

(b) $\mathbf{D} = \mathbf{BA}$

$\mathbf{D}$ is a 2×2 matrix.

$$d_{11} = (3)(-3) + (4)(1) + (-5)(2) = -15$$
$$d_{12} = (3)(-1) + (4)(0) + (-5)(4) = -23$$
$$d_{21} = (1)(-3) + (-2)(1) + (0)(2) = -5$$
$$d_{22} = (1)(-1) + (-2)(0) + (0)(4) = -1$$

$$\mathbf{D} = \boxed{\begin{bmatrix} -15 & -23 \\ -5 & -1 \end{bmatrix}}$$

Note: $\mathbf{AB} \neq \mathbf{BA}$.

8. $\mathbf{AB} = \begin{bmatrix} 2 & 1 & -4 \\ 0 & 1 & 3 \\ 2 & 3 & 2 \end{bmatrix} \begin{bmatrix} 2 & 1 & -4 \\ 0 & 1 & 3 \\ 0 & 0 & 5 \end{bmatrix}$

$= \begin{bmatrix} 4 & 3 & -25 \\ 0 & 1 & 18 \\ 4 & 5 & 11 \end{bmatrix}$

$\mathbf{BA} = \begin{bmatrix} 2 & 1 & -4 \\ 0 & 1 & 3 \\ 0 & 0 & 5 \end{bmatrix} \begin{bmatrix} 2 & 1 & -4 \\ 0 & 1 & 3 \\ 2 & 3 & 2 \end{bmatrix}$

$= \begin{bmatrix} -4 & -9 & -13 \\ 6 & 10 & 9 \\ 10 & 15 & 10 \end{bmatrix}$

$\mathbf{AB} - \mathbf{BA} = \begin{bmatrix} 4 & 3 & -25 \\ 0 & 1 & 18 \\ 4 & 5 & 11 \end{bmatrix} - \begin{bmatrix} -4 & -9 & -13 \\ 6 & 10 & 9 \\ 10 & 15 & 10 \end{bmatrix}$

$= \boxed{\begin{bmatrix} 8 & 12 & -12 \\ -6 & -9 & 9 \\ -6 & -10 & 1 \end{bmatrix}}$

Note: $\mathbf{AB} - \mathbf{BA} \neq 0$.

MATHEMATICS
Linear Alg

9.
$$\mathbf{A}^t = \begin{bmatrix} 1 & -2 & -3 \\ -1 & 5 & 0 \\ 3 & 4 & 1 \\ 2 & 0 & -4 \end{bmatrix}$$

10.
$$\mathbf{A} = \begin{bmatrix} a & c \\ b & d \end{bmatrix}$$

$$\mathbf{A}^t = \begin{bmatrix} a & b \\ c & d \end{bmatrix}$$

$$\mathbf{A}\mathbf{A}^t = \begin{bmatrix} a^2 + c^2 & ab + cd \\ ab + cd & b^2 + d^2 \end{bmatrix}$$

$$\mathbf{A}^t\mathbf{A} = \begin{bmatrix} a^2 + b^2 & ac + bd \\ ac + bd & c^2 + d^2 \end{bmatrix}$$

$$\mathbf{A}\mathbf{A}^t = \mathbf{A}^t\mathbf{A} \longrightarrow \begin{cases} a^2 + c^2 = a^2 + b^2 \\ ab + cd = ac + bd \\ b^2 + d^2 = c^2 + d^2 \end{cases}$$

$$\longrightarrow \begin{cases} b^2 = c^2 \\ ab + cd = ac + bd \end{cases}$$

These require

$$\{b = c\} \text{ or } \begin{cases} b \neq 0 \\ b = -c \\ a = d \end{cases}$$

When a, b, and d are arbitrary, the set of matrices is

$$\mathbf{A} = \left\{ \begin{bmatrix} a & b \\ b & d \end{bmatrix} \right\} \cup \left\{ \begin{bmatrix} a & -b \\ b & a \end{bmatrix} \right\}$$

Note: These are normal 2×2 matrices.

11. The determinant is zero and the sum of the first two rows is equal to two times the third row. Consequently, the matrix is singular; therefore, its rank is less than 3.

One 2×2 submatrix is

$$\begin{bmatrix} 6 & 3 \\ -4 & 1 \end{bmatrix}$$

Its determinant is

$$(6)(1) - (3)(-4) = 18$$

The determinant is nonzero; therefore, the matrix is nonsingular.

The rank of the original 3×3 matrix is 2.

12. (a) Expand by cofactors using the first column.

$$(bc^2 - cb^2) - (ac^2 - ca^2) + (ab^2 - ba^2)$$
$$= bc^2 - cb^2 - ac^2 + ca^2 + ab^2 - ba^2$$

$$(a - b)(b - c)(c - a) = (a - b)(bc - c^2 - ab + ac)$$
$$= abc - ac^2 - ba^2 + ca^2$$
$$- cb^2 + bc^2 + ab^2 - abc$$
$$= bc^2 - cb^2 - ac^2 + ca^2 + ab^2 - ba^2$$

Since these two results are the same, the equality has been proved.

(b) If $a, b,$ and c are all distinctly different, the determinant $\neq 0$ and the rank is 3. If two of the parameters are equal and the third one is different, the matrix has two identical rows, with the remaining row independent from the other two. Therefore, the rank is 2. If $a = b = c$, the three rows are equal and the rank is 1.

13. The cofactor matrix is

$$\begin{bmatrix} 1 & -3 & 2 \\ 0 & -1 & 1 \\ -2 & 6 & -5 \end{bmatrix}$$

The adjoint matrix is

$$\begin{bmatrix} 1 & 0 & -2 \\ -3 & -1 & 6 \\ 2 & 1 & -5 \end{bmatrix}$$

14.
$$\mathbf{A} = \begin{bmatrix} a & c \\ b & d \end{bmatrix}$$

$$\mathbf{A}_{\text{adj}} = \begin{bmatrix} d & -c \\ -b & a \end{bmatrix}$$

$$\mathbf{A} = \mathbf{A}_{\text{adj}} \longrightarrow \begin{cases} a = d \\ b = -b \\ c = -c \end{cases} \text{ or } \begin{cases} a = d \\ b = 0 \\ c = 0 \end{cases}$$

When a is arbitrary, the set of matrices is

$$\mathbf{A} = \begin{bmatrix} a & 0 \\ 0 & a \end{bmatrix}$$

Note: These are scalar matrices.

15. (a) $|\mathbf{A}| = (6)(4) - (1)(2) = 22$

$$\mathbf{A}^{-1} = \frac{1}{|\mathbf{A}|} \begin{bmatrix} 4 & -1 \\ -2 & 6 \end{bmatrix}$$

$$= \left(\frac{1}{22}\right) \begin{bmatrix} 4 & -1 \\ -2 & 6 \end{bmatrix}$$

$$= \boxed{\begin{bmatrix} \frac{2}{11} & -\frac{1}{22} \\ -\frac{1}{11} & \frac{3}{11} \end{bmatrix}}$$

(b) Expand by cofactors using the third column.

$$|\mathbf{A}| = (1) \begin{vmatrix} 1 & 3 \\ -1 & 4 \end{vmatrix} - (1) \begin{vmatrix} 2 & 1 \\ -1 & 4 \end{vmatrix}$$

$$= [(1)(4) - (3)(-1)] - [(2)(4) - (1)(-1)]$$

$$= \boxed{-2}$$

The cofactor matrix with appropriate signs is

$$\begin{bmatrix} -4 & -1 & 7 \\ 4 & 1 & -9 \\ -2 & -1 & 5 \end{bmatrix}$$

$$\mathbf{A}_{\text{adj}} = \begin{bmatrix} -4 & 4 & -2 \\ -1 & 1 & -1 \\ 7 & -9 & 5 \end{bmatrix}$$

$$\mathbf{A}^{-1} = \frac{\mathbf{A}_{\text{adj}}}{|\mathbf{A}|} = \left(-\frac{1}{2}\right) \begin{bmatrix} -4 & 4 & -2 \\ -1 & 1 & -1 \\ 7 & -9 & 5 \end{bmatrix}$$

$$= \boxed{\begin{bmatrix} 2 & -2 & 1 \\ \frac{1}{2} & -\frac{1}{2} & \frac{1}{2} \\ -\frac{7}{2} & \frac{9}{2} & -\frac{5}{2} \end{bmatrix}}$$

(c) Compute the classical adjoint of the matrix. The cofactor matrix is

$$\begin{bmatrix} -1 & -1 & 2 \\ -1 & 1 & -2 \\ 1 & -1 & 0 \end{bmatrix}$$

The adjoint is

$$\begin{bmatrix} -1 & -1 & 1 \\ -1 & 1 & -1 \\ 2 & -2 & 0 \end{bmatrix}$$

The determinant is

$$|\mathbf{A}| = -1 - 1 = -2$$

$$\mathbf{A}^{-1} = \frac{\mathbf{A}_{\text{adj}}}{|\mathbf{A}|}$$

$$= \begin{bmatrix} \frac{1}{2} & \frac{1}{2} & -\frac{1}{2} \\ \frac{1}{2} & -\frac{1}{2} & \frac{1}{2} \\ -1 & 1 & 0 \end{bmatrix}$$

(d) The cofactor matrix is

$$\begin{bmatrix} 1 & 0 & 0 \\ 1 & 1 & 0 \\ -3 & -1 & 1 \end{bmatrix}$$

The adjoint is

$$\begin{bmatrix} 1 & 1 & -3 \\ 0 & 1 & -1 \\ 0 & 0 & 1 \end{bmatrix}$$

Expanding by cofactors using the first column, the determinant is found to be 1.

The inverse is

$$\mathbf{A}^{-1} = \frac{\mathbf{A}_{\text{adj}}}{|\mathbf{A}|} = \begin{bmatrix} 1 & 1 & -3 \\ 0 & 1 & -1 \\ 0 & 0 & 1 \end{bmatrix}$$

Note: This is a general result. The inverse of an upper triangular matrix is another upper triangular matrix.

16. Since $|\mathbf{AB}| = |\mathbf{A}||\mathbf{B}|$,

$$\left|\mathbf{AA}^{-1}\right| = |\mathbf{A}|\left|\mathbf{A}^{-1}\right|$$

Solving for $|\mathbf{A}^{-1}|$,

$$\left|\mathbf{A}^{-1}\right| = \frac{\left|\mathbf{AA}^{-1}\right|}{|\mathbf{A}|}$$

But $\mathbf{AA}^{-1} = \mathbf{I} = 1$. Therefore,

$$\left|\mathbf{A}^{-1}\right| = \frac{\left|\mathbf{AA}^{-1}\right|}{|\mathbf{A}|} = \frac{1}{|\mathbf{A}|}$$

17. $$\mathbf{A} = \begin{bmatrix} a & c \\ b & d \end{bmatrix} \text{ and } \mathbf{A}^t = \begin{bmatrix} a & b \\ c & d \end{bmatrix}$$

$$\mathbf{A}^{-1} = \frac{\mathbf{A}_{\text{adj}}}{|\mathbf{A}|} = \frac{1}{ad - bc} \begin{bmatrix} d & -c \\ -b & a \end{bmatrix}$$

$$\mathbf{A}^{-1} = \mathbf{A}^t \longrightarrow \left\{ \begin{array}{l} a(ad - bc) = d \\ c(ad - bc) = -b \\ b(ad - bc) = -c \\ d(ad - bc) = a \end{array} \right\}$$

$$\longrightarrow \left\{ \begin{array}{l} (ad - bc)^2 = 1 \\ a(ad - bc) = d \\ b(ad - bc) = -c \end{array} \right\}$$

If $ad - bc = 1$,

$$\left\{\begin{array}{l} a = d \\ b = -c \end{array}\right\}$$

$$ad - bc = a^2 + b^2 = 1$$
$$a = \cos\theta$$
$$b = \sin\theta$$
$$c = -\sin\theta$$
$$d = \cos\theta$$

If $ad - bc = -1$,

$$\left\{\begin{array}{l} a = -d \\ b = c \end{array}\right\}$$

$$ad - bc = -a^2 - b^2 = -1$$
$$a = \cos\theta$$
$$b = \sin\theta$$
$$c = \sin\theta$$
$$d = -\cos\theta$$

When θ is an arbitrary real number, the set of matrices is

$$\left\{\begin{bmatrix} \cos\theta & -\sin\theta \\ \sin\theta & \cos\theta \end{bmatrix}\right\} \cup \left\{\begin{bmatrix} \cos\theta & \sin\theta \\ \sin\theta & -\cos\theta \end{bmatrix}\right\}$$

18.

$$\Delta = \begin{vmatrix} 2 & -3 \\ 3 & 7 \end{vmatrix} = 23$$

$$\Delta_x = \begin{vmatrix} 5 & -3 \\ -2 & 7 \end{vmatrix} = 29$$

$$\Delta_y = \begin{vmatrix} 2 & 5 \\ 3 & -2 \end{vmatrix} = -19$$

$$x = \frac{\Delta_x}{\Delta} = \boxed{\frac{29}{23}}$$

$$y = \frac{\Delta_y}{\Delta} = \boxed{\frac{-19}{23}}$$

19. Rewrite the equations in standard form.

$$x - 2y - 3z = -1$$
$$2x + 3y + 5z = 8$$
$$3x - y + 2z = 1$$

$$\Delta = \begin{vmatrix} 1 & -2 & -3 \\ 2 & 3 & 5 \\ 3 & -1 & 2 \end{vmatrix}$$
$$= (1)(6+5) - (2)(-4-3) + (3)(-10+9) = 22$$

$$\Delta_x = \begin{vmatrix} -1 & -2 & -3 \\ 8 & 3 & 5 \\ 1 & -1 & 2 \end{vmatrix} = 44$$

$$\Delta_y = \begin{vmatrix} 1 & -1 & -3 \\ 2 & 8 & 5 \\ 3 & 1 & 2 \end{vmatrix} = 66$$

$$\Delta_z = \begin{vmatrix} 1 & -2 & -1 \\ 2 & 3 & 8 \\ 3 & -1 & 1 \end{vmatrix} = -22$$

$$\boxed{\begin{aligned} x &= \frac{\Delta_x}{\Delta} = \frac{44}{22} = 2 \\ y &= \frac{\Delta_y}{\Delta} = \frac{66}{22} = 3 \\ z &= \frac{\Delta_z}{\Delta} = \frac{-22}{22} = -1 \end{aligned}}$$

20. Arrange the system in standard form.

$$x - 2y - 3z = -1$$
$$2x + 3y + 5z = 8$$
$$3x - y + 2z = 1$$

The augmented matrix is

$$\begin{bmatrix} 1 & -2 & -3 & | & -1 \\ 2 & 3 & 5 & | & 8 \\ 3 & -1 & 2 & | & 1 \end{bmatrix}$$

Multiply the first row by 3 and subtract from the third row.

$$\begin{bmatrix} 1 & -2 & -3 & -1 \\ 2 & 3 & 5 & 8 \\ 0 & 5 & 11 & 4 \end{bmatrix}$$

Proceed with row operations.

$$\begin{bmatrix} 1 & -2 & -3 & -1 \\ 0 & 7 & 11 & 10 \\ 0 & 5 & 11 & 4 \end{bmatrix} \longrightarrow \begin{bmatrix} 1 & -2 & -3 & -1 \\ 0 & 7 & 11 & 10 \\ 0 & 0 & 22 & -22 \end{bmatrix}$$

$$\longrightarrow \begin{bmatrix} 7 & 0 & 1 & 13 \\ 0 & 7 & 11 & 10 \\ 0 & 0 & 22 & -22 \end{bmatrix} \longrightarrow \begin{bmatrix} 7 & 0 & 1 & 13 \\ 0 & 7 & 11 & 10 \\ 0 & 0 & 1 & -1 \end{bmatrix}$$

$$\longrightarrow \begin{bmatrix} 7 & 0 & 0 & 14 \\ 0 & 7 & 11 & 10 \\ 0 & 0 & 1 & -1 \end{bmatrix} \longrightarrow \begin{bmatrix} 7 & 0 & 0 & 14 \\ 0 & 7 & 0 & 21 \\ 0 & 0 & 1 & -1 \end{bmatrix}$$

$$\longrightarrow \begin{bmatrix} 1 & 0 & 0 & 2 \\ 0 & 1 & 0 & 3 \\ 0 & 0 & 1 & -1 \end{bmatrix}$$

$$\boxed{\begin{aligned} x &= 2 \\ y &= 3 \\ z &= -1 \end{aligned}}$$

21. $k\mathbf{I} - \mathbf{A} = \begin{vmatrix} k-1 & -4 \\ -2 & k-3 \end{vmatrix}$

$|k\mathbf{I} - \mathbf{A}| = (k-1)(k-3) - (-2)(-4) = 0$

$$k^2 - 4k - 5 = 0$$
$$(k-5)(k+1) = 0$$

The eigenvalues are

$$\boxed{-1, \; 5}$$

Substitute $k = -1$ to find the eigenvector corresponding to -1.

$$-\mathbf{I} - \mathbf{A} = \begin{bmatrix} -2 & -4 \\ -2 & -4 \end{bmatrix}$$

$$|-\mathbf{I} - \mathbf{A}|\mathbf{X} = \begin{bmatrix} 0 \\ 0 \end{bmatrix}$$

$$\mathbf{X} = \boxed{\begin{bmatrix} 2 \\ -1 \end{bmatrix}}$$

For $k = 5$,

$$5\mathbf{I} - \mathbf{A} = \begin{bmatrix} 4 & -4 \\ -2 & 2 \end{bmatrix}$$

$$|5\mathbf{I} - \mathbf{A}|\mathbf{X} = \begin{bmatrix} 0 \\ 0 \end{bmatrix}$$

$$\mathbf{X} = \boxed{\begin{bmatrix} 1 \\ 1 \end{bmatrix}}$$

22. $k\mathbf{I} - \mathbf{A} = \begin{bmatrix} k-1 & -2 & -2 \\ -1 & k-2 & 1 \\ 1 & -1 & k-4 \end{bmatrix}$

$|k\mathbf{I} - \mathbf{A}| = (k-1)[(k-2)(k-4)+1]$
$\qquad\qquad + [(-2)(k-4) - 2] + [-2 + (2)(k-2)] = 0$

$$(k-1)[(k-2)(k-4)+1] = 0$$
$$(k-1)(k^2 - 6k + 9) = (k-1)(k-3)^2 = 0$$

The eigenvalues are

$$\boxed{1, \; 3}$$

For $k = 1$,

$$\mathbf{I} - \mathbf{A} = \begin{bmatrix} 0 & -2 & -2 \\ -1 & -1 & 1 \\ 1 & -1 & -3 \end{bmatrix}$$

If $\mathbf{X} = \begin{bmatrix} x \\ y \\ z \end{bmatrix}$,

$$|\mathbf{I} - \mathbf{A}|\mathbf{X} = 0 \longrightarrow \begin{cases} -2y - 2z = 0 \\ -x - y + z = 0 \\ x - y - 3z = 0 \end{cases}$$

$$\mathbf{X} = \boxed{\begin{bmatrix} 2 \\ -1 \\ 1 \end{bmatrix}}$$

For $k = 3$,

$$3\mathbf{I} - \mathbf{A} = \begin{bmatrix} 2 & -2 & -2 \\ -1 & 1 & 1 \\ 1 & -1 & -1 \end{bmatrix}$$

$$|\mathbf{I} - \mathbf{A}|\mathbf{X} = 0 \longrightarrow \begin{cases} 2x - 2y - 2z = 0 \\ -x + y + z = 0 \\ x - y - z = 0 \end{cases}$$

All three equations are equivalent to $x - y - z = 0$, and result in two independent eigenvectors for the eigenvalue $k = 3$.

$$\mathbf{X} = \boxed{\begin{bmatrix} 1 \\ 1 \\ 0 \end{bmatrix}} \text{ and } \boxed{\begin{bmatrix} 1 \\ 0 \\ 1 \end{bmatrix}}$$

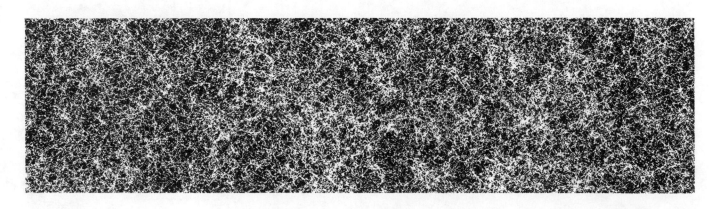

VECTORS

1. (a) $\mathbf{V}_1 \cdot \mathbf{V}_2 = \mathbf{V}_{1x}\mathbf{V}_{2x} + \mathbf{V}_{1y}\mathbf{V}_{2y}$

$= (2)(5) + (3)(-2)$

$= \boxed{4}$

(b) $\mathbf{V}_1 \cdot \mathbf{V}_2 = (1)(9) + (4)(-3)$

$= \boxed{-3}$

(c) $\mathbf{V}_1 \cdot \mathbf{V}_2 = (7)(3) + (-3)(4)$

$= \boxed{9}$

(d) $\mathbf{V}_1 \cdot \mathbf{V}_2 = \mathbf{V}_{1x}\mathbf{V}_{2x} + \mathbf{V}_{1y}\mathbf{V}_{2y} + \mathbf{V}_{1z}\mathbf{V}_{2z}$

$= (2)(8) + (-3)(2) + (6)(-3)$

$= \boxed{-8}$

(e) $\mathbf{V}_1 \cdot \mathbf{V}_2 = (6)(1) + (2)(0) + (3)(1)$

$= \boxed{9}$

2. (a) $\cos\phi = \dfrac{\mathbf{V}_1 \cdot \mathbf{V}_2}{|\,\mathbf{V}_1\,|\,|\,\mathbf{V}_2\,|}$

$= \dfrac{4}{\left(\sqrt{2^2 + 3^2}\right)\left(\sqrt{5^2 + 2^2}\right)} = 0.206$

$\phi = \cos^{-1}(0.206) = \boxed{78.1°}$

(b) $\cos\phi = \dfrac{\mathbf{V}_1 \cdot \mathbf{V}_2}{|\,\mathbf{V}_1\,|\,|\,\mathbf{V}_2\,|}$

$= \dfrac{-3}{\left(\sqrt{1^2 + 4^2}\right)\left(\sqrt{9^2 + (-3)^2}\right)}$

$= -0.077$

$\phi = \boxed{94.4°}$

(c) $\cos\phi = \dfrac{\mathbf{V}_1 \cdot \mathbf{V}_2}{|\,\mathbf{V}_1\,|\,|\,\mathbf{V}_2\,|}$

$= \dfrac{9}{\left(\sqrt{7^2 + (-3)^2}\right)\left(\sqrt{3^2 + 4^2}\right)} = 0.236$

$\phi = \boxed{76.3°}$

3. (a) $\mathbf{V}_1 \times \mathbf{V}_2 = \begin{vmatrix} \mathbf{i} & \mathbf{V}_{1x} & \mathbf{V}_{2x} \\ \mathbf{j} & \mathbf{V}_{1y} & \mathbf{V}_{2y} \\ \mathbf{k} & \mathbf{V}_{1z} & \mathbf{V}_{2z} \end{vmatrix}$

$= \begin{vmatrix} \mathbf{i} & 2 & 5 \\ \mathbf{j} & 3 & -2 \\ \mathbf{k} & 0 & 0 \end{vmatrix}$

Expand by the third row.

$= \mathbf{k}\begin{vmatrix} 2 & 5 \\ 3 & -2 \end{vmatrix} = \boxed{-19\,\mathbf{k}}$

(b) $\mathbf{V}_1 \times \mathbf{V}_2 = \begin{vmatrix} \mathbf{i} & 1 & 9 \\ \mathbf{j} & 4 & -3 \\ \mathbf{k} & 0 & 0 \end{vmatrix}$

Expand by the third row.

$= \mathbf{k}\begin{vmatrix} 1 & 9 \\ 4 & -3 \end{vmatrix} = \boxed{-39\,\mathbf{k}}$

(c) $\mathbf{V}_1 \times \mathbf{V}_2 = \begin{vmatrix} \mathbf{i} & 7 & 3 \\ \mathbf{j} & -3 & 4 \\ \mathbf{k} & 0 & 0 \end{vmatrix}$

Expand by the third row.

$\mathbf{k}\begin{vmatrix} 7 & 3 \\ -3 & 4 \end{vmatrix} = \boxed{37\,\mathbf{k}}$

(d) $\mathbf{V}_1 \times \mathbf{V}_2 = \begin{vmatrix} \mathbf{i} & 2 & 8 \\ \mathbf{j} & -3 & 2 \\ \mathbf{k} & 6 & -3 \end{vmatrix}$

Expand by the first column.

$= \mathbf{i}\begin{vmatrix} -3 & 2 \\ 6 & -3 \end{vmatrix} - \mathbf{j}\begin{vmatrix} 2 & 8 \\ 6 & -3 \end{vmatrix}$

$+\,\mathbf{k}\begin{vmatrix} 2 & 8 \\ -3 & 2 \end{vmatrix}$

$= \boxed{-3\mathbf{i} + 54\mathbf{j} + 28\mathbf{k}}$

(e) $\mathbf{V}_1 \times \mathbf{V}_2 = \begin{vmatrix} \mathbf{i} & 6 & 1 \\ \mathbf{j} & 2 & 0 \\ \mathbf{k} & 3 & 1 \end{vmatrix}$

Expand by the second row.

$= -\mathbf{j}\begin{vmatrix} 6 & 1 \\ 3 & 1 \end{vmatrix} + (2)\begin{vmatrix} \mathbf{i} & 1 \\ \mathbf{k} & 1 \end{vmatrix}$

$= \boxed{2\mathbf{i} - 3\mathbf{j} - 2\mathbf{k}}$

TRIGONOMETRY

1. (a) $\theta = \tan^{-1}\left(\dfrac{9.6}{17.2}\right) = \boxed{29.17°}$

(b) $\theta = \sin^{-1}\left(\dfrac{3.1}{5.4}\right) = \boxed{35.03°}$

(c) $\theta = \cos^{-1}\left(\dfrac{2.9}{14.1}\right) = \boxed{78.13°}$

2.

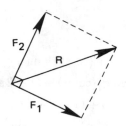

$$R = \sqrt{F_1^2 + F_2^2} = \sqrt{(20)^2 + (30)^2} = \boxed{36}$$

3.

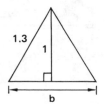

$$(1.3)^2 = (1)^2 + \left(\dfrac{b}{2}\right)^2$$

$$b = 2\sqrt{(1.3)^2 - (1)^2} = \boxed{1.66}$$

4. Construct a right triangle, ABC.

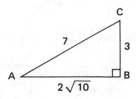

From the Pythagorean theorem, the adjacent side is

$$AB = \sqrt{(7)^2 - (3)^2} = \sqrt{40} = 2\sqrt{10}$$

$\sin A = \boxed{\dfrac{3}{7}}$

$\cos A = \boxed{\dfrac{2\sqrt{10}}{7}}$

$\tan A = \dfrac{3}{2\sqrt{10}} = \boxed{\dfrac{3\sqrt{10}}{20}}$

$\cot A = \boxed{\dfrac{2\sqrt{10}}{3}}$

$\sec A = \dfrac{7}{2\sqrt{10}} = \boxed{\dfrac{7\sqrt{10}}{20}}$

$\csc A = \boxed{\dfrac{7}{3}}$

5. Construct an isosceles right triangle with the two equal sides measuring 1.

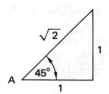

$\sin 45° = \dfrac{1}{\sqrt{2}} = \boxed{\dfrac{\sqrt{2}}{2}}$

$\cos 45° = \dfrac{1}{\sqrt{2}} = \boxed{\dfrac{\sqrt{2}}{2}}$

$\tan 45° = \dfrac{1}{1} = \boxed{1}$

$\cot 45° = \boxed{1}$

$\sec 45° = \boxed{\sqrt{2}}$

$\csc 45° = \boxed{\sqrt{2}}$

PROFESSIONAL PUBLICATIONS, INC. ● Belmont, CA

6. An angle with a positive sine and a negative cosine will be located in the second quadrant ($90° \leq \theta \leq 180°$).

7. When k is any integer, the angle is

$$\boxed{\dfrac{5\pi}{4} + 2k\pi \text{ or } 225° + k360°}$$

8. $\sin(270° - \theta) = \sin[180° + (90° - \theta)]$

$$= -\sin(90° - \theta) = \boxed{-\cos\theta}$$

$\cos(270° - \theta) = \cos[180° + (90° - \theta)]$

$$= -\cos(90° - \theta) = \boxed{-\sin\theta}$$

$\tan(270° - \theta) = \tan[180° + (90° - \theta)]$

$$= \tan(90° - \theta) = \boxed{\cot\theta}$$

9. $\sin(270° + \theta) = \sin[180° + (90° + \theta)]$

$$= -\sin(90° + \theta) = \boxed{-\cos\theta}$$

$\cos(270° + \theta) = \cos[180° + (90° + \theta)]$

$$= -\cos(90° + \theta) = \boxed{\sin\theta}$$

$\tan(270° + \theta) = \tan[180° + (90° + \theta)]$

$$= \tan(90° + \theta) = \boxed{-\cot\theta}$$

10.   (a) $\sec\theta - (\sec\theta)(\sin^2\theta) = \dfrac{1}{\cos\theta} - \dfrac{\sin^2\theta}{\cos\theta}$

$$= \dfrac{1 - \sin^2\theta}{\cos\theta} = \dfrac{\cos^2\theta}{\cos\theta}$$

$$= \boxed{\cos\theta}$$

(b) $\sin^2\theta(1 + \cot^2\theta) = (\sin^2\theta)(\csc^2\theta)$

$$= (\sin^2\theta)\left(\dfrac{1}{\sin^2\theta}\right)$$

$$= \boxed{1}$$

(c) $(\tan^2\theta)(\cos^2\theta) + (\cot^2\theta)(\sin^2\theta)$

$$= \left(\dfrac{\sin^2\theta}{\cos^2\theta}\right)(\cos^2\theta) + \left(\dfrac{\cos^2\theta}{\sin^2\theta}\right)(\sin^2\theta)$$

$$= \sin^2\theta + \cos^2\theta$$

$$= \boxed{1}$$

11. For the equality to hold, the products of the diagonal terms must be equal.

$$(\cos x)(\cos x) = (1 + \sin x)(1 - \sin x)$$

This can be shown as follows.

$$\cos^2 x + \sin^2 x = 1$$

$$\cos^2 x = 1 - \sin^2 x$$

$$(\cos x)(\cos x) = (1 - \sin x)(1 + \sin x)$$

12. $\sin 3\alpha = \sin(2\alpha + \alpha)$

$$= (\sin 2\alpha)(\cos\alpha) + (\cos 2\alpha)(\sin\alpha)$$

$$= [(2\sin\alpha)(\cos\alpha)](\cos\alpha) + (1 - 2\sin^2\alpha)(\sin\alpha)$$

$$= (2\sin\alpha)(\cos^2\alpha) + (1 - 2\sin^2\alpha)(\sin\alpha)$$

$$= (2\sin\alpha)(1 - \sin^2\alpha) + (1 - 2\sin^2\alpha)(\sin\alpha)$$

$$= \boxed{3\sin\alpha - 4\sin^3\alpha}$$

13. $\cosh^2\theta - \sinh^2\theta = \left(\dfrac{e^\theta + e^{-\theta}}{2}\right)^2 - \left(\dfrac{e^\theta - e^{-\theta}}{2}\right)^2$

$$= \dfrac{e^{2\theta} + e^{-2\theta} + 2 - e^{2\theta} - e^{-2\theta} + 2}{4}$$

$$= \dfrac{4}{4} = 1$$

14. $\sinh 2\theta = \dfrac{e^{2\theta} - e^{-2\theta}}{2} = \dfrac{(e^\theta - e^{-\theta})(e^\theta + e^{-\theta})}{2}$

$$= (2)\left(\dfrac{e^\theta - e^{-\theta}}{2}\right)\left(\dfrac{e^\theta + e^{-\theta}}{2}\right)$$

$$= 2\sinh\theta\cosh\theta$$

15. (a) $C = 180° - 70° - 32° = \boxed{78°}$

$$\dfrac{\sin A}{a} = \dfrac{\sin B}{b} = \dfrac{\sin C}{c} = \dfrac{\sin 78°}{27}$$

$$a = \dfrac{(27)(\sin 32°)}{\sin 78°} = \boxed{14.63}$$

$$b = \dfrac{(27)(\sin 70°)}{\sin 78°} = \boxed{25.94}$$

(b) $C = 180° - 25° - 40° = \boxed{115°}$

$$a = \frac{(63)(\sin 25°)}{\sin 115°} = \boxed{29.38}$$

$$b = \frac{(63)(\sin 40°)}{\sin 115°} = \boxed{44.68}$$

18.

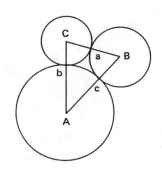

$a = 110 + 140 = 250$

$b = 110 + 220 = 330$

$c = 140 + 220 = 360$

$$\cos A = \frac{b^2 + c^2 - a^2}{2bc} = \frac{(330)^2 + (360)^2 - (250)^2}{(2)(330)(360)}$$

$$= 0.7407$$

$A = \boxed{42.2°}$

$$\cos B = \frac{a^2 + c^2 - b^2}{2ac} = \frac{(250)^2 + (360)^2 - (330)^2}{(2)(250)(360)}$$

$$= 0.4622$$

$B = \boxed{62.5°}$

$$\cos C = \frac{a^2 + b^2 - c^2}{2ab} = \frac{(250)^2 + (330)^2 - (360)^2}{(2)(250)(330)}$$

$$= 0.2533$$

$C = \boxed{75.3°}$

16. $c^2 = a^2 + b^2 - 2ab \cos C$

$\quad = (132)^2 + (224)^2 - (2)(132)(224)(\cos 28.7°)$

$c = \boxed{125.4}$

$$\sin A = \frac{a \sin C}{c} = \frac{(132)(\sin 28.7°)}{125.4} = 0.5055$$

$A = \boxed{30.4°}$

$$\sin B = \frac{b \sin C}{c} = \frac{(224)(\sin 28.7°)}{125.4} = 0.8578$$

$B = \boxed{120.9°}$

17. (a) In parallelogram $ABCD$,

$A + B = C + D = 180°$

$\quad B = 180° - A = 180° - 50.8° = 129.2°$

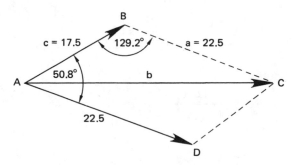

In $\triangle ABC$, the magnitude of the resultant is diagonal b.

$b^2 = c^2 + a^2 - 2ac \cos B$

$\quad = (17.5)^2 + (22.5)^2 - (2)(17.5)(22.5)(\cos 129.2°)$

$b = \boxed{36.2}$

(b) The required angle is BAC. From the law of sines,

$$\sin BAC = \frac{a \sin B}{b} = \frac{(22.5)(\sin 129.2°)}{36.2} = 0.4817$$

$BAC = \boxed{28.8°}$

19. There are two possible solutions to this problem.

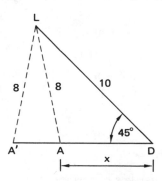

$x = DA$ and $x = DA'$ solve the equation

$$d^2 = a^2 + x^2 - 2ax \cos D$$

$$(8)^2 = (10)^2 + x^2 - (2)(10)(x)(\cos 45°)$$

For $\triangle DLA$ and $\triangle DLA'$,

$$x^2 - 10\sqrt{2}x + 36 = 0$$

Use the quadratic equation with $a = 1$, $b = -10\sqrt{2}$, and $c = 36$,

$$x_1 = DA = \frac{10\sqrt{2} - 2\sqrt{14}}{2} = 3.33$$

$$x_2 = DA' = \frac{10\sqrt{2} + 2\sqrt{14}}{2} = 10.8$$

$$t_1 = \frac{x_1}{v} = \frac{3.33}{12} = 0.2775 \text{ hr} = 16.7 \text{ min}$$

The ship will reach A at 8:17 a.m.

$$t_2 = \frac{x_2}{v} = \frac{10.8}{12} = 0.9 \text{ hr} = 54 \text{ min}$$

The ship will reach A' at 8:54 a.m.

20.

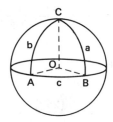

$$c = 60°$$
$$b = a = 90°$$

If planes OAC and OBC are perpendicular to the equator, $C = 60°$ and angles A and B are right angles.

$$A = B = 90°$$
$$\epsilon = A + B + C - 180°$$
$$= 90° + 90° + 60° - 180°$$
$$= 60°$$

21. From the first law of cosines,

$$\cos A = \frac{\cos a - (\cos b)(\cos c)}{(\sin b)(\sin c)}$$
$$= \frac{\cos 56.4° - (\cos 65.9°)(\cos 78.5°)}{(\sin 65.9°)(\sin 78.5°)} = 0.52765$$

$$A = 58.15°$$

$$\cos B = \frac{\cos b - (\cos a)(\cos c)}{(\sin a)(\sin c)}$$
$$= \frac{\cos 65.9° - (\cos 56.4°)(\cos 78.5°)}{(\sin 56.4°)(\sin 78.5°)}$$
$$= 0.36511$$

$$B = 68.59°$$

$$\cos C = \frac{\cos c - (\cos a)(\cos b)}{(\sin a)(\sin b)}$$
$$= \frac{\cos 78.5° - (\cos 56.4°)(\cos 65.9°)}{(\sin 56.4°)(\sin 65.9°)}$$
$$= -0.03498$$

$$C = 92.00°$$

Use the law of sines to check the angles.

$$\frac{\sin A}{\sin a} = \frac{\sin 58.15°}{\sin 56.4°} = 1.020$$

$$\frac{\sin B}{\sin b} = \frac{\sin 68.59°}{\sin 65.9°} = 1.020$$

$$\frac{\sin C}{\sin c} = \frac{\sin 92.00°}{\sin 78.5°} = 1.020$$

$A + B + C = 218.61°$ so that $180° < A + B + C < 540°$.

22.

$$\omega = \frac{A_{\text{earth}}}{r^2} = \frac{\pi (r_{\text{earth}})^2}{r^2}$$

$$= \frac{(\pi)(3800 \text{ mi})^2}{185{,}000 \text{ mi}^2} = 0.00133 \text{ sr}$$

23.

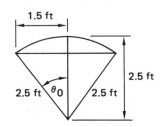

The surface area of a sphere of radius r cut out by an angle, θ_0, rotated from the center about a radius, r, is

$$A = 2\pi r^2 (1 - \cos \theta_0)$$

$$\omega = \frac{A}{r^2} = 2\pi (1 - \cos \theta_0)$$

$$\sin \theta_0 = \frac{1.5 \text{ ft}}{2.5 \text{ ft}} = 0.6$$

$$\cos \theta_0 = \sqrt{1 - (0.6)^2} = 0.8$$

$$\omega = 2\pi (1 - 0.8) = 1.26 \text{ sr}$$

ANALYTIC GEOMETRY

1.

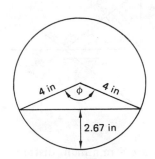

$$\frac{\phi}{2} = \arccos\left(\frac{4 - 2.67}{4}\right)$$

$$= 1.23 \text{ rad}$$

$$\phi = (2)(1.23 \text{ rad}) = 2.46 \text{ rad}$$

$$A = \frac{1}{2}r^2(\phi - \sin\phi) \quad [\phi \text{ in radians}]$$

$$= \left(\frac{1}{2}\right)(4 \text{ in})^2[2.46 - \sin(2.46 \text{ rad})]$$

$$= \boxed{14.64 \text{ in}^2}$$

2. area of base $= \pi r^2 = \pi\left(\frac{3}{2} \text{ ft}\right)^2$

$$= 7.07 \text{ ft}^2$$

area of cone $= \pi r\sqrt{r^2 + h^2}$

$$= \pi\left(\frac{3}{2} \text{ ft}\right)\sqrt{\left(\frac{3}{2} \text{ ft}\right)^2 + (4 \text{ ft})^2}$$

$$= 20.13 \text{ ft}^2$$

total area $= 7.07 \text{ ft}^2 + 20.13 \text{ ft}^2$

$$= \boxed{27.20 \text{ ft}^2}$$

3.

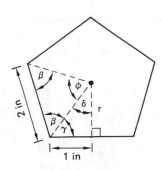

$$\phi = \frac{2\pi}{n} = \frac{360°}{5} = 72°$$

$$\beta = \left(\frac{1}{2}\right)(180° - 72°) = 54°$$

$$\gamma = \beta = 54°$$

$$\delta = 180° - 90° - 54° = 36°$$

$$\frac{1 \text{ in}}{r} = \tan(36°)$$

$$r = 1.376 \text{ in}$$

largest diameter $= 2r = (2)(1.376 \text{ in})$

$$= \boxed{2.752 \text{ in}}$$

4.

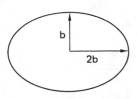

$$18 \text{ in} = p = 2\pi\sqrt{\frac{1}{2}(b^2 + 4b^2)}$$

$$= \pi b\sqrt{10}$$

$$b = \frac{18 \text{ in}}{\pi\sqrt{10}}$$

$$A = \pi b(2b) = 2\pi b^2$$

$$= 2\pi\left(\frac{18 \text{ in}}{\pi\sqrt{10}}\right)^2 = \boxed{20.63 \text{ in}^2}$$

5. (a) $\quad m = \dfrac{y_2 - y_1}{x_2 - x_1}$

$$= \frac{9.5 - 3.4}{8.3 - 1.7} = \frac{6.1}{6.6} = 0.924$$

Substitute any point into the slope form to find b.

$$y = mx + b$$

$$3.4 = (0.924)(1.7) + b$$

$$b = 1.830$$

$$\boxed{y = 0.924x + 1.830}$$

(b) $\quad m = \dfrac{y_2 - y_1}{x_2 - x_1} = \dfrac{-7.2 - (-3.8)}{5.1 - (-6)} = -0.306$

Substitute any point into the slope form to find b.

$$-3.8 = (-0.306)(-6) + b$$

$$b = -5.636$$

$$\boxed{y = -0.306x - 5.636}$$

6. For $y = 0.924x + 1.830$,

(a)
$$\boxed{0.924x - y + 1.830 = 0}$$

(b)
$$y - y_1 = m(x - x_1)$$
$$\boxed{y - 3.4 = 0.924(x - 1.7)}$$

(c)
$$\frac{y}{1.830} = \frac{0.924x}{1.830} + 1$$
$$\boxed{-\frac{x}{1.98} + \frac{y}{1.83} = 1}$$

For $y = -0.306x - 5.636$,

(a)
$$\boxed{0.306x + y + 5.636 = 0}$$

(b)
$$\boxed{y + 3.8 = -0.306(x + 6)}$$

(c)
$$-\frac{0.306}{5.636}x - \frac{1}{5.636}y = 1$$
$$\boxed{-\frac{x}{18.42} - \frac{y}{5.636} = 1}$$

7. $\phi = \arctan(m) = \arctan(0.4) = \boxed{21.8°}$

8. $d = \sqrt{L^2 + M^2 + N^2}$
$$= \sqrt{(x_2 - x_1)^2 + (y_2 - y_1)^2 + (z_2 - z_1)^2}$$
$$= \sqrt{(-4 - 1)^2 + [5 - (-3)]^2 + (-7 - 9)^2}$$
$$= \boxed{18.57}$$

9. The line passing through $(3,-2)$ and $(-6,5)$ has the slope
$$m = \frac{y_2 - y_1}{x_2 - x_1} = \frac{5 - (-2)}{-6 - 3} = \frac{7}{-9}$$
The perpendicular line has a slope of
$$m' = -\frac{1}{m} = \frac{9}{7}$$
Substitute $(3,1)$ into the form
$$y = m'x + b$$
$$1 = \left(\frac{9}{7}\right)(3) + b$$
$$b = -\frac{20}{7}$$
$$\boxed{y = \frac{9}{7}x - \frac{20}{7}}$$

10.
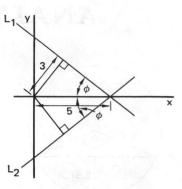

Both L_1 and L_2 are a minimum distance of 3 from the origin,
$$\phi = \pm \arcsin\left(\frac{3}{5}\right) = \pm36.87°$$
$$\tan\phi = \pm0.75$$
$$m = \pm0.75$$

Substitute $(5,0)$ and $m = \pm0.75$ into the form
$$y = mx + b$$
$$0 = (\pm0.75)(5) + b$$
$$b = \mp3.75$$
$$\boxed{y = \pm0.75x \mp 3.75}$$

11.
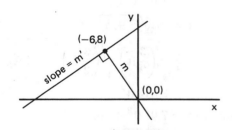

$$m = \frac{y_2 - y_1}{x_2 - x_1} = \frac{8}{-6} = -\frac{4}{3}$$
The desired line must have the slope
$$m' = -\frac{1}{m} = \frac{3}{4}$$
Substitute $(-6,8)$ and $m' = \frac{3}{4}$ into the form
$$y = m'x + b$$
$$8 = \left(\frac{3}{4}\right)(-6) + b$$
$$b = \frac{50}{4}$$
$$\boxed{y = \frac{3}{4}x + \frac{50}{4}}$$

12. (a) The direction numbers are

$$L = 9 \qquad M = 7 \qquad d = \sqrt{(9)^2 + (7)^2} = 11.40$$

The direction cosines are

$$\cos\alpha = \frac{L}{d} = \frac{9}{11.4} = 0.789$$

$$\cos\beta = \frac{M}{d} = \frac{7}{11.4} = 0.614$$

The direction angles are

$$\alpha = \arccos\left(\frac{L}{d}\right) = 37.86°$$

$$\beta = \arccos\left(\frac{M}{d}\right) = 52.12°$$

(b) The direction numbers are

$$L = 3 \qquad M = 2 \qquad N = -6$$

$$d = \sqrt{(3)^2 + (2)^2 + (-6)^2} = 7$$

The direction cosines are

$$\cos\alpha = \frac{L}{d} = \frac{3}{7} = 0.429$$

$$\cos\beta = \frac{M}{d} = \frac{2}{7} = 0.286$$

$$\cos\gamma = \frac{N}{d} = \frac{-6}{7} = -0.857$$

The direction angles are

$$\alpha = \arccos\left(\frac{L}{d}\right) = 64.62°$$

$$\beta = \arccos\left(\frac{M}{d}\right) = 73.4°$$

$$\gamma = \arccos\left(\frac{N}{d}\right) = 149.00°$$

(c) The direction numbers are

$$L = -2 \qquad M = 1 \qquad N = -3$$

$$d = \sqrt{(-2)^2 + (1)^2 + (-3)^2} = 3.74$$

The direction cosines are

$$\cos\alpha = \frac{L}{d} = \frac{-2}{3.74} = -0.535$$

$$\cos\beta = \frac{M}{d} = \frac{1}{3.74} = 0.267$$

$$\cos\gamma = \frac{N}{d} = \frac{-3}{3.74} - 0.802$$

The direction angles are

$$\alpha = \arccos\left(\frac{L}{d}\right) = 122.33°$$

$$\beta = \arccos\left(\frac{M}{d}\right) = 74.49°$$

$$\gamma = \arccos\left(\frac{N}{d}\right) = 143.33°$$

13. Assume the plane is of the form

$$A(x - x_0) + B(y - y_0) + C(z - z_0) = 0$$

Since it is normal to vector $(1,2,-3)$,

$$A = 1 \qquad B = 2 \qquad C = -3$$

Substitute point $(3,1,2)$ into the form

$$(x - 3) + 2(y - 1) - 3(z - 2) = 0$$

$$\boxed{x + 2y - 3z + 1 = 0}$$

14. Let

$$(x_1, y_1, z_1) = (0, 0, 2)$$
$$(x_2, y_2, z_2) = (2, 4, 1)$$
$$(x_3, y_3, z_3) = (-2, 3, 3)$$

$$\mathbf{V}_1 = (x_2 - x_1)\mathbf{i} + (y_2 - y_1)\mathbf{j} + (z_2 - z_1)\mathbf{k}$$
$$= 2\mathbf{i} + 4\mathbf{j} - \mathbf{k}$$

$$\mathbf{V}_2 = (x_3 - x_1)\mathbf{i} + (y_3 - y_1)\mathbf{j} + (z_3 - z_1)\mathbf{k}$$
$$= -2\mathbf{i} + 3\mathbf{j} + \mathbf{k}$$

$$\mathbf{N} = \begin{vmatrix} \mathbf{i} & \mathbf{j} & \mathbf{k} \\ 2 & 4 & -1 \\ -2 & 3 & 1 \end{vmatrix}$$

Expand by the top row.

$$\mathbf{N} = \mathbf{i}[(4)(1) - (3)(-1)] - \mathbf{j}[(2)(1) - (-2)(-1)]$$
$$+ \mathbf{k}[(2)(3) - (-2)(4)]$$
$$= 7\mathbf{i} + 14\mathbf{k}$$

Substitute any point. Use $(0,0,2)$.

$$7(x - x_o) + 0(y - y_o) + 14(z - z_o) = 0$$
$$7(x - 0) + 14(z - 2) = 0$$
$$7x + 14z - 28 = 0$$

$$\boxed{x + 2z - 4 = 0}$$

15. $\cos\phi = \dfrac{|\mathbf{N}_1 \cdot \mathbf{N}_2|}{|\mathbf{N}_1||\mathbf{N}_2|}$

$= \dfrac{|(2)(6) + (-1)(-2) + (-2)(3)|}{\sqrt{(2)^2 + (-1)^2 + (-2)^2}\sqrt{(6)^2 + (-2)^2 + (3)^2}}$

$= \boxed{0.381}$

16. $d = \dfrac{|Ax + By + Cz + D|}{\sqrt{A^2 + B^2 + C^2}}$

$= \dfrac{|(2)(2) + (2)(2) + (-1)(-4) - 6|}{\sqrt{(2)^2 + (2)^2 + (-1)^2}} = \boxed{2}$

17. (a) Since the vertex is at the origin, $p = 4 - 0 = 4$.
Substitute $p = 4$ into the form

$$y^2 = 4px = (4)(4)x$$

$$\boxed{y^2 = 16x}$$

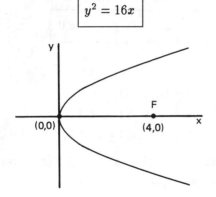

(0,0)   F   (4,0)

(b) $x = h - p$

$5 = 2 - p$

$p = -3$

Substitute $h = 2$, $k = 3$, $p = -3$ into the form

$$(y - k)^2 = 4p(x - h)$$

$$(y - 3)^2 = (4)(-3)(x - 2)$$

$$(y - 3)^2 = -12(x - 2)$$

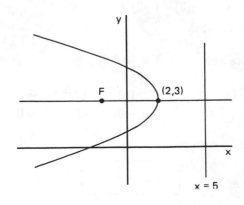

F   (2,3)

x = 5

(c) $(h, p + k) = (-1, 2)$

$h = -1$

$p + k = 2 \qquad\qquad [1]$

$y = k - p$

$-2 = k - p \qquad\qquad [2]$

Solving Eqs. [1] and [2] simultaneously gives

$$k = 0 \qquad p = 2$$

Substitute $h = -1$, $k = 0$, $p = 2$ into the form

$$(x - h)^2 = 4p(y - k)$$

$$\boxed{(x + 1)^2 = 8y}$$

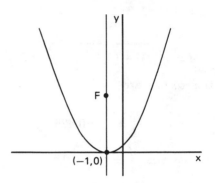

F

(−1,0)

18. $9x^2 + 16y^2 + 54x - 64y = -1$

$9x^2 + 54x + 81 + 16y^2 - 64y + 64 = -1 + 81 + 64$

$(3x + 9)^2 + (4y - 8)^2 = 144$

$\dfrac{9(x + 3)^2}{144} + \dfrac{16(y - 2)^2}{144} = 1$

$\dfrac{(x + 3)^2}{16} + \dfrac{(y - 2)^2}{9} = 1$

$$\boxed{\text{ellipse}}$$

19. (a) $a = \dfrac{8}{2} = 4$

$c = 3$

$b^2 = a^2 - c^2 = (4)^2 - (3)^2 = 16 - 9 = 7$

$b = \sqrt{7}$

Since the ellipse is centered at (0,0), the equation is

$$\boxed{\dfrac{x^2}{16} + \dfrac{y^2}{7} = 1}$$

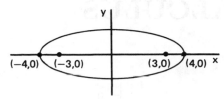

(b)
$$b = 4$$
$$\sqrt{\frac{33}{49}} = \epsilon = \frac{\sqrt{a^2 - b^2}}{a} = \frac{\sqrt{a^2 - 16}}{a}$$
$$a = 7$$

Since the ellipse is centered at the origin, the equation is

$$\boxed{\frac{x^2}{49} + \frac{y^2}{16} = 1}$$

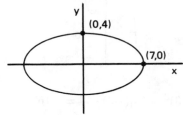

(c) Since the ellipse is centered at the origin and $b = 3$, substitute point $(\sqrt{2}, 1.5\sqrt{2})$ into the form

$$\frac{x^2}{a^2} + \frac{y^2}{9} = 1$$
$$\frac{(\sqrt{2})^2}{a^2} + \frac{(1.5\sqrt{2})^2}{9} = 1$$
$$a = 2$$

$$\boxed{\frac{x^2}{4} + \frac{y^2}{9} = 1}$$

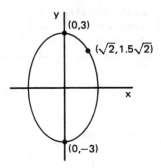

20. Substitute $(h, k, l) = (3, 2, 5)$ and the point $(5,6,4)$ into the form

$$(x - h)^2 + (y - k)^2 + (z - l)^2 = r^2$$
$$(5 - 3)^2 + (6 - 2)^2 + (4 - 5)^2 = r^2$$
$$r = \sqrt{21}$$
$$\boxed{(x - 3)^2 + (y - 2)^2 + (z - 5)^2 = 21}$$

21. (a)
$$A = 4\pi r^2 = 4\pi (4 \text{ m})^2$$
$$= \boxed{201.06 \text{ m}^2}$$

(b)
$$V = \frac{4\pi r^3}{3} = \frac{4\pi (4 \text{ m})^3}{3}$$
$$= \boxed{268.08 \text{ m}^3}$$

22. In the x-y plane, $z = 0$, and the x-intercept is where $y = 0$.

$$(x - 2)^2 + (0 - 2)^2 + (0)^2 = 9$$
$$(x - 2)^2 = 5$$
$$x = 2 \pm \sqrt{5}$$

The sphere intercepts the x-axis at

$$\boxed{(2 \pm \sqrt{5}, 0, 0)}$$

23. (a) old volume $= \dfrac{4\pi \left(\dfrac{d_1}{2}\right)^3}{3} = 4\pi \left(\dfrac{3 \text{ cm}}{2}\right)^3$
$$= 14.14 \text{ cm}^3$$

new volume $= \dfrac{4\pi \left(\dfrac{d_2}{2}\right)^3}{3} = 4\pi \left(\dfrac{4 \text{ cm}}{2}\right)^3$
$$= 33.51 \text{ cm}^3$$

percent increased $= \dfrac{\text{new volume} - \text{old volume}}{\text{old volume}} \times 100\%$

$$= \frac{33.51 - 14.14}{14.14} \times 100\%$$

$$= \boxed{137\%}$$

(b) old surface $= 4\pi \left(\dfrac{d_1}{2}\right)^2 = 4\pi \left(\dfrac{3 \text{ cm}}{2}\right)^2$
$$= 28.27 \text{ cm}^2$$

new surface $= 4\pi \left(\dfrac{d_2}{2}\right)^2 = 4\pi \left(\dfrac{4 \text{ cm}}{2}\right)^2$
$$= 50.26 \text{ cm}^2$$

percent increased $= \dfrac{50.26 - 28.27}{28.27}$

$$= \boxed{78\%}$$

DIFFERENTIAL CALCULUS

1. (a) $\mathbf{D}f(x) = \boxed{4x + 8}$

(b) $\mathbf{D}f(x) = \dfrac{(x^2 - 1)\mathbf{D}(x + 5) - \mathbf{D}(x^2 - 1)(x + 5)}{(x^2 - 1)^2}$

$= \dfrac{(x^2 - 1)(1) - 2x(x + 5)}{(x^2 - 1)^2}$

$= \boxed{\dfrac{-x^2 - 10x - 1}{(x^2 - 1)^2}}$

(c) $\mathbf{D}f(x) = (-1)(2)(x^2 - 1)^{-2}\mathbf{D}(x^2 - 1)$

$= -2(x^2 - 1)^{-2}(2x)$

$= \boxed{\dfrac{-4x}{(x^2 - 1)^2}}$

(d) $\mathbf{D}f(x) = \dfrac{1}{2}(2 - 3x^2)^{-\frac{1}{2}}\mathbf{D}(2 - 3x^2)$

$= \dfrac{1}{2}(2 - 3x^2)^{-\frac{1}{2}}(-6x)$

$= \boxed{\dfrac{-3x}{\sqrt{2 - 3x^2}}}$

(e) $\mathbf{D}f(x) = 2\sin(x^2 + 3x)\mathbf{D}[\sin(x^2 + 3x)]$

$= 2\sin(x^2 + 3x)\cos(x^2 + 3x)\mathbf{D}(x^2 + 3x)$

$= 2\sin(x^2 + 3x)\cos(x^2 + 3x)(2x + 3)$

$= \boxed{(4x + 6)\sin(x^2 + 3x)\cos(x^2 + 3x)}$

2. (a) $\dfrac{\partial f}{\partial x} = \boxed{8x - 3y}$

(b) $\dfrac{\partial f}{\partial x} = \boxed{y^2}$

(c) $\dfrac{\partial f}{\partial x} = \boxed{2x}$

3. (a)

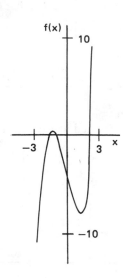

$f(-3) = (-3)^3 - (5)(-3) - 4 = -16$

$f(-1) = (-1)^3 - (5)(-1) - 4 = 0$

$f'(x) = 3x^2 - 5$

$f'(x) = 0$ at $x = \pm\sqrt{\dfrac{5}{3}} = \pm 1.291$

$f(-1.291) = (-1.291)^3 - (5)(-1.291) - 4 = 0.303$

$\boxed{\begin{array}{l}\text{On the interval } [-3, -1], \\ \text{maximum value is } 0.303. \\ \text{Minimum value is } -16.\end{array}}$

(b)

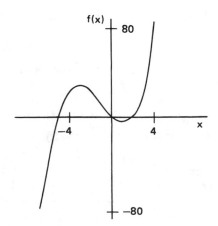

$f(-4) = (-4)^3 + (3)(-4)^2 - (9)(-4) = 20$

$f(4) = (4)^3 + (3)(4)^2 - (9)(4) = 76$

$f'(x) = 3x^2 + 6x - 9 = 3(x + 3)(x - 1)$

$f'(x) = 0$ at $x = -3$ or $x = 1$

$f(-3) = (-3)^3 + (3)(-3) - (9)(-3) = 27$

$f(1) = (1)^3 + (3)(1) - (9)(1) = -5$

> On the interval $[-4, 4]$,
> maximum value is 76.
> Minimum value is -5.

(d)

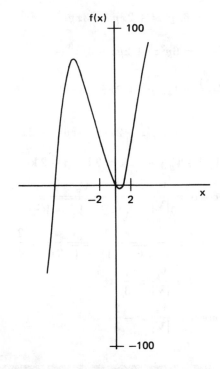

$f(-2) = (-2)^3 + (7)(-2)^2 - (5)(-2) = 30$

$f(2) = (2)^3 + (7)(2)^2 - (5)(2) = 26$

$f'(x) = 3x^2 + 14x - 5 = (3x - 1)(x + 5)$

$f'(x) = 0$ if $x = \frac{1}{3}$ or -5

$f(\frac{1}{3}) = \left(\frac{1}{3}\right)^3 + (7)\left(\frac{1}{3}\right)^2 - (5)\left(\frac{1}{3}\right) = -0.85$

> On the interval $[-2, 2]$,
> maximum value is 30.
> Minimum value is -0.85.

(c)

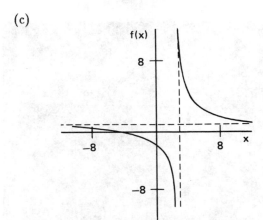

$f(-5) = \dfrac{-5 + 5}{-5 - 3} = 0$

$f(2) = \dfrac{2 + 5}{2 - 3} = -7$

$f'(x) = \dfrac{(x - 3) - (x + 5)}{(x - 3)^2}$

$\quad = \dfrac{-8}{(x - 3)^2} \neq 0$ for any x in $[-5, 2]$

> On the interval $[-5, 2]$,
> maximum value is 0.
> Minimum value is -7.

4. $\left.\dfrac{\partial f(x, y, z)}{\partial x}\right|_{(8,2,2)} = 6x\bigg|_{(8,2,2)} = 48$

$\left.\dfrac{\partial f(x, y, z)}{\partial y}\right|_{(8,2,2)} = 10y\bigg|_{(8,2,2)} = 20$

$\left.\dfrac{\partial f(x, y, z)}{\partial z}\right|_{(8,2,2)} = -4z\bigg|_{(8,2,2)} = -8$

(a) The normal vector at $f(8, 2, 2)$ is

$$\boxed{\mathbf{N} = 48\,\mathbf{i} + 20\,\mathbf{j} - 8\,\mathbf{k}}$$

(b) $(48)(x - 8) + (20)(y - 2) - (8)(z - 2) = 0$

$$\boxed{12x + 5y - 2z - 102 = 0}$$

5. $f(x, y, z) = 6xy^2z^2 + 2xyz - 3xyz^3 - xz^2$

$$\frac{\partial f(x, y, z)}{\partial x} = 6y^2z^2 + 2yz - 3yz^3 - z^2$$

$$\frac{\partial f(x, y, z)}{\partial y} = 12xyz^2 + 2xz - 3xz^3$$

$$\frac{\partial f(x, y, z)}{\partial z} = 12xy^2z + 2xy - 9xyz^2 - 2xz$$

$$V_x\,\mathbf{i} + V_y\,\mathbf{j} + V_z\,\mathbf{k} = 2\,\mathbf{i} + \mathbf{j} - 2\,\mathbf{k}$$

$$\cos\alpha = \frac{V_x}{|\mathbf{V}|} = \frac{V_x}{\sqrt{V_x^2 + V_y^2 + V_z^2}}$$

$$= \frac{2}{\sqrt{(2)^2 + (1)^2 + (-2)^2}} = \frac{2}{3}$$

$$\cos\beta = \frac{V_y}{|\mathbf{V}|} = \frac{1}{3}$$

$$\cos\gamma = \frac{V_z}{|\mathbf{V}|} = -\frac{2}{3}$$

The slope at $(1,1,1)$ is

$$\nabla_u f(1,1,1) = \frac{\partial f(1,1,1)}{\partial x}\cos\alpha + \frac{\partial f(1,1,1)}{\partial y}\cos\beta$$

$$+ \frac{\partial f(1,1,1)}{\partial z}\cos\gamma$$

$$= (6 - 3 + 2 - 1)\left(\frac{2}{3}\right) + (12 - 3 + 2)\left(\frac{1}{3}\right)$$

$$+ (12 - 9 + 2 - 2)\left(-\frac{2}{3}\right)$$

$$= (4)\left(\frac{2}{3}\right) + (11)\left(\frac{1}{3}\right) + (3)\left(-\frac{2}{3}\right)$$

$$= \boxed{\frac{13}{3}}$$

INTEGRAL CALCULUS

1. (a) $\int \sqrt{1-x}\,dx = \int (1-x)^{\frac{1}{2}}\,dx$

$$= \boxed{\left(-\frac{2}{3}\right)(1-x)^{\frac{3}{2}} + C}$$

(b) $\int \dfrac{x}{x^2+1}\,dx = \dfrac{1}{2}\int \dfrac{2x}{x^2+1}\,dx$

$$= \boxed{\frac{1}{2}\ln(x^2+1) + C}$$

(c) $\dfrac{x^2}{x^2+x-6} = 1 - \dfrac{x-6}{x^2+x-6}$

$$= 1 - \frac{x-6}{(x+3)(x-2)}$$

$$= 1 - \frac{\frac{9}{5}}{x+3} + \frac{\frac{4}{5}}{x-2}$$

$\int \dfrac{x^2}{x^2+x-6}\,dx = \int \left(1 - \dfrac{\frac{9}{5}}{x+3} + \dfrac{\frac{4}{5}}{x-2}\right)dx$

$$= \int dx - \int \frac{\frac{9}{5}}{x+3}\,dx + \int \frac{\frac{4}{5}}{x-2}\,dx$$

$$= \boxed{x - \frac{9}{5}\ln(x+3) + \frac{4}{5}\ln(x-2) + C}$$

2. (a) $\displaystyle\int_1^3 (x^2+4x)\,dx = \left[\dfrac{x^3}{3} + 2x^2\right]_1^3 = \boxed{24\tfrac{2}{3}}$

(b) $\displaystyle\int_{-2}^2 (x^3+1)\,dx = \left[\dfrac{x^4}{4} + x\right]_{-2}^2 = \boxed{4}$

(c) $\displaystyle\int_1^2 (4x^3 - 3x^2)\,dx = \left[x^4 - x^3\right]_1^2 = \boxed{8}$

3.

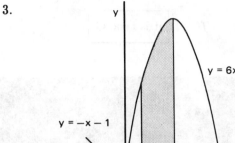

$\text{area} = \displaystyle\int_1^3 \left((6x - x^2) - (-x - 1)\right)dx$

$$= \int_1^3 (-x^2 + 7x + 1)\,dx$$

$$= \left[-\frac{x^3}{3} + \frac{7}{2}x^2 + x\right]_1^3 = \boxed{21\tfrac{1}{3}}$$

4. (a) $a_0 = \dfrac{1}{2\pi}\displaystyle\int_0^{2\pi} f(t)\,dt$

$$= \frac{1}{\pi}\int_0^{\pi} f(t)\,dt$$

$$= \frac{1}{\pi}\left[(r)\left(\frac{\pi}{2}\right) + (-3r)\left(\frac{\pi}{2}\right)\right] = \boxed{-r}$$

(b) $a_0 = \dfrac{1}{2\pi}\displaystyle\int_0^{2\pi} f(t)\,dt$

$$= \frac{1}{\pi}\int_0^{\pi} f(t)\,dt$$

$$= \left(\frac{1}{\pi}\right)\left(\frac{1}{2}\pi h\right)$$

$$= \boxed{h/2}$$

5. (a) Since $f(t) = -f(-t)$, $\boxed{\text{it is type B}}$

   (b) Since $f(t) = f(-t)$, $\boxed{\text{it is type C}}$

DIFFERENTIAL EQUATIONS

MATHEMATICS
Diff Eq

1. The equation is third-order, nonlinear, and non-homogeneous.

2. This is a second-order, linear, homogeneous equation with nonconstant coefficients.

3. $3x^3 + 7 = 0$

4. The Cauchy equation is an nth-order linear equation with nonconstant coefficients. It is nonhomogeneous unless the function, f, is zero.

5. (a)
$$y' = x^2 - 2x - 4$$
$$y = \int (x^2 - 2x - 4)dx$$
$$= \frac{x^3}{3} - x^2 - 4x + C$$

Since $y(3) = -6$, $C = 6$

$$y = \frac{x^3}{3} - x^2 - 4x + 6$$

(b)
$$T' = 2\sin 4t$$
$$T = \int 2\sin 4t\, dt$$
$$= \frac{-\cos 4t}{2} + C$$

Since $T(0) = \frac{1}{2}$, $C = 1$

$$T = \frac{-\cos 4t}{2} + 1$$

6. The differential equation can be rewritten as
$$y' + \frac{1}{\tau}y = 0$$

The solution is
$$y(t) = y_0 e^{\frac{-t}{\tau}}$$
At $t = \tau$,
$$y(\tau) = y_0 e^{-1} = \boxed{\frac{y_0}{e}}$$

7. For $x \neq 0$,
$$y' + \frac{2}{x}y = \frac{\sin x}{x}$$

Let $p(x) = \frac{2}{x}$ and $g(x) = \frac{\sin x}{x}$.

The integrating factor is

$$u(x) = \exp\left[\int p(x)dx\right]$$
$$= \exp\left[\int \frac{2}{x}dx\right] = e^{2\ln x} = x^2$$

The solution is
$$y(x) = \frac{1}{u(x)}\left[\int u(x)g(x)dx + C\right]$$
$$= \frac{1}{x^2}\left[\int (x^2)\left(\frac{\sin x}{x}\right)dx + C\right]$$
$$= \frac{1}{x^2}\int x\sin x\, dx$$

Using integration by parts,
$$\int x\sin x\, dx = (-x\cos x) - \int -\cos x\, dx$$
$$= -x\cos x + \sin x + C$$
$$y(x) = -\frac{\cos x}{x} + \frac{\sin x}{x^2} + \frac{C}{x^2}$$

Since $y(\pi) = \frac{1}{\pi}$,
$$\frac{1}{\pi} = \frac{1}{\pi} + \frac{C}{\pi^2}$$
$$C = 0$$

$$y = -\frac{\cos x}{x} + \frac{\sin x}{x^2}$$

Note: The solution is undefined for $x = 0$.

8. Let $p(x) = \cot x$ and $g(x) = 2\csc(x) = \frac{2}{\sin x}$.

The integrating factor is

$$u(x) = \exp\left[\int \cot x\, dx\right]$$
$$= \exp[\ln\sin x] = \sin x$$

$$y(x) = \frac{1}{u(x)}\left[\int u(x)g(x)dx + C\right]$$

$$= \frac{1}{\sin x}\left[\int (\sin x)\left(\frac{2}{\sin x}\right)dx + C\right]$$

$$= \frac{1}{\sin x}(2x + C)$$

$$y\left(\frac{\pi}{2}\right) = 1 = (2)\left(\frac{\pi}{2}\right) + C$$

$$C = 1 - \pi$$

$$\boxed{y(x) = \frac{2x + 1 - \pi}{\sin x}}$$

9.
$$y' = \frac{dy}{dx} = \frac{2x}{y + x^2 y} = \frac{2x}{y(1 + x^2)}$$

$$y\,dy = \frac{2x}{1 + x^2}dx$$

Integrating both sides of the equation,

$$\frac{y^2}{2} = \ln(1 + x^2) + C$$

C is determined by the initial value, $y(0) = -2$.

$$\frac{(-2)^2}{2} = 0 + C$$

$$C = 2$$

$$y^2 = 2\ln(1 + x^2) + 4$$

$$\boxed{y = \sqrt{2\ln(1 + x^2) + 4}}$$

10. By inspection, this is a first-order exact differential equation.

$$f_x(x, y) = \cos x \ln y$$

$$f_y(x, y) = \frac{\sin x}{y}$$

$$f(x, y) = \int f_x(x, y)dx$$

$$= \ln y \sin x$$

Using the general solution $F(x, y) - C = 0$,

$$C = \ln y \sin x$$

$$y\left(\frac{\pi}{2}\right) = e$$

$$C = (1)(1) = 1$$

$$\ln y \sin x = 1$$

$$\boxed{y = e^{1/\sin x}}$$

Note: The solution is not defined for $x = 2k\pi$, where k is an integer.

11. (a) The auxiliary equation is

$$r^2 + 4r + 4 = 0$$

$$(r + 2)(r + 2) = 0$$

There is a double root at $r = -2$.

$$y = A_1 e^{-2t} + A_2 t e^{-2t}$$

Since $y(0) = 1$, $A_1 = 1$.

$$y' = -2e^{-2t} + A_2 e^{-2t} - 2A_2 t e^{-2t}$$

Since $y'(0) = 1$, $A_2 = 3$.

$$\boxed{y = e^{-2t}(1 + 3t)}$$

(b) The auxiliary equation is

$$r^2 + 3r + 2 = 0$$

$$(r + 1)(r + 2) = 0$$

$$r = -1, -2$$

$$y = A_1 e^{-t} + A_2 e^{-2t}$$

Since $y(0) = 1$,

$$1 = A_1 + A_2 \qquad \text{[Eq. 1]}$$

$$y' = -A_1 e^{-t} - 2A_2 e^{-2t}$$

Since $y'(0) = 0$,

$$0 = -A_1 - 2A_2 \qquad \text{[Eq. 2]}$$

Solving Eqs. 1 and 2 simultaneously,

$$A_1 = 2$$

$$A_2 = -1$$

$$\boxed{y = 2e^{-t} - e^{-2t}}$$

(c) The auxiliary equation is

$$r^2 + 2r + 2 = 0$$

$$r = -1 \pm i$$

$$y_c = e^{-t}(A_1 \cos t + A_2 \sin t)$$

Let $y_p = Ce^{-t}$.

$$y_p'' + 2y_p' + 2y_p = Ce^{-t} = e^{-t}$$
$$C = 1$$
$$y = y_c + y_p$$
$$= e^{-t} + e^{-t}(A_1 \cos t + A_2 \sin t)$$

Since $y(0) = 0$, $A_1 = -1$.

$$y' = -e^{-t} - e^{-t}(A_1 \cos t + A_2 \sin t)$$
$$+ e^{-t}(-A_1 \sin t + A_2 \cos t)$$

Since $y'(0) = 1$, $A_2 = 1$.

$$\boxed{y = e^{-t}(1 + \sin t - \cos t)}$$

(d) The auxiliary equation is

$$4r^2 + 4r - 3 = 0$$
$$(2r + 3)(2r - 1) = 0$$
$$r = -\frac{3}{2}, \frac{1}{2}$$

$$x = A_1 e^{-\frac{3}{2}t} + A_2 e^{\frac{1}{2}t}$$

Since $x(0) = 3$,

$$3 = A_1 + A_2 \qquad \text{[Eq. 3]}$$
$$x' = -\frac{3}{2}A_1 e^{-\frac{3}{2}t} + \frac{1}{2}A_2 e^{\frac{1}{2}t}$$

Since $x'(0) = 1$,

$$1 = -\frac{3}{2}A_1 + \frac{1}{2}A_2 \qquad \text{[Eq. 4]}$$

By solving Eqs. 3 and 4 simultaneously,

$$A_1 = \frac{1}{4}$$
$$A_2 = \frac{11}{4}$$

$$\boxed{x = \frac{1}{4}e^{-\frac{3}{2}t} + \frac{11}{4}e^{\frac{1}{2}t}}$$

(e) The auxiliary equation is

$$4r^2 - 7r - 2 = 0$$
$$(4r + 1)(r - 2) = 0$$
$$r = -\frac{1}{4}, 2$$

$$x_c = A_1 e^{-\frac{1}{4}t} + A_2 e^{2t}$$

Let $x_p = B_1 te^{-t} + B_2 e^{-t}$.

$$x_p' = B_1 e^{-t} - B_1 te^{-t} - B_2 e^{-t}$$
$$x_p'' = -2B_1 e^{-t} + B_1 te^{-t} + B_2 e^{-t}$$
$$4x_p'' - 7x_p' - 2x_p = -8B_1 e^{-t} + 4B_1 te^{-t} + 4B_2 e^{-t}$$
$$- 7B_1 e^{-t} + 7B_1 te^{-t} + 7B_2 e^{-t}$$
$$- 2B_1 te^{-t} - 2B_2 e^{-t}$$
$$= -15B_1 e^{-t} + 9B_1 te^{-t} + 9B_2 e^{-t}$$
$$= te^{-t}$$

$$-15B_1 + 9B_2 = 0$$
$$9B_1 = 1$$
$$B_1 = \frac{1}{9}$$
$$B_2 = \frac{15}{81}$$
$$x_p = \frac{1}{9}te^{-t} + \frac{15}{81}e^{-t}$$
$$x = x_c + x_p$$
$$= A_1 e^{-\frac{1}{4}t} + A_2 e^{2t} + \frac{1}{9}te^{-t} + \frac{15}{81}e^{-t}$$

Since $x(0) = 0$,

$$0 = A_1 + A_2 + \frac{15}{81} \qquad \text{[Eq. 5]}$$
$$x' = -\frac{1}{4}A_1 e^{-\frac{1}{4}t} + 2A_2 e^{2t}$$
$$+ \frac{1}{9}e^{-t} - \frac{1}{9}te^{-t} - \frac{15}{81}e^{-t}$$

Since $x'(0) = 1$,

$$1 = -\frac{1}{4}A_1 + 2A_2 + \frac{1}{9} - \frac{15}{81} \qquad \text{[Eq. 6]}$$

Solving Eqs. 5 and 6 simultaneously,

$$A_1 = -\frac{52}{81}$$
$$A_2 = \frac{37}{81}$$

$$x = -\frac{52}{81}e^{-\frac{1}{4}t} + \frac{37}{81}e^{2t} + \frac{1}{9}te^{-t} + \frac{15}{81}e^{-t}$$

$$\boxed{= \left(\frac{1}{81}\right)\left(-52e^{-\frac{1}{4}t} + 37e^{2t} + 9te^{-t} + 15e^{-t}\right)}$$

(f) The auxiliary equation is

$$r^2 - r - 2 = 0$$
$$(r + 1)(r - 2) = 0$$
$$r = -1, 2$$
$$x_c = A_1 e^{2t} + A_2 e^{-t}$$

A particular solution is

$$x_p = -5$$

The general solution is

$$x = x_p + x_c = C_1 e^{2t} + C_2 e^{-t} - 5$$
$$x(0) = C_1 + C_2 - 5 = -5$$
$$C_1 + C_2 = 0 \qquad \text{[Eq. 7]}$$
$$x'(0) = 2C_1 - C_2 = 3$$
$$2C_1 - C_2 = 3 \qquad \text{[Eq. 8]}$$

Solving Eqs. 7 and 8 simultaneously,

$$C_1 = 1$$
$$C_2 = -1$$

$$\boxed{x = e^{2t} - e^{-t} - 5}$$

12. The complementary solution satisfies

$$y'' - 2y' + y = 0$$

The auxiliary equation is

$$r^2 - 2r + 1 = 0$$
$$r_1, r_2 = 1$$

The complementary solution is

$$y_c = A_1 e^x + A_2 x e^x$$

The forcing function is $xe^x + 4$. Look for a particular solution of the form

$$y_p = (B_0 x + B_1)e^x + B_2$$

The solution to the homogeneous equation is xe^x.

$$y_p = x^2(B_0 x + B_1)e^x + B_2 \quad [s = 1 \text{ will not work}]$$
$$y_p' = (3B_0 x^2 + 2B_1 x)e^x + x^2(B_0 x + B_1)e^x$$
$$y_p'' = (6B_0 x + 2B_1)e^x + (3B_0 x^2 + 2B_1 x)e^x$$
$$\quad + (3B_0 x^2 + 2B_1 x)e^x + x^2(B_0 x + B_1)e^x$$
$$y_p'' - 2y_p' + y_p = (6B_0 x + 2B_1)e^x + B_2$$
$$= xe^x + 4$$

Therefore, this requires

$$6B_0 = 1 \text{ or } B_0 = \frac{1}{6}$$
$$2B_1 = 0 \text{ or } B_0 = 0$$
$$B_2 = 4$$
$$y_p = \frac{x^3}{6}e^x + 4$$

The general solution is

$$y = y_c + y_p$$
$$= A_1 e^x + A_2 x e^x + \frac{x^3}{6}e^x + 4$$

The initial conditions require

$$1 = A_1 + 4 \text{ for } y(0) = 1$$
$$1 = A_1 + A_2 \text{ for } y'(0) = 1$$

Therefore,

$$A_1 = -3, \ A_2 = 4$$

$$\boxed{y = \left(\frac{x^3}{6} + 4x - 3\right)e^x + 4}$$

13. The complementary solution satisfies

$$y'' + 2y' + y = 0$$

The auxiliary equation is

$$r^2 + 2r + 1 = 0$$

The auxiliary equation has a double root at $r = -1$.

$$y_c = A_1 e^{-x} + A_2 x e^{-x}$$

From the forcing function form, the particular solution is of the form

$$y_p = B_1 e^x \cos x + B_2 e^x \sin x$$
$$y_p' = B_1 e^x \cos x - B_1 e^x \sin x + B_2 e^x \sin x$$
$$\quad + B_2 e^x \cos x$$
$$y_p'' = B_1 e^x \cos x - B_1 e^x \sin x - B_1 e^x \sin x - B_1 e^x \cos x$$
$$\quad + B_2 e^x \sin x + B_2 e^x \cos x + B_2 e^x \cos x - B_2 e^x \sin x$$
$$= -2B_1 e^x \sin x + 2B_2 e^x \cos x$$

$$y_p'' + 2y_p' + y_p = -4B_1 e^x \sin x + 4B_2 e^x \cos x$$
$$\quad + 3B_1 e^x \cos x + 3B_2 e^x \sin x$$
$$= e^x [\cos x(4B_2 + 3B_1)$$
$$\quad + \sin x(-4B_1 + 3B_2)]$$

But, $y_p'' + 2y_p' + y_p = e^x \cos x$ (the forcing function). This requires

$$-4B_1 + 3B_2 = 0 \text{ and } 4B_2 + 3B_1 = 1$$
$$B_1 = \frac{3}{25} \quad \text{and} \quad B_2 = \frac{4}{25}$$

The general solution is

$$\boxed{y = (A_1 + A_2 x)e^{-x} + \left(\frac{3}{25}\cos x + \frac{4}{25}\sin x\right)e^x}$$

14. (a) $\mathcal{L}(2p'') + \mathcal{L}(p) = \mathcal{L}(\cosh(3t))$

$$2s^2\mathcal{L}(p) - 2sp(0) - 2p'(0) + \mathcal{L}(p) = \frac{s}{s^2-9}$$

$$(2s^2+1)\mathcal{L}(p) - 2 = \frac{s}{s^2-9}$$

$$\mathcal{L}(p) = \frac{s}{(s^2-9)(2s^2+1)} + \frac{2}{2s^2+1}$$

By partial fractions,

$$\frac{s}{(s^2-9)(2s^2+1)} = \frac{A_1s+B_1}{s^2-9} + \frac{A_2s+B_2}{2s^2+1}$$

$$= \frac{s^3(2A_1+A_2)+s^2(2B_1+B_2)+s(A_1-9A_2)+(B_1-9B_2)}{(s^2-9)(2s^2+1)}$$

$$A_1 = \frac{1}{19} \qquad B_1 = 0$$
$$\text{and}$$
$$A_2 = -\frac{2}{19} \qquad B_2 = 2$$

$$\mathcal{L}(p) = \frac{\frac{s}{19}}{s^2-9} - \frac{\frac{2s}{19}}{2s^2+1} + \frac{2}{2s^2+1}$$

$$= \left(\frac{1}{19}\right)\left(\frac{s}{s^2-9}\right) - \left(\frac{1}{19}\right)\left(\frac{s}{s^2+\frac{1}{2}}\right)$$

$$+ (\sqrt{2})\left(\frac{\frac{1}{\sqrt{2}}}{s^2+\frac{1}{2}}\right)$$

$$\boxed{p = \frac{1}{19}\cosh(3t) - \frac{1}{19}\cos\left(\frac{t}{\sqrt{2}}\right) + \sqrt{2}\sin\left(\frac{t}{\sqrt{2}}\right)}$$

(b) $\mathcal{L}(x'') - 2\mathcal{L}(x') - 3\mathcal{L}(x) = 0$

$$s^2\mathcal{L}(x) - sx(0) - x'(0) - 2s\mathcal{L}(x) - 2x(0) - 3\mathcal{L}(x) = 0$$
$$(s^2-2s-3)\mathcal{L}(x) = 1$$

$$\mathcal{L}(x) = \frac{1}{s^2-2s-3}$$

$$= \frac{1}{(s+1)(s-3)}$$

$$x = \frac{e^{-(-3)t}-e^{-t}}{1-(-3)}$$

$$= \boxed{\frac{e^{3t}-e^{-t}}{4}}$$

(c) $\mathcal{L}(x'') - \mathcal{L}(x) = 0$

$$s^2\mathcal{L}(x) - sx(0) - x'(0) - \mathcal{L}(x) = 0$$
$$(s^2-1)\mathcal{L}(x) - 4s - 2 = 0$$

$$\mathcal{L}(x) = \frac{4s}{s^2-1} + \frac{2}{s^2-1}$$

$$= (4)\left(\frac{s}{s^2-1}\right) + (2)\left(\frac{1}{s^2-1}\right)$$

$$\boxed{x = 4\cosh t + 2\sinh t}$$

(d) $\mathcal{L}(x'') + 4\mathcal{L}(x') + 4\mathcal{L}(x) = \mathcal{L}(\sin 2t)$
$$+ 9\mathcal{L}(\cos 2t)$$

$$s^2\mathcal{L}(x) - sx(0) - x'(0)$$
$$+ 4s\mathcal{L}(x) + 4x(0) + 4\mathcal{L}(x) = \frac{2}{s^2+4} + \frac{9s}{s^2+4}$$

$$(s^2+4s+4)\mathcal{L}(x) - 2 = \frac{9s+2}{s^2+4}$$

$$\mathcal{L}(x) = \frac{9s+2+2s^2+8}{(s+2)^2(s^2+4)}$$

$$= \frac{(2s+5)(s+2)}{(s+2)^2(s^2+4)}$$

$$= \frac{2s+5}{(s+2)(s^2+4)}$$

$$= \frac{A}{s+2} + \frac{B_1s+B_2}{s^2+4}$$

$$A = \frac{1}{8}$$
$$B_1 = -\frac{1}{8}$$
$$B_2 = \frac{9}{4}$$

$$\mathcal{L}(x) = \frac{\frac{1}{8}}{s+2} - \frac{\frac{s}{8}}{s^2+4} + \frac{\frac{9}{4}}{s^2+4}$$

$$= \left(\frac{1}{8}\right)\left(\frac{1}{s+2}\right) - \left(\frac{1}{8}\right)\left(\frac{s}{s^2+4}\right)$$

$$+ \left(\frac{9}{8}\right)\left(\frac{2}{s^2+4}\right)$$

$$\boxed{x = \frac{1}{8}e^{-2t} - \frac{1}{8}\cos 2t + \frac{9}{8}\sin 2t}$$

15. $$\mathcal{L}(f(t)) = F(s) = \int_0^\infty e^{-st}t^n\,dt$$

Integrating by parts,

$$\mathcal{L}(f(t)) = \left[\frac{e^{-st}}{-s}t^n\right]_0^\infty - \int_0^\infty \left(\frac{e^{-st}}{-s}\right)(nt^{n-1})dt$$

Integrating by parts, if $s > 0$,

$$\left[\frac{e^{-st}}{-s}t^n\right] = 0$$

$$F_n(s) = \frac{n}{s}F_{n-1}(s)$$

$$F_p(s) = \mathcal{L}(t^p)$$

$$F_n(s) = \frac{n!}{s^n}F_0(s)$$

$$F_0(s) = \int_0^\infty e^{-st}dt = \left[\frac{e^{-st}}{-s}\right]_0^\infty = \frac{1}{s}$$

$$F_n(s) = \left(\frac{n!}{s^n}\right)\left(\frac{1}{s}\right) = \boxed{\frac{n!}{s^{n+1}}}$$

16.
$$e^{at}\sin bt = e^{at}\left(\frac{e^{ibt} - e^{-ibt}}{2i}\right)$$

$$= \frac{e^{(a+ib)t} - e^{(a-ib)t}}{2i}$$

$$F(s) = \int_0^\infty e^{-st}\left(\frac{e^{(a+ib)t} - e^{(a-ib)t}}{2i}\right)$$

$$= \left(\frac{1}{2i}\right)\left[\int_0^\infty e^{(a+ib-s)t}dt - \int_0^\infty e^{(a-ib-s)t}dt\right]$$

$$= \left(\frac{1}{2i}\right)\left(\left[\frac{e^{(a+ib-s)t}}{a+ib-s}\right]_0^\infty - \left[\frac{e^{(a-ib-s)t}}{a-ib-s}\right]_0^\infty\right)$$

$$= \left(\frac{1}{2i}\right)\left(\frac{-1}{a+ib-s} - \frac{-1}{a-ib-s}\right)$$

This is valid if $s > a$.

$$F(s) = \left(\frac{-1}{2i}\right)\left[\frac{a-ib-s-a-ib+s}{(a-s)^2+b^2}\right]$$

$$= \left(\frac{-1}{2i}\right)\left[\frac{-2ib}{(s-a)^2+b^2}\right]$$

$$= \boxed{\frac{b}{(s-a)^2+b^2}}$$

17.
$$\mathcal{L}\left(\frac{1}{t}f(t)\right) = \int_s^\infty F(u)du$$

$$f(t) = \sin t$$

From a table of transforms,

$$F(u) = \frac{1}{1+u^2}$$

$$\mathcal{L}\left(\frac{\sin t}{t}\right) = \int_s^\infty\left(\frac{du}{1+u^2}\right)$$

$$= \left[\arctan u\right]_s^\infty = \boxed{\frac{\pi}{2} - \arctan s}$$

18.

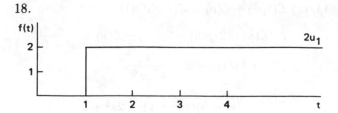

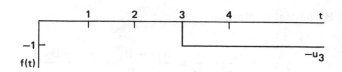

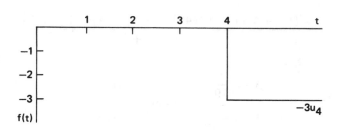

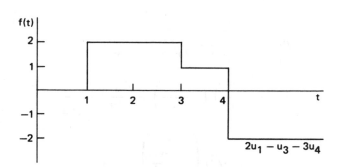

19. (a)
$$\mathcal{L}(\delta(t)) = \int_0^\infty e^{-st}\delta(t)dt$$

$$= e^{-st0} = \boxed{1}$$

(b)
$$\mathcal{L}(\delta(t-\tau)) = \int_0^\infty e^{-st}\delta(t-\tau)dt$$

Use the time-shifting theorem.

$$\mathcal{L}(\delta(t-\tau)) = \int_{-\tau}^\infty e^{-s(t+\tau)}\delta(t)dt$$

$$= e^{-s(0+\tau)} = \boxed{e^{-s\tau}}$$

20. Let $x(t)$ be the amount of salt in the tank at time t, where x is measured in lbm and t is measured in min.

$$x'(t) = 2\frac{\text{lbm}}{\text{min}} - \left(3\frac{\text{gal}}{\text{min}}\right)\left(\frac{x}{100\,\text{gal} - t\frac{\text{gal}}{\text{min}}}\right)$$

After eliminating the units for clarity and rearranging,

$$x'(t) + \frac{3}{100 - t}x(t) = 2$$

Since this is first order linear,

$$u(t) = \exp\left[\int \frac{3}{100 - t}dt\right] = (100 - t)^{-3}$$

$$x(t) = \frac{1}{u(t)}\left[\int u(t)g(t)dt + C\right]$$

$$= (100 - t)^3\left[\int \frac{2}{(100 - t)^3}dt + C\right]$$

$$= 100 - t + C(100 - t)^3$$

Since $x(0) = 60$, $C = -0.00004$.

$$x(t) = 100 - t - 0.00004\,(100 - t)^3$$

$$\boxed{x(60) = 100 - 60 - (0.00004)(100 - 60)^3 = 37.4\,\text{lbm}}$$

21. The differential equation is

$$L\left(\frac{di(t)}{dt}\right) + Ri(t) = V_0 u_1(t)$$

$$i(0) = 0$$

Taking the Laplace transform of each term,

$$sL\mathcal{L}(i(s)) - i(0) + R\mathcal{L}(i(s)) = \frac{V_0}{s}$$

$$sLI(s) + RI(s) = \frac{V_0}{s}$$

$$I(s) = \frac{V_0}{s(sL + R)} = \frac{\dfrac{V_0}{R}}{s\left(\dfrac{sL}{R} + 1\right)}$$

$$= \frac{V_0}{R}\left(\frac{1}{s} - \frac{1}{s + \dfrac{R}{L}}\right)$$

Taking the inverse Laplace transform of each term,

$$\boxed{i(t) = \mathcal{L}^{-1}(I(s)) = \frac{V_0}{R}\left(1 - e^{-\frac{R}{L}t}\right)}$$

22. Let x be the amount of sugar in the tank at time t, where x is measured in lbm and t is expressed in min.

$$x'(t) = 0.5\,\frac{\text{lbm}}{\text{min}} - \left(1.5\,\frac{\text{gal}}{\text{min}}\right)\left(\frac{x(t)}{100\,\text{gal}}\right)$$

$$x'(t) + 0.015x(t) = 0.5$$

For a differential equation of the form $y' + ay + b$, the solution is

$$y = \frac{b}{a}(1 - e^{ax}) + y_0\,(e^{-ax})$$

Since $x(0) = 0$,

$$x(t) = \left(\frac{0.5}{0.015}\right)(1 - e^{-0.015t})$$

$$= \left(\frac{100}{3}\right)(1 - e^{-0.015t})$$

For the sugar concentration in the tank to equal 0.3 lbm/gal,

$$\frac{x}{100} = 0.3$$

$$x = 30\,\text{lbm}$$

$$30 = \left(\frac{100}{3}\right)(1 - e^{-0.015t})$$

$$t = \left(-\frac{1}{0.015}\right)\ln\left(1 - \frac{90}{100}\right)$$

$$= \boxed{153.5\,\text{min}}$$

23. Let $m(t)$ be the quantity of radioactive material at time t.

$$m(t) = m_0 e^{kt}$$

If $m(t) = 0.1m_0$,

$$e^{kt} = 0.1$$

$$kt = \ln 0.1$$

$$kt_{\frac{1}{2}} = \ln\left(\frac{1}{2}\right)$$

$$t = \left(\frac{\ln 0.1}{\ln 0.5}\right)t_{\frac{1}{2}} = \left(\frac{\ln 0.1}{\ln 0.5}\right)(100{,}000\,\text{years})$$

$$= \boxed{332{,}193\,\text{years}}$$

MATHEMATICS
Diff Eq

PROBABILITY AND ANALYSIS OF STATISTICAL DATA

1. $P(n,r) = \dfrac{n!}{(n-r)!}$

$P(6,3) = \dfrac{6!}{(6-3)!} = \dfrac{6 \cdot 5 \cdot 4 \cdot 3 \cdot 2 \cdot 1}{3 \cdot 2 \cdot 1} = 6 \cdot 5 \cdot 4$

$= \boxed{120}$

2. The first seat can be filled 7 ways, the second 6 ways, etc.

(a) $P(n,n) = \dfrac{n!}{(n-n)!}$

$P(7,7) = \dfrac{7!}{(7-7)!} = \dfrac{7 \cdot 6 \cdot 5 \cdot 4 \cdot 3 \cdot 2 \cdot 1}{0!}$

$= 7 \cdot 6 \cdot 5 \cdot 4 \cdot 3 \cdot 2 \cdot 1 = \boxed{5040}$

(b) This is the same as part (a) except there is no "first" seat.

$P_{\text{ring}}(n,n) = \dfrac{P(n,n)}{n} = (n-1)!$

$P_{\text{ring}}(7,7) = (7-1)! = \boxed{720}$

3. The number of sequences, if all flags were distinguishable, would be

$$(n_{\text{r}} + n_{\text{g}})!$$

To account for the indistinguishability of the red flags, divide by $n_{\text{r}}!$

$$\dfrac{(n_{\text{r}} + n_{\text{g}})!}{n_{\text{r}}!}$$

To account for the indistinguishability of the green flags, divide by $n_{\text{g}}!$

$$\dfrac{(n_{\text{r}} + n_{\text{g}})!}{n_{\text{r}}! \, n_{\text{g}}!} = \dfrac{(4+2)!}{(4!)(2!)} = \boxed{15}$$

4. (a) This is a combination since the order is not relevant.

$C(n,r) = \dfrac{n!}{(n-r)! \, r!}$

$C(8,6) = \dfrac{8!}{(8-6)! \, 6!} = \boxed{28}$

(b) $C(n,r) = \dfrac{n!}{(n-r)! \, r!}$

$n = 8 - 2 = 6$

$r = 6 - 2 = 4$

$C(6,4) = \dfrac{6!}{(6-4)! \, (4)!} = \boxed{15}$

5. $p\{\text{heads}\} = 2p\{\text{tails}\}$

$p\{\text{heads}\} + p\{\text{tails}\} = 1$

$p\{\text{heads}\} = \dfrac{2}{3}, \quad p\{\text{tails}\} = \dfrac{1}{3}$

(a) $p\{\text{heads}\} = \dfrac{2}{3} = \boxed{0.667}$

(b) $p\{\text{heads}\}^3 = \left(\dfrac{2}{3}\right)^3 = \boxed{0.296}$

(c) $p\{x\} = \dbinom{4}{x} \left(\dfrac{2}{3}\right)^x \left(1 - \dfrac{2}{3}\right)^{4-x}$

$p\{2\} = \dbinom{4}{2} \left(\dfrac{2}{3}\right)^2 \left(1 - \dfrac{2}{3}\right)^{4-2}$

$= \dfrac{4!}{(4-2)! \, 2!} \left(\dfrac{2}{3}\right)^2 \left(\dfrac{1}{3}\right)^2$

$= \boxed{0.296}$

6. (a) $p\{\spadesuit_1\} p\{\spadesuit_2\} = \left(\dfrac{13}{52}\right)\left(\dfrac{12}{51}\right) = \boxed{0.059}$

(b) Either a spade or diamond may be first.

$p\{\spadesuit_1\} p\{\blacklozenge_2\} + p\{\blacklozenge_1\} p\{\spadesuit_2\}$

$= \left(\dfrac{13}{52}\right)\left(\dfrac{13}{51}\right) + \left(\dfrac{13}{52}\right)\left(\dfrac{13}{51}\right)$

$= \boxed{0.127}$

(c) $p\{\spadesuit_1\} p\{\spadesuit_2\} + p\{\blacklozenge_1\} p\{\blacklozenge_2\}$

$= \left(\dfrac{13}{52}\right)\left(\dfrac{12}{51}\right) + \left(\dfrac{13}{52}\right)\left(\dfrac{12}{51}\right)$

$= \boxed{0.118}$

(d) $p\{\spadesuit\}p\{\heartsuit\} + p\{\spadesuit\}p\{\diamondsuit\}$
$\quad + p\{\heartsuit\}p\{\spadesuit\} + p\{\diamondsuit\}p\{\spadesuit\}$

$\quad = \left(\dfrac{13}{52}\right)\left(\dfrac{13}{51}\right) + \left(\dfrac{13}{52}\right)\left(\dfrac{13}{51}\right)$

$\quad\quad + \left(\dfrac{13}{52}\right)\left(\dfrac{13}{51}\right) + \left(\dfrac{13}{52}\right)\left(\dfrac{13}{51}\right)$

$\quad = \boxed{0.255}$

7. $p\{x\} = \dbinom{7}{x}\left(\dfrac{1}{4}\right)^{x}\left(1-\dfrac{1}{4}\right)^{7-x}$

(a) $p\{0\} = \dbinom{7}{0}\left(\dfrac{1}{4}\right)^{0}\left(1-\dfrac{1}{4}\right)^{7-0} = \boxed{0.133}$

(b) $p\{1\} = \dbinom{7}{1}\left(\dfrac{1}{4}\right)^{1}\left(1-\dfrac{1}{4}\right)^{7-1} = \boxed{0.311}$

(c) $p\{2\} = \dbinom{7}{2}\left(\dfrac{1}{4}\right)^{2}\left(1-\dfrac{1}{4}\right)^{7-2} = \boxed{0.311}$

8. Assume that the probability distribution of the number of customers serviced is a Poisson distribution.

$\quad p\{x\} = \dfrac{e^{-38}38^{x}}{x!}$

$\quad p\{34\} = \dfrac{(e^{-38})(38^{34})}{34!} = \boxed{0.0549}$

9. $p\{x\} = \dbinom{5}{x}\left(\dfrac{1}{10}\right)^{x}\left(1-\dfrac{1}{10}\right)^{5-x}$

$\quad p\{2\} = \dbinom{5}{2}\left(\dfrac{1}{10}\right)^{2}\left(1-\dfrac{1}{10}\right)^{5-2}$

$\quad = \boxed{0.0729}$

10. $p\{x\} = \dbinom{20}{x}\left(\dfrac{15}{100}\right)^{x}\left(1-\dfrac{15}{100}\right)^{20-x}$

$\quad p\{2\} = \dbinom{20}{2}\left(\dfrac{15}{100}\right)^{2}\left(1-\dfrac{15}{100}\right)^{20-2}$

$\quad = \boxed{0.229}$

11. $z_n = \dfrac{x_n - \mu}{\sigma}$

$\quad z_1 = \dfrac{x_1 - \mu}{\sigma} = \dfrac{15 - 12}{3} = 1$

$\quad z_2 = \dfrac{x_2 - \mu}{\sigma} = \dfrac{18 - 12}{3} = 2$

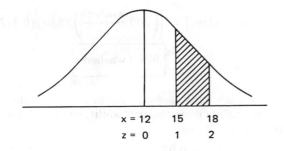

From the normal table,

$$p\{15 < x < 18\} = p\{x < 18\} - p\{x < 15\}$$
$$= p\{z < 2\} - p\{z < 1\}$$
$$= 0.4772 - 0.3413$$
$$= \boxed{0.1359}$$

12. $z_n = \dfrac{x_n - \mu}{\sigma}$

$\quad z_1 = \dfrac{x_1 - \mu}{\sigma} = \dfrac{0.497 - 0.502}{0.005} = -1$

$\quad z_2 = \dfrac{x_2 - \mu}{\sigma} = \dfrac{0.507 - 0.502}{0.005} = 1$

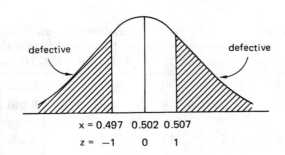

(a) $p\{\text{defective}\} = p\{x < 0.497 \text{ and } x > 0.507\}$
$\quad = 2p\{x > 0.507\}$
$\quad = 2\left(0.500 - p\{0.502 < x < 0.507\}\right)$
$\quad = 2\left(0.500 - 0.3413\right)$
$\quad = \boxed{0.3174}$

(b) $p\{x\} = \dbinom{3}{x}(0.3174)^{x}(1-0.3174)^{3-x}$

$\quad p\{2\} = \dbinom{3}{2}(0.3174)^{2}(1-0.3174)^{1}$

$\quad = \boxed{0.2063}$

(c) $np\{\text{defective}\} = \left[\left(200\ \dfrac{\text{washers}}{\text{hr}}\right)(8\ \text{hr})\right](0.3174)$

$= \boxed{507.8\ \text{washers}}$

13. Since each bank is a one-out-of-n system, the reliability of each type is the probability that at least one is operating.

$$R_A = 0.97$$
$$R_B = 1 - (1 - 0.85)^2 = 0.9775$$
$$R_C = 1 - (1 - 0.60)^3 = 0.936$$
$$R_D = 1 - (1 - 0.80)^3 = 0.992$$

The probability of a serial system operating correctly is

$$R_A R_B R_C R_D = (0.97)(0.9775)(0.936)(0.992)$$

$$= \boxed{0.880}$$

14. $\text{MTBF} = \dfrac{(2)(50\ \text{hr}) + (5)(100\ \text{hr}) + (18)(150\ \text{hr})}{50}$

$$+ \dfrac{(22)(200\ \text{hr}) + (2)(250\ \text{hr}) + (1)(300\ \text{hr})}{50}$$

$$= \boxed{170\ \text{hr}}$$

elapsed time t_n (hr)	failures $F(t_n)$	survivors $S(t_n)$	failure probability $F(t_n)/50$	conditional distribution $F(t_n)/S(t_{n-1})$
0	0	50	–	–
50	2	48	0.04	0.040
100	5	43	0.10	0.104
150	18	25	0.36	0.419
200	22	3	0.44	0.880
250	2	1	0.04	0.667
300	1	0	0.02	1.000

15. (a) $R\{t\} = e^{-\lambda t}$

$$\lambda = \frac{\ln R\{t\}}{t} = -\frac{\ln 0.99}{8\ \text{hr}} = 0.001256\ \text{hr}^{-1}$$

$$\text{MTBF} = \frac{1}{\lambda} = \frac{1}{0.00126} = \boxed{796\ \text{hr}}$$

(b) $F(t) = 1 - R\{t\} = 1 - e^{-(0.001256\ \text{hr}^{-1})t}$

$$F(20\ \text{hr}) = 1 - e^{-(0.001256\ \text{hr}^{-1})(20\ \text{hr})}$$

$$= \boxed{0.025}$$

(c) The exponential distribution is "memoryless". It does not know the system has already operated for 10 hours. Therefore, the probability is the same as in part(b).

$$\boxed{0.025}$$

Proof: This is a conditional probability.

$$p\{T < 30 | T > 10\} = \frac{p\{10 < T < 30\}}{p\{T > 10\}}$$

$$= \frac{p\{T < 30\} - p\{T < 10\}}{p\{T > 10\}}$$

$$= \frac{1 - e^{-30\lambda} - (1 - e^{-10\lambda})}{e^{-10\lambda}}$$

$$= \frac{e^{-10\lambda} - e^{-30\lambda}}{e^{-10\lambda}}$$

$$= \frac{e^{-10\lambda}(1 - e^{-20\lambda})}{e^{-10\lambda}}$$

$$= 1 - e^{-20\lambda} = 0.025$$

(d) $$z(t) = \lambda$$

$$\boxed{z(t) = 0.001256\ \text{hr}^{-1}\ \text{(independent of } t\text{)}}$$

16. system A:

$$R_1 = 1 - (1 - 0.4)(1 - 0.61)(1 - 0.52)$$
$$= 0.888$$
$$R_2 = 0.99$$

$$R = R_1 R_2 = (0.888)(0.99) = \boxed{0.879}$$

system B:

$$R_1 = (0.4)(0.61)(0.52) = 0.127$$
$$R_2 = 0.99$$
$$R = 1 - (1 - R_1)(1 - R_2)$$
$$= 1 - (1 - 0.127)(1 - 0.99)$$
$$= \boxed{0.991}$$

17.
$$\bar{x} = \frac{\sum x_i}{n} = \frac{114}{12} = \boxed{9.5}$$

$$\sigma = \sqrt{\frac{\sum x_i^2}{n} - x^2}$$

$$= \sqrt{\frac{1428}{12} - 9.5^2} = \boxed{5.36}$$

$$s = \sqrt{\frac{\sum (x_i - \bar{x})^2}{n-1}}$$

$$= \sqrt{\frac{\sum x_i^2 - 2\bar{x}\sum x_i + n\bar{x}^2}{n-1}}$$

$$= \sqrt{\frac{1428 - (2)(9.5)(114) + (12)(9.5)^2}{12-1}}$$

$$= \boxed{5.60}$$

$$\sigma^2 = 5.36^2 = \boxed{28.73}$$

$$s^2 = 5.60^2 = \boxed{31.36}$$

18.
$$n = 3 + 8 + 18 + 12 + 9 = 50$$

$$\sum x_i = (3)(1.5) + (8)(2.5) + (18)(3.5)$$
$$+ (12)(4.5) + (9)(5.5) = 191$$

$$\bar{x} = \frac{\sum x_i}{n} = \frac{191}{50} = 3.82$$

$$\sum x_i^2 = (3)(1.5)^2 + (8)(2.5)^2 + (18)(3.5)^2$$
$$+ (12)(4.5)^2 + (9)(5.5)^2 = 792.5$$

$$\sigma = \sqrt{\frac{\sum x_i^2}{n} - \bar{x}^2}$$

$$= \sqrt{\frac{792.5}{50} - 3.82^2} = \boxed{1.121}$$

$$s = \sqrt{\frac{\sum (x_i - \bar{x})^2}{n-1}} = \sqrt{\frac{\sum x_i^2 - 2\bar{x}\sum x_i + n\bar{x}^2}{n-1}}$$

$$= \sqrt{\frac{792.5 - (2)(3.82)(191) + (50)(3.82)^2}{50-1}}$$

$$= \boxed{1.133}$$

MATHEMATICS
Prob/Stat

NUMBERING SYSTEMS

1. (a)
$$\begin{array}{r} 101 \\ +\ 011 \\ \hline \boxed{1000} \end{array}$$

(b)
$$\begin{array}{r} 101 \\ +\ 110 \\ \hline \boxed{1011} \end{array}$$

(c)
$$\begin{array}{r} 101 \\ +\ 100 \\ \hline \boxed{1001} \end{array}$$

(d)
$$-\left(\begin{array}{r} 1100 \\ -\ 0100 \\ \hline 1000 \end{array}\right) = \boxed{-1000}$$

(e)
$$\begin{array}{r} 1110 \\ -\ 1000 \\ \hline \boxed{0110} \end{array}$$

(f)
$$-\left(\begin{array}{r} 101 \\ -\ 010 \\ \hline 011 \end{array}\right) = \boxed{-011}$$

(g)
$$\begin{array}{r} 111 \\ \times\ \ 11 \\ \hline 111 \\ 111 \\ \hline \boxed{10101} \end{array}$$

(h)
$$\begin{array}{r} 100 \\ \times\ \ 11 \\ \hline 100 \\ 100 \\ \hline \boxed{1100} \end{array}$$

(i)
$$\begin{array}{r} 1011 \\ \times\ 1101 \\ \hline 1011 \\ 1011 \\ 1011 \\ \hline \boxed{10001111} \end{array}$$

2. (a)
$$\begin{array}{r} 466 \\ +\ 457 \\ \hline \boxed{1145} \end{array}$$

(b)
$$\begin{array}{r} 1007 \\ +\ 6661 \\ \hline \boxed{7670} \end{array}$$

(c)
$$\begin{array}{r} 321 \\ +\ 465 \\ \hline \boxed{1006} \end{array}$$

(d)
$$\begin{array}{r} 71 \\ -\ 27 \end{array} = \begin{array}{r} 6\ \ 11 \\ -\ 2\ \ \ 7 \\ \hline \boxed{4\ \ 2} \end{array}$$

(e)
$$\begin{array}{r} 1143 \\ -\ 367 \end{array} = \begin{array}{r} 113\ \ 13 \\ -\ 36\ \ \ 7 \end{array} = \begin{array}{r} 10\ \ 13\ \ 13 \\ -\ 3\ \ \ 6\ \ \ 7 \\ \hline \boxed{5\ \ 5\ \ 4} \end{array}$$

(f)
$$-\left(\begin{array}{r} 677 \\ -\ 646 \\ \hline 31 \end{array}\right) = \boxed{-31}$$

(g)
$$\begin{array}{r} 77 \\ \times\ 66 \\ \hline 572 \\ 572 \\ \hline \boxed{6512} \end{array}$$

(h)
$$\begin{array}{r} 325 \\ \times\ \ 36 \\ \hline 2376 \\ 1177 \\ \hline \boxed{14366} \end{array}$$

(i)
$$\begin{array}{r} 3251 \\ \times\ 16.1 \\ \hline 3251 \\ 23766 \\ 3251 \\ \hline \boxed{57023.1} \end{array}$$

3. (a)
$$\begin{array}{r} BA \\ +\ \ C \\ \hline \boxed{C6} \end{array}$$

(b)
$$\begin{array}{r} BB \\ +\ \ A \\ \hline \boxed{C5} \end{array}$$

(c)
$$\begin{array}{r} BE \\ 10 \\ +\ 1A \\ \hline \boxed{E8} \end{array}$$

(d)
$$\begin{array}{r} FF \\ -\ \ E \\ \hline \boxed{F1} \end{array}$$

(e)
$$\begin{array}{cc} \overset{\displaystyle 74}{\ } & \overset{\displaystyle 6\ 14}{\ } \\ -\ 4A & =\quad -\ 4\ \ A \\ \hline & \boxed{2\ A} \end{array}$$

(f)
$$\begin{array}{ccc} FB & & E\ 1B \\ -\ BF & = & -\ B\ \ F \\ \hline & & \boxed{3\ \ C} \end{array}$$

(g)
$$\begin{array}{r} 4\,A \\ \times\ \ \ 3\,E \\ \hline 40\,C \\ DE \\ \hline \boxed{11\,EC} \end{array}$$

(h)
$$\begin{array}{r} FE \\ \times\ \ EF \\ \hline EE\,2 \\ DE\,4 \\ \hline \boxed{ED\,22} \end{array}$$

(i)
$$\begin{array}{r} 17 \\ \times\ \ 7A \\ \hline E\,6 \\ A\,1 \\ \hline \boxed{AF\,6} \end{array}$$

4. (a) $(674)_8 = (6)(8)^2 + (7)(8)^1 + (4)(8)^0$

   $= \boxed{444}$

(b) $(101101)_2 = (1)(2)^5 + (0)(2)^4 + (1)(2)^3$
   $+ (1)(2)^2 + (0)(2)^1 + (1)(2)^0$

   $= \boxed{45}$

(c) $(734.262)_8 = (7)(8)^2 + (3)(8)^1 + (4)(8)^0$
   $+ (2)(8)^{-1} + (6)(8)^{-2} + (2)(8)^{-3}$

   $= \boxed{476.348}$

(d) $(1011.11)_2 = (1)(2)^3 + (0)(2)^2 + (1)(2)^1$
   $+ (1)(2)^0 + (1)(2)^{-1} + (1)(2)^{-2}$

   $= \boxed{11.75}$

5. (a)
$$75 \div 8 = 9 \quad \text{remainder } 3$$
$$9 \div 8 = 1 \quad \text{remainder } 1$$
$$1 \div 8 = 0 \quad \text{remainder } 1$$

$$(75)_{10} = \boxed{(113)_8}$$

(b)
$$0.375 \div \tfrac{1}{8} = 3 \quad \text{remainder } 0$$

$$(0.375)_{10} = \boxed{(0.3)_8}$$

(c)
$$121 \div 8 = 15 \quad \text{remainder } 1$$
$$15 \div 8 = 1 \quad \text{remainder } 7$$
$$1 \div 8 = 0 \quad \text{remainder } 1$$
$$0.875 \div \tfrac{1}{8} = 7 \quad \text{remainder } 0$$

$$(121.875)_{10} = \boxed{(171.7)_8}$$

(d) Since $(2)^3 = 8$, group the bits into groups of three starting at the decimal point and working outward in both directions.

$$001011100.011100 = 001\ 011\ 100.011\ 100$$

Convert each of the groups into its octal equivalent.

$$001\ 011\ 100.011\ 100 = 1\ 3\ 4.3\ 4$$

$$\boxed{(134.34)_8}$$

6. (a)
$$83 \div 2 = 41 \quad \text{remainder } 1$$
$$41 \div 2 = 20 \quad \text{remainder } 1$$
$$20 \div 2 = 10 \quad \text{remainder } 0$$
$$10 \div 2 = 5 \quad \text{remainder } 0$$
$$5 \div 2 = 2 \quad \text{remainder } 1$$
$$2 \div 2 = 1 \quad \text{remainder } 0$$
$$1 \div 2 = 0 \quad \text{remainder } 1$$

$$(83)_{10} = \boxed{(1010011)_2}$$

(b)
$$100 \div 2 = 50 \quad \text{remainder } 0$$
$$50 \div 2 = 25 \quad \text{remainder } 0$$
$$25 \div 2 = 12 \quad \text{remainder } 1$$

$$12 \div 2 = 6 \quad \text{remainder } 0$$
$$6 \div 2 = 3 \quad \text{remainder } 0$$
$$3 \div 2 = 1 \quad \text{remainder } 1$$
$$1 \div 2 = 0 \quad \text{remainder } 1$$
$$0.3 \div \tfrac{1}{2} = 0 \quad \text{remainder } 0.3$$
$$0.3 \div \left(\tfrac{1}{2}\right)^2 = 1 \quad \text{remainder } 0.05$$
$$0.05 \div \left(\tfrac{1}{2}\right)^3 = 0 \quad \text{remainder } 0.05$$
$$0.05 \div \left(\tfrac{1}{2}\right)^4 = 0 \quad \text{remainder } 0.05$$
$$0.05 \div \left(\tfrac{1}{2}\right)^5 = 1 \quad \text{remainder } 0.01875$$
$$0.01875 \div \left(\tfrac{1}{2}\right)^6 = 1 \quad \text{remainder } 0.003125$$
$$\vdots$$

$$(100.3)_{10} = \boxed{(1100100.010011\cdots)_2}$$

(c)
$$0.97 \div \tfrac{1}{2} = 1 \quad \text{remainder } 0.47$$
$$0.47 \div \left(\tfrac{1}{2}\right)^2 = 1 \quad \text{remainder } 0.22$$
$$0.22 \div \left(\tfrac{1}{2}\right)^3 = 1 \quad \text{remainder } 0.095$$
$$0.095 \div \left(\tfrac{1}{2}\right)^4 = 1 \quad \text{remainder } 0.0325$$
$$0.0325 \div \left(\tfrac{1}{2}\right)^5 = 1 \quad \text{remainder } 0.00125$$
$$0.00125 \div \left(\tfrac{1}{2}\right)^6 = 0 \quad \text{remainder } 0.00125$$
$$\vdots$$

$$(0.97)_{10} = \boxed{(0.111110\cdots)_2}$$

(d) Since $8 = (2)^3$, convert each octal digit into its binary equivalent.

$$
\begin{aligned}
321.422: \quad &3 = 011 \\
&2 = 010 \\
&1 = 001 \\
&4 = 100 \\
&2 = 010 \\
&2 = 010
\end{aligned}
$$

$$\boxed{(321.422)_8 = 011010001.100010010}$$

ENGINEERING ECONOMIC ANALYSIS

1. $F = P(F/P, 6\%, 9) = (\$250)(1.6895)$

$= \boxed{\$422}$

2. $P = F(P/F, 6\%, 5) = (\$2000)(0.7473)$

$= \boxed{\$1495}$

3. $P = F(P/F, 6\%, 7) = (\$2000)(0.6651)$

$= \boxed{\$1330}$

4. $A = P(A/P, 6\%, 10) = (\$50)(0.1359)$

$= \boxed{\$6.80}$

5. $F = A(F/A, 6\%, 10) = (\$20,000)(13.1808)$

$= \boxed{\$263,600}$

6. $A = F(A/F, 6\%, 7) = (\$5000)(0.1191)$

$= \boxed{\$595}$

7. $P = A(P/A, 6\%, 7) = (\$400)(5.5824)$

$= \boxed{\$2233}$

8. $F = (\$500)(F/P, 4\%, 10) + (\$700)(F/P, 4\%, 8)$
$\qquad + (\$900)(F/P, 4\%, 6)$

$= (\$500)(1.4802) + (\$700)(1.3686)$
$\qquad + (\$900)(1.2653)$

$= \boxed{\$2837}$

9. $i = \dfrac{r}{k} = \dfrac{6\%}{2} = 3\% \qquad n = (2)(8) = 16$

$F = P(F/P, 3\%, 16) = (\$550)(1.6047)$

$= \boxed{\$883}$

10. $P = F(P/F, i\%, 5)$

$(P/F, i\%, 5) = \dfrac{P}{F} = \dfrac{50}{75} = 0.6667$

From the table,

$(P/F, 8\%, 5) = 0.6806$

$(P/F, 9\%, 5) = 0.6499$

$i = 8\% + (9\% - 8\%)\left(\dfrac{0.6667 - 0.6806}{0.6499 - 0.6806}\right)$

$= \boxed{8.45\%}$

$(F/P, i, 5)$ can also be solved directly for i.

$(1 + i)^5 = 1.5 \qquad i = 0.08447 \ (8.447\%)$

11. $i = \dfrac{r}{k} = \dfrac{4\%}{12} = 0.33\% \quad n = (30)(12) = 360$

$A = F(A/F, 0.33\%, 360) = F\left[\dfrac{i}{(1+i)^n - 1}\right]$

$= (\$50,000)\left[\dfrac{0.0033}{(1 + 0.0033)^{360} - 1}\right] = \boxed{\$72.56}$

12. $n = (18)(12) = 216$

$F = A(F/A, 0.33\%, 216) = A\left[\dfrac{(1+i)^n - 1}{i}\right]$

$= (\$72.56)\left[\dfrac{(1 + 0.0033)^{216} - 1}{0.0033}\right]$

$= (\$72.56)(314.33)$

$= \boxed{\$22,808}$

13. $A = F(A/F, 5\%, 19) = (\$20,000)(0.0327)$

$= \boxed{\$654}$

14. It helps to draw a cash flow diagram. Present worth as of January 1, year 0 of the deposits is

$P_{\text{deposits}} = (\$50)(P/A, 6\%, 10)$

$= (\$50)(7.3601) = \$368.00$

Present worth as of January 1, year 0 of the withdrawal is

$P_{\text{withdrawal}} = (A_w)\big[(P/A, 6\%, 19) - (P/A, 6\%, 14)\big]$

$(A_w)(11.1581 - 9.2950) = 1.8631A_w$

Since the last withdrawal exhausts the fund,

$$P_{\text{deposits}} = P_{\text{withdrawals}}$$
$$\$368.00 = 1.8631 A_w$$
$$A_w = \boxed{\$197.52}$$

15. (a) Number of compounding periods:

$$n = \frac{\$2000}{\$400} = 5$$

Effective interest rate:

$$i = \frac{r}{k} = \frac{0.10}{12} = 0.0083$$

$$\text{payment} = \frac{\text{monthly}}{\text{payment}} + \frac{\text{interest on}}{\text{unpaid balance}}$$

first month:
$$\text{payment} = \$400 + (\$2000)(0.0083) = \boxed{\$417}$$

second month:
$$\text{payment} = \$400 + (\$2000 - \$400)(0.0083) = \boxed{\$413}$$

third month:
$$\text{payment} = \$400 + (\$2000 - \$800)(0.0083) = \boxed{\$410}$$

fourth month:
$$\text{payment} = \$400 + (\$2000 - \$1200)(0.0083) = \boxed{\$407}$$

fifth month:
$$\text{payment} = \$400 + (\$2000 - \$1600)(0.0083) = \boxed{\$403}$$

(b) Principal remaining after the third payment is
$$\$2000 - (3)(\$400) = \boxed{\$800}$$

(c) Interest on the fourth payment is $\$407 - \$400 = \boxed{\$7}$

16. $P = \$12,000 - (\$2000)(P/F, 10\%, 10)$
$\qquad + (\$1000)(P/A, 10\%, 10)$
$\qquad + (\$200)(P/G, 10\%, 10)$
$\quad = \$12,000 - (\$2000)(0.3855) + (\$1000)(6.1446)$
$\qquad + (\$200)(22.8913)$
$\quad = \boxed{\$21,950}$

17. Amount of money saved per year:

$$A = \left(40,000 \, \frac{\text{pieces}}{\text{yr}}\right)\left(\frac{7 \text{ sec}}{\text{piece}}\right)\left(\frac{\text{hr}}{3600 \text{ sec}}\right)\left(\frac{\$15.00}{\text{hr}}\right)$$
$$= \$1166.67/\text{yr}$$

For a 3-year life, $n = 3$, the maximum purchase price, P, is

$$P = A(P/A, 8\%, 3) = (\$1166.67)(2.57466)$$
$$= \boxed{\$3003.78}$$

18. The service has an infinite life; the capitalized cost, CC, is

$$CC = \text{initial cost} + \frac{\text{actual cost}}{i}$$
$$= \$100,000 + \frac{\$18,000}{0.08} = \boxed{\$325,000}$$

19. (a) Brand A:

$$CC = \$120 + \frac{(\$400)(1 - 0.93)}{0.10} = \boxed{\$400}$$

Brand B:

$$CC = \$70 + \frac{(\$400)(1 - 0.87)}{0.10} = \boxed{\$590}$$

(b) $\boxed{\text{Brand A is superior}}$

20. $\text{EUAC} = C_1(A/P, 6\%, 5) - S_1(A/F, 6\%, 5)$
$\qquad + C_2(A/P, 6\%, 5) - S_2(A/F, 6\%, 5)$
$\qquad + \text{maintenance}$
$\quad = (\$17,000)(0.2374) - (\$14,000)(0.1774)$
$\qquad + (\$5000)(0.2374) - (\$2500)(0.1774)$
$\qquad + \$200$
$\quad = \boxed{\$2495.70}$

21. Aluminum:

$$\text{EUAC} = (\$6000)(A/P, 10\%, 50)$$
$$= (\$6000)(0.1009) = \$605$$

Shingles:

$$\text{EUAC} = (\$3500)(A/P, 10\%, 15)$$
$$= (\$3500)(0.1315) = \$460$$

$$\boxed{\text{shingles are superior}}$$

22. Alternative A:
$$\text{EUAC} = C(A/P, 12\%, 20) + \text{other annual costs}$$
$$- S(A/F, 12\%, 20)$$
$$= (\$90,000)(0.1339)$$
$$+ (\$3000 + \$2200 + \$400)$$
$$- (\$10,000)(0.0139) = \$17,512$$

Alternative B:
$$\text{EUAC} = (\$60,000)(0.1339) + (\$5000 + \$3000)$$
$$- (\$6000)(0.139) = \$15,950$$

$$\boxed{\text{alternative B is best}}$$

23. $A = \left(\begin{array}{c}\text{amount}\\\text{of loan}\end{array}\right)(A/P, i\%, 12)$
$$= (\$14,000 - \$4000)(A/P, i\%, 12)$$

$$(A/P, i\%, 12) = \frac{A}{\$10,000} = \frac{\$1200}{\$10,000} = 0.1200$$

From the table,
$$(A/P, 6\%, 12) = 0.1193$$
$$(A/P, 7\%, 12) = 0.1259$$

$$i = 6\% + (7\% - 6\%)\left(\frac{0.1200 - 0.1193}{0.1259 - 0.1193}\right) = \boxed{6.1\%}$$

24. $F = (\$14,000 + \$1000)(F/P, 10\%, 10)$
$$+ [\$150 + \$250 - (\$75)(12)](F/A, 10\%, 10)$$
$$= (\$15,000)(2.5937) - (\$500)(15.9374)$$

$$= \boxed{\$30,937}$$

25. $P_A(0.08) = P_B(0.15) + (P_A - P_B)i$
$$i = \frac{P_A(0.08) - P_B(0.15)}{P_A - P_B}$$
$$= \frac{(\$40,000)(0.08) - (\$10,000)(0.15)}{\$40,000 - \$10,000}$$
$$= 0.0567 = \boxed{5.67\%}$$

26. (a) $B - C = \$500,000 - \$175,000 - \$50,000$
$$= \boxed{\$275,000}$$

(b) $\dfrac{B}{C} = \dfrac{\$500,000 - \$50,000}{\$175,000} = \boxed{2.57}$

27. (a) $\dfrac{B}{C} = \dfrac{\$1,500,000 - \$300,000}{\$1,000,000} = \boxed{1.2}$

(b) $B - C = \$1,500,000 - \$300,000 - \$1,000,000$
$$= \boxed{\$200,000}$$

28. $\text{EUAC of capital} = (P - \text{salvage of defender})$
$$\times (A/P, 8\%, n) + Si$$
$\text{EUAC of maintenance} = \text{EUAC of operating costs}$

year 1:
$$\text{total EUAC} = (\$10,000 - \$6000)(1.0800)$$
$$+ (\$6000)(0.08) + \$2300$$
$$= \$7100$$

year 2:
$$\text{total EUAC} = (\$10,000 - \$4000)(0.5608)$$
$$+ (\$4000)(0.08) + [(\$2300)(P/F, 8\%, 1)$$
$$+ (\$2500)(P/F, 8\%, 2)](A/P, 8\%, 2)$$
$$= \$3685 + [(\$2300)(0.9259)$$
$$+ (\$2500)(0.8573)](0.5608)$$
$$= \$6081$$

year 3:
$$\text{total EUAC} = (\$10,000 - \$3200)(0.3880)$$
$$+ (\$3200)(0.08) + [(\$2300)(0.9259)$$
$$+ (\$2500)(0.8573)$$
$$(\$3300)(P/F, 8\%, 3)](A/P, 8\%, 3)$$
$$= \$2894 + [\$2130 + \$2143$$
$$+ (\$3300)(0.7938)](0.3880)$$
$$= \$5568$$

year 4:
$$\text{total EUAC} = (\$10,000 - \$2500)(0.3019)$$
$$+ (\$2500)(0.08) + [\$2130 + \$2143$$
$$+ \$2620 + (\$4800)(P/F, 8\%, 4)]$$
$$\times (A/P, 8\%, 4)$$
$$= \$2464 + [\$6893 + (\$4800)(0.7350)]$$
$$\times (0.3019)$$
$$= \$5610$$

$$\boxed{\text{Replace the asset at the end of year 3.}}$$

29.   EUAC of capital $= P(A/P, 10\%, n)$

EUAC of maintenance $= A + (\$100)(A/G, 10\%, n)$

year	EUAC of capital	EUAC of maintenance	total EUAC
1	($6000)(1.1000)	$400 + ($100)(0)	$7000
2	($6000)(0.5762)	$400 + ($100)(0.4762)	$3905
3	($6000)(0.4021)	$400 + ($100)(0.9366)	$2906
4	($6000)(0.3155)	$400 + ($100)(1.3812)	$2431
5	($6000)(0.2638)	$400 + ($100)(1.8101)	$2164
6	($6000)(0.2296)	$400 + ($100)(2.2236)	$2000
7	($6000)(0.2054)	$400 + ($100)(2.6216)	$1895
8	($6000)(0.1874)	$400 + ($100)(3.0045)	$1825
9	($6000)(0.1736)	$400 + ($100)(3.3724)	$1779
10	($6000)(0.1627)	$400 + ($100)(3.7255)	$1749
11	($6000)(0.1540)	$400 + ($100)(4.0641)	$1730
12	($6000)(0.1468)	$400 + ($100)(4.3884)	$1720
13	($6000)(0.1408)	$400 + ($100)(4.6988)	$1715
14	($6000)(0.1357)	$400 + ($100)(4.9955)	$1714
15	($6000)(0.1315)	$400 + ($100)(5.2789)	$1717

> The car should be replaced at the end of year 14.

30. SL:

$$\text{annual depreciation} = \frac{\$500,000 - \$100,000}{25 \text{ yr}}$$

$$= \$16,000/\text{yr}$$

$$D_1 = D_2 = D_3 = \boxed{\$16,000}$$

DDB:

$$D_j = \left(\frac{2C}{n}\right)\left(1 - \frac{2}{n}\right)^{j-1}$$

$$D_1 = \left[\frac{(2)(\$500,000)}{25}\right]\left(1 - \frac{2}{25}\right)^{1-1} = \boxed{\$40,000}$$

$$D_2 = \left[\frac{(2)(\$500,000)}{25}\right]\left(1 - \frac{2}{25}\right)^{2-1} = \boxed{\$36,800}$$

$$D_3 = \left[\frac{(2)(\$500,000)}{25}\right]\left(1 - \frac{2}{25}\right)^{3-1} = \boxed{\$33,860}$$

SOYD: $T = \frac{1}{2}n(n+1) = \left(\frac{1}{2}\right)(25)(25+1) = 325$

Depreciation in year j:

$$D_j = \frac{(c - S_n)(n - J + 1)}{T}$$

$$D_1 = \frac{(\$500,000 - \$100,000)(25 - 1 + 1)}{325} = \boxed{\$30,770}$$

$$D_2 = \frac{(\$500,000 - \$100,000)(25 - 2 + 1)}{325} = \boxed{\$29,540}$$

$$D_3 = \frac{(\$500,000 - \$100,000)(25 - 3 + 1)}{325} = \boxed{\$28,310}$$

31. SOYD: $T = \frac{1}{2}n(n+1) = \left(\frac{1}{2}\right)(6)(6+1) = 21$

year 5:

$$D_5 = \frac{(C - S_n)(n - j + 1)}{T}$$

$$= \frac{(\$12,000 - \$2000)(6 - 5 + 1)}{21}$$

$$= \boxed{\$952}$$

DDB:

$$D_j = \left(\frac{2C}{n}\right)\left(1 - \frac{2}{n}\right)^{j-1}$$

$$D_5 = \left[\frac{(2)(\$12,000)}{6}\right]\left(1 - \frac{2}{6}\right)^{5-1} = \$790$$

Check to see if the book value has dropped below the salvage value (which is not permitted).

$$BV_5 = (\$12,000)\left(1 - \frac{2}{6}\right)^5 = 1580$$

Since the book value would be less than $2000, the maximum depreciation allowed is the difference between the previous book value and $2000.

$$BV_4 = (\$12,000)\left(1 - \frac{2}{6}\right)^4 = \$2370$$

$$D_5 = BV_4 - S_5 = \$2370 - \$2000 = \boxed{\$370}$$

SL: $D_5 = \frac{C - S_n}{n} = \frac{\$12,000 - \$2000}{6} = \boxed{\$1667}$

32. SL: $D = \frac{C - S}{n} = \frac{\$2500 - \$1100}{6} = \$233$

$$BV_1 = \$2500 - \$233 = \boxed{\$2267}$$

$$BV_2 = \$2267 - \$233 = \boxed{\$2034}$$

$$BV_3 = \$2034 - \$233 = \boxed{\$1801}$$

$$BV_4 = \$1801 - \$233 = \boxed{\$1568}$$

$$BV_5 = \$1568 - \$233 = \boxed{\$1335}$$

$$BV_6 = \$1335 - \$233 = \boxed{\$1102}$$

ECONOMICS

DDB:

$$D_j = \frac{2C}{n}\left(1 - \frac{2}{n}\right)^{j-1}$$

$$D_1 = \left[\frac{(2)(\$2500)}{6}\right]\left(1 - \frac{2}{6}\right)^{1-1} = \$833$$

$$BV_1 = \$2500 - \$833 = \boxed{\$1667}$$

$$D_2 = \left[\frac{(2)(\$2500)}{6}\right]\left(1 - \frac{2}{6}\right)^{2-1} = \$556$$

$$BV_2 = \$1667 - \$556 = \boxed{\$1111}$$

$$D_3 = \left(\frac{\$2500}{3}\right)\left(1 - \frac{2}{6}\right)^{3-1} = \$370$$

$$BV_3 = \$1111 - \$370 = \$741$$

Since book value cannot be less than salvage value,

$$BV_3 = \boxed{\$1100}$$

$$D_3 = \$1111 - \$1100 = \$11$$

SOYD: $T = \frac{1}{2}n(n+1) = \left(\frac{1}{2}\right)(6)(6+1) = 21$

$$D_j = \frac{(C - S_n)(n - j + 1)}{T}$$

$$C - S_n = \$2500 - \$1100 = \$1400$$

$$D_1 = \frac{(\$1400)(6 - 1 + 1)}{21} = \$400$$

$$BV_1 = \$2500 - \$400 = \boxed{\$2100}$$

$$D_2 = \frac{(\$1400)(6 - 2 + 1)}{21} = \$333$$

$$BV_2 = \$2100 - \$333 = \boxed{\$1767}$$

$$D_3 = \frac{(\$1400)(6 - 3 + 1)}{21} = \$267$$

$$BV_3 = \$1767 - \$267 = \boxed{\$1500}$$

$$D_4 = \frac{(\$1400)(6 - 4 + 1)}{21} = \$200$$

$$BV_4 = \$1500 - \$200 = \boxed{\$1300}$$

$$D_5 = \frac{(\$1400)(6 - 5 + 1)}{21} = \$133$$

$$BV_5 = \$1300 - \$133 = \boxed{\$1167}$$

$$D_6 = \frac{(\$1400)(6 - 6 + 1)}{21} = \$67$$

$$BV_6 = \$1167 - \$67 = \boxed{\$1100}$$

sinking fund:

$$D_j = (C - S_n)(A/F, i\%, n)(F/P, i\%, j - 1)$$
$$C - S_n = \$2500 - \$1100 = \$1400$$

$$D_1 = (\$1400)(A/F, 6\%, 6)(F/P, 6\%, 0)$$
$$= (\$1400)(0.1434)(1) = \$201$$

$$BV_1 = \$2500 - \$201 = \boxed{\$2299}$$

$$D_2 = (\$201)(F/P, 6\%, 1) = (\$201)(1.0600) = \$213$$

$$BV_2 = \$2299 - \$213 = \boxed{\$2086}$$

$$D_3 = (\$201)(F/P, 6\%, 2) = (\$201)(1.1236) = \$226$$

$$BV_3 = \$2086 - \$226 = \boxed{\$1860}$$

$$D_4 = (\$201)(F/P, 6\%, 3) = (\$201)(1.1910) = \$239$$

$$BV_4 = \$1860 - \$239 = \boxed{\$1621}$$

$$D_5 = (\$201)(F/P, 6\%, 4) = (\$201)(1.2625) = \$254$$

$$BV_5 = \$1621 - \$254 = \boxed{\$1367}$$

$$D_6 = (\$201)(F/P, 6\%, 5) = (\$201)(1.3382) = \$269$$

$$BV_6 = \$1367 - \$269 = \boxed{\$1098}$$

33. SL: Annual depreciation, $D = \$233$

$$DR = (0.53)D(P/A, 6\%, 6)$$
$$= (0.53)(\$233)(4.9173) = \$607$$

DDB: $DR = (0.53)[D_1(P/F, 6\%, 1) + D_2(P/F, 6\%, 2)$
$$+ D_3(P/F, 6\%, 3)]$$
$$= (0.53)[(\$833)(0.9434) + (\$556)(0.8900)$$
$$+ (\$11)(0.8396)]$$
$$= \boxed{\$684}$$

SOYD: The depreciation begins at \$400 per year and decreases \$67 per year after the first year.

$$A = \$400 \qquad G = \$67$$

$$DR = (0.53)[A(P/A, 6\%, 6) - G(P/G, 6\%, 6)]$$
$$= (0.53)[(\$400)(4.9173) - (\$67)(11.4594)]$$
$$= \boxed{\$635}$$

sinking fund: The present worth at the end of each year of depreciation is equal to the depreciation at the end of the first year ($201).

$$DR = (0.53)[(\$201)(6)(P/F, 6\%, 1)]$$

$$= (0.53)[(\$201)(6)(0.9434)] = \boxed{\$603}$$

34. Annual depreciation

$$= \frac{\$80,000}{25} = \$3200$$

annual taxable income

$$= \$22,500 - \$12,000 - \$3200$$

$$= \$7300$$

annual income (net)

$$= \$22,500 - \$12,000 - (0.53)(\$7300)$$

$$= \$6631$$

$$P_t = -\$80,000 + A(P/A, 10\%, 25)$$

$$= -\$80,000 + (\$6631)(9.0770) = \boxed{-\$19,810}$$

35. $\quad T = \frac{1}{2}n(n+1) = \left(\frac{1}{2}\right)(25)(25+1) = 325$

End of the first year depreciation:

$$D_1 = \frac{Cn}{T} = \frac{(\$80,000)(25)}{325} = \$6154$$

Decreasing depreciation gradient:

$$G = \frac{C}{T} = \frac{\$80,000}{325} = \$246$$

$$P_t = -\$80,000 + (\$22,500 - \$12,000)(P/A, 10\%, 25)$$
$$\times (0.47) + (\$6154)(P/A, 10\%, 25)(0.53)$$
$$- (\$246)(P/G, 10\%, 25)(0.53)$$
$$= -\$80,000 + (\$10,500)(9.0770)(0.47)$$
$$+ (\$6154)(9.0770)(0.53)$$
$$- (\$246)(67.6964)(0.53)$$

$$= \boxed{-\$14,430}$$

36. SL: $\quad\dfrac{\text{annual}}{\text{depreciation}} = \dfrac{C}{n} = \dfrac{\$2000}{4} = \$500$

$$P_t = -\$2000 + (\$1200)(1 - 0.30)(P/A, 8\%, 4)$$
$$+ D(0.30)(P/A, 8\%, 4)$$
$$= -\$2000 + (\$1200)(0.70)(3.3121)$$
$$+ (\$500)(0.30)(3.3121)$$

$$= \boxed{\$1279}$$

SOYD: $\quad T = \frac{1}{2}n(n+1) = \left(\frac{1}{2}\right)(4)(4+1) = 10$

End of the first year depreciation:

$$D_1 = \frac{Cn}{T} = \frac{(\$2000)(4)}{10} = \$800$$

Decreasing depreciation gradient:

$$G = \frac{C}{T} = \frac{\$2000}{10} = \$200$$

$$P_t = -\$2000 + (\$1200)(0.70)(P/A, 8\%, 4)$$
$$+ (\$800)(0.30)(P/A, 8\%, 4)$$
$$- (\$200)(0.30)(P/G, 8\%, 4)$$
$$= -\$2000 + (\$1200)(0.70)(3.3121)$$
$$+ (\$800)(0.30)(3.3121)$$
$$- (\$200)(0.30)(4.6501)$$

$$= \boxed{\$1298}$$

37. Effective annual rate:

$$i = (1 + \phi)^k - 1 = (1 + 0.015)^{12} - 1$$

$$= 0.1956 = \boxed{19.56\%}$$

38. $P = A(P/A, i\%, 12)$

$$(P/A, i\%, 12) = \frac{P}{A} = \frac{\$100}{\$9.46} = 10.5708$$

At $n = 12$ in the (P/A) column, $\phi = 2\%$

Effective annual rate:

$$i = (1 + \phi)^k - 1 = (1 + 0.02)^{12} - 1$$

$$= 0.2682 = \boxed{26.82\%}$$

39. $P = A(P/A, i\%, n)$

$$(P/A, i\%, 30) = \frac{\$2000}{\$89.30} = 22.3964$$

From the table, monthly effective interest rate is 2%.

$$\frac{\text{effective annual}}{\text{interest rate}} = (1 + 0.02)^{12} - 1 = 0.2682 = \boxed{26.82\%}$$

40. (a) average unit cost of first 240 units

$$= \frac{\$3400}{240 \text{ units}} = \boxed{\$14.17/\text{unit}}$$

(b) incremental cost

$$= \frac{\$4000 - \$3400}{360 \text{ units} - 240 \text{ units}} = \frac{\$600}{120 \text{ units}}$$

$$= \boxed{\$5.00/\text{unit}}$$

(c) fixed cost

$$= \$3400 - (\$5.00/\text{unit})(240 \text{ units})$$

$$= \boxed{\$2200}$$

(d) Without the fixed cost:

$$\text{profit} = R - C = (10)(\$10.47 - \$5) = \boxed{\$54.70}$$

With the fixed cost:

$$\text{profit} = (10)(\$10.47) - (10)\left(\$5.00 + \frac{\$2200}{249}\right)$$

$$= -\$33.65 = \boxed{\text{loss of } \$33.65}$$

41. Annual power cost:

$$A_1 = \left[\frac{(2000 \text{ units})\left(48 \frac{\text{min}}{\text{unit}}\right)}{60 \frac{\text{min}}{\text{hr}}}\right]\left(\frac{\$2.15}{\text{hr}}\right) = \$3440$$

Annual labor cost:

$$A_2 = (1600 \text{ hr})\left(\frac{\$12.90}{\text{hr}}\right) = \$20,640$$

unit cost =

$$\frac{C(A/P,10\%,8) - S_8(A/F,10\%,8) + A_1 + A_2 + \overset{\text{operation}}{\underset{\text{cost}}{}}}{2000 \text{ units}}$$

$$= \frac{(\$40,000)(0.1874) - (\$5000)(0.0874) + \$3440 + \$20,640 + \$800}{2000 \text{ units}}$$

$$= \boxed{\$15.97/\text{unit}}$$

42. (a) total cost $= (700,000)(0.62)(\$0.348)$

$$+ \$190,000$$

$$= \$341,032$$

$$\text{profit} = R - C = \$430,000 - \$341,032$$

$$= \boxed{\$88,968}$$

(b) $\quad p = \dfrac{\$430,000}{(700,000 \text{ units})(0.62)} = \$0.991/\text{unit}$

break-even point:

$$Q^* = \frac{f}{p-a} = \frac{\$190,000}{\dfrac{\$0.991}{\text{unit}} - \dfrac{\$0.348}{\text{unit}}} = \boxed{295,500 \text{ units}}$$

43. annual amount saved

$$= (3500)(\$0.06) = \$210$$

pay-back period (traditional definition):

$$= \frac{\text{initial investment}}{\text{net annual profit}} = \frac{\$700}{\dfrac{\$210}{\text{yr}} - \dfrac{\$40}{\text{yr}}}$$

$$= \boxed{4.12 \text{ yr}}$$

pay-back period at 10% interest:

$$P = A(P/A, 10\%, n)$$

$$\$700 = (\$210 - \$40)(P/A, 10\%, n)$$

$$(P/A, 10\%, n) = \frac{\$700}{\$170} = 4.1176 = \frac{(1+i)^n - 1}{i(1+i)^n}$$

$$(4.1176)(0.10)(1.10)^n - (1.10)^n = -1$$

$$n = \frac{\ln(1.6999)}{\ln(1.10)} = \boxed{5.57 \text{ yr}}$$

44. $\text{EUAC}_{\text{hand tool}} = (\$200)(A/P, 5\%, n)$

$$+ (4000)(\$1.21)$$

$$\text{EUAC}_{\text{machine tool}} = (\$3600)(A/P, 5\%, n)$$

$$+ (4000)(\$0.75)$$

$$(\$3400)(A/P, 5\%, n) = \$1840$$

$$(A/P, 5\%, n) = \frac{\$1840}{\$3400} = 0.5412$$

$$0.5412 = \frac{i(1+i)^n}{(1+i)^n - 1}$$

$$= \frac{(0.05)(1+0.05)^n}{(1+0.05)^n - 1}$$

$$(0.05)(1.05)^n = (0.5412)(1.05)^n - 0.5412$$

$$(1.05)^n = \frac{0.5412}{0.4912} = 1.1018$$

Take the log of both sides of the equation.

$$n = \frac{\log(1.1018)}{\log(1.05)} = \boxed{1.99 \text{ yr}}$$

ECONOMICS

SUPPLEMENTARY PRACTICE PROBLEMS

1. Present worth of the second tank and pumping system 10 years from now:

$$P_{10} = \$60,000 + \$7000 + (\$800)$$
$$\times (P/A, 10\%, 8) + (\$100)(P/G, 10\%, 8)$$
$$= \$67,000 + (\$800)(5.3349)$$
$$+ (\$100)(16.0287)$$
$$= \$72,870$$

$$CC = \$40,000 + P_{10}(P/F, 10\%, 10)$$
$$= \$40,000 + (\$72,870)(0.3855)$$
$$= \boxed{\$68,090}$$

2. Kept for 3 years:

$$EUAC(3) = (\$28,000)(A/P, 10\%, 3)$$
$$+ \$600 + (\$300)(A/G, 10\%, 3)$$
$$- (\$15,000)(A/F, 10\%, 3)$$
$$= (\$28,000)(0.4021) + \$600$$
$$+ (\$300)(0.9366)$$
$$- (\$15,000)(0.3021)$$
$$= \boxed{\$7608}$$

Kept for 6 years:

$$EUAC(6) = [(\$28,000)(A/P, 10\%, 6)$$
$$- (\$10,000)(A/F, 10\%, 6)]$$
$$+ [(\$600)(P/A, 10\%, 3)$$
$$+ (\$300)(P/G, 10\%, 3)$$
$$+ (\$7000)(P/F, 10\%, 3)$$
$$+ (\$900)(P/F, 10\%, 4)$$
$$+ (\$1150)(P/F, 10\%, 5)$$
$$+ (\$1300)(P/F, 10\%, 6)](A/P, 10\%, 6)$$
$$= [(\$28,000)(0.2296) - (\$10,000)(0.1296)]$$
$$+ [(\$600)(2.4869) + (\$300)(2.3291)$$
$$+ (\$7000)(0.7513) + (\$900)(0.6830)$$
$$+ (\$1150)(0.6209)$$
$$+ (\$1300)(0.5645)](0.2296)$$
$$= \$5132 + (\$9512)(0.2296)$$
$$= \boxed{\$7317}$$

Kept for 8 years:

$$EUAC(8) = (\$28,000)(A/P, 10\%, 8)$$
$$+ [(\$600)(P/A, 10\%, 3)$$
$$+ (\$300)(P/G, 10\%, 3)$$
$$+ (\$7000)(P/F, 10\%, 3)$$
$$+ (\$900)(P/F, 10\%, 4)$$
$$+ (\$1150)(P/F, 10\%, 5)$$
$$+ (\$1300 + \$14,000)(P/F, 10\%, 6)$$
$$+ (\$9900)(P/F, 10\%, 7)$$
$$+ (\$9900)(P/F, 10\%, 8)](A/P, 10\%, 8)$$
$$= (\$28,000)(0.1874) + [(\$600)(2.4869)$$
$$+ (\$300)(2.3291)$$
$$+ (\$7000)(0.7513) + (\$900)(0.6830)$$
$$+ (\$1150)(0.6209) + (\$15,300)(0.5645)$$
$$+ (\$9900)(0.5132)$$
$$+ (\$9900)(0.4665)](0.1874)$$
$$= \$5247 + (\$27,115)(0.1874)$$
$$= \boxed{\$10,328}$$

3. After-tax revenue per year

$$= (\$1.20)\left(\frac{3000}{\text{month}}\right)\left(\frac{12\,\text{month}}{\text{yr}}\right)(1 - 0.45)$$
$$= \$23,760/\text{yr}$$

For annual depreciation, assume straight-line method.

$$D = \frac{C}{n} = \frac{\$200,000}{20}$$
$$= \$10,000/\text{yr}$$

annual production cost

$$= (\$1.10)\left(\frac{3000}{\text{month}}\right)\left(\frac{12\,\text{month}}{\text{yr}}\right)$$
$$= \$39,600\,\text{yr}$$

total after-tax annual cost

$$= (-\$10,000)(0.45)$$
$$+ (\$20,000 + \$10,000 + \$39,600)(1 - 0.45)$$
$$= \$33,780/\text{yr}$$

$$\text{cost} > \text{revenue}$$

$$\boxed{\text{The factory should be shut down.}}$$

4. Assume an 8-hour workday.

$$\text{time required} = \frac{200\,\text{units}}{\left(6\,\dfrac{\text{unit}}{\text{hr}}\right)\left(\dfrac{8\,\text{hr}}{\text{day}}\right)}$$
$$= 4.17\,\text{days or 5 days}$$

$$\text{cost per item} = \frac{\$20 + (5)(\$100) + (200)(\$1.00)}{200}$$

$$= \boxed{\$3.60}$$

5. Notice in this problem that no interest rate is given, so a rigorous (time value of money) analysis cannot be performed.

$$\text{first year: } A_1 = \text{annual cost} + \frac{\text{average drop}}{\text{in value}}$$

$$= \$1000 + \frac{\$8000 - \$3900}{1}$$

$$= \$5100$$

$$\text{second year: } A_2 = \$1000 + \frac{\$8000 - (\$3900 - \$550)}{2}$$

$$= \$3325$$

$$\text{third year: } A_3 = \$1000 + \frac{\$8000 - (\$3900 - \$550 \times 2)}{3}$$

$$= \$2733$$

$$\text{fourth year: } A_4 = \$1000 + \frac{\$8000 - (\$3900 - \$550 \times 3)}{4}$$

$$= \$2437$$

$$\text{fifth year: } A_5 = \$1000 + \frac{\$8000 - (\$3900 - \$550 \times 4)}{5}$$

$$= \$2260$$

$$\text{sixth year: } A_6 = \$1000 + \frac{\$8000 - (\$3900 - \$550 \times 5)}{6}$$

$$= \$2142$$

$$\text{seventh year: } A_7 = \$1000 + \frac{\$8000 - (\$3900 - \$550 \times 6)}{7}$$

$$= \$2057$$

$$\text{eighth year: } A_8 = \$1000 + \frac{\$8000 - (\$3900 - \$550 \times 7)}{8}$$

$$= \$1994$$

In the ninth year, the salvage value is zero.

$$A_9 = \$1000 + \frac{\$8,000}{9} = \$1899$$

$$A_{10} = \$1000 + \frac{\$8,000}{10} = \$1800$$

$$\boxed{\text{Keep the car for at least 10 years.}}$$

6. $P = -\$50,000 - (\$10,000)(P/A, 4\%, 20)$

$\quad - (\$1,000,000)(0.04)(P/A, 4\%, 20)$

$\quad - (\$1,000,000)(P/F, 4\%, 20) + \$970,000$

$= -\$50,000 - (\$10,000)(13.5903)$

$\quad - (\$1,000,000)(0.04)(13.5903)$

$\quad - (\$1,000,000)(0.4564) + \$970,000$

$= \boxed{-\$215,900}$

7. Using the present worth method,

$\$1000 = (\$140)(P/F, i\%, 1) + (\$150)(P/F, i\%, 2)$
$\quad + (\$170)(P/F, i\%, 3) + (\$50)(P/F, i\%, 4)$
$\quad + (\$730)(P/F, i\%, 5)$

Try $i = 6\%$

$P = (\$140)(0.9434) + (\$150)(0.8900)$
$\quad + (\$170)(0.8396) + (\$50)(0.7921)$
$\quad + (\$730)(0.7473)$
$= \$993$

Try $i = 5\%$

$P = (\$140)(0.9524) + (\$150)(0.9070)$
$\quad + (\$170)(0.8638) + (\$50)(0.8227)$
$\quad + (\$730)(0.7835)$
$= \$1029$

The rate of return is about 6%. The exact rate of return can be computed as follows: (x is the incremental rate)

$A_1[(P/F, 5\%, 1) - (P/F, 6\%, 1)]x$
$+A_2[(P/F, 5\%, 2) - (P/F, 6\%, 2)]x$
$+A_3[(P/F, 5\%, 3) - (P/F, 6\%, 3)]x$
$+A_4[(P/F, 5\%, 4) - (P/F, 6\%, 4)]x$
$+A_5[(P/F, 5\%, 5) - (P/F, 6\%, 5)]x$
$\qquad = P(5\%) - \text{investment}$

$(\$140)[(0.9524 - 0.9434)]x$
$+(\$150)[(0.9070 - 0.8900)]x$
$+(\$170)[(0.8638 - 0.8396)]x$
$+(\$50)[(0.8227 - 0.7921)]x$
$+(\$730)[(0.7835 - 0.7473)]x$
$\qquad = (\$35.88)x = \$1029 - \$1000$
$\qquad = \$29$
$\qquad x = \frac{\$29}{\$35.88} = 0.81$

PROFESSIONAL PUBLICATIONS, INC. ● Belmont, CA

$$i = 0.05 + \frac{0.81}{100}$$
$$= 0.0581 \text{ or } \boxed{5.81\%}$$

8. Unit of time is a year. $a = 1000$ item/yr

Fixed cost: $K = \$40$

Inventory cost: $h = \dfrac{\$0.80}{\text{item-yr}} + \dfrac{(0.12)(\$5.20)}{\text{item-yr}}$

$$= \$1.424/\text{item-yr}$$

$$Q^* = \sqrt{2\frac{aK}{h}}$$

$$= \sqrt{(2)\dfrac{\left(1000\,\dfrac{\text{item}}{\text{yr}}\right)(\$40)}{\dfrac{\$1.424}{\text{item-yr}}}}$$

$$= \boxed{237 \text{ items}}$$

9. Year-end compounding:

P_j is the present worth of year j return.

$$P_1 = (\$10{,}000)(F/P, 15\%, 0)(P/F, 5\%, 1)$$
$$= (\$10{,}000)(1)(0.9524)$$
$$= \$9524$$

$$P_2 = (\$10{,}000)(F/P, 15\%, 1)(P/F, 5\%, 2)$$
$$= (\$10{,}000)(1.1500)(0.9070)$$
$$= \$10{,}430$$

$$P_3 = (\$10{,}000)(F/P, 15\%, 2)(P/F, 5\%, 3)$$
$$= (\$10{,}000)(1.3225)(0.8638)$$
$$= \$11{,}424$$

$$P_4 = (\$10{,}000)(F/P, 15\%, 3)(P/F, 5\%, 4)$$
$$= (\$10{,}000)(1.5209)(0.8227)$$
$$= \$12{,}512$$

$$P_5 = (\$10{,}000)(F/P, 15\%, 4)(P/F, 5\%, 5)$$
$$= (\$10{,}000)(1.7490)(0.7835)$$
$$= \$13{,}703$$

$$P = \boxed{\$57{,}593}$$

Month-end compounding:

$$\phi_1 = \frac{r_1}{k} = \frac{15\%}{12}$$
$$= 1.25\%$$

$$\phi_2 = \frac{r_2}{k} = \frac{5\%}{12}$$
$$= 0.4167\%$$

Effective annual rate:

$$i_1 = (1 + \phi_1)^k - 1 = (1 + 0.0125)^{12} - 1$$
$$= 0.1608$$
$$i_2 = (1 + \phi_2)^k - 1 = (1 + 0.004167)^{12} - 1$$
$$= 0.0512$$

As in year-end compounding,

$$(F/P, 16.08\%, 0)(P/F, 5.12\%, 1) = \frac{(1 + 0.1608)^0}{(1 + 0.0512)^1}$$
$$= 0.9513$$
$$P_1 = (\$10{,}000)(0.9513) = \$9513$$

$$(F/P, 16.08\%, 1)(P/F, 5.12\%, 2) = \frac{(1 + 0.1608)^1}{(1 + 0.0512)^2}$$
$$= 1.0505$$
$$P_2 = (\$10{,}000)(1.0505) = \$10{,}505$$

$$(F/P, 16.08\%, 2)(P/F, 5.12\%, 3) = \frac{(1.1608)^2}{(1 + 0.0512)^3}$$
$$= 1.1600$$
$$P_3 = (\$10{,}000)(1.1600) = \$11{,}600$$

$$(F/P, 16.08\%, 3)(P/F, 5.12\%, 4) = \frac{(1.1608)^3}{(1.0512)^4}$$
$$= 1.2809$$
$$P_4 = (\$10{,}000)(1.2809) = \$12{,}809$$

$$(F/P, 16.08\%, 4)(P/F, 5.12\%, 5) = \frac{(1.1608)^4}{(1.0512)^5}$$
$$= 1.4145$$
$$P_5 = (\$10{,}000)(1.4145) = \$14{,}145$$

$$P = P_1 + P_2 + P_3 + P_4 + P_5$$
$$= \$9513 + \$10{,}505 + \$11{,}600$$
$$+ \$12{,}809 + \$14{,}145$$
$$= \boxed{\$58{,}572}$$

10. The corrected inflation rate is i'.

$$i' = i + e + ie = 0.10 + 0.08 + (0.10)(0.08)$$
$$= 0.188$$

$$P = A(P/A, 18.8\%, 10) + G(P/G, 18.8\%, 10)$$

$$= A\left[\frac{(1+i)^n - 1}{i(1+i)^n}\right]$$

$$+ G\left[\frac{(1+i)^n - 1}{i^2(1+i)^n} - \frac{n}{i(1+i)^n}\right]$$

$$= (\$100,000)\left[\frac{(1+0.188)^{10} - 1}{(0.188)(1+0.188)^{10}}\right]$$

$$+ (\$10,000)\left[\frac{(1+0.188)^{10} - 1}{(0.188)^2(1+0.188)^{10}}\right.$$

$$\left. - \frac{10}{(0.188)(1+0.188)^{10}}\right]$$

$$= \boxed{\$574,300}$$

11. DDB: Depreciation in any year $j = \frac{2C}{n}\left(1 - \frac{2}{n}\right)^{j-1}$

$$\frac{2C}{n} = \frac{(2)(\$40,000)}{12} = \$6667$$

$$\left(1 - \frac{2}{n}\right) = \left(1 - \frac{2}{12}\right) = 0.8333$$

$$D_1 = (\$6667)(0.8333)^{1-1} = \$6667$$
$$BV_1 = \$40,000 - \$6667 = \$33,333$$
$$D_2 = (\$6667)(0.8333)^{2-1} = \$5556$$
$$BV_2 = \$33,333 - \$5556 = \$27,777$$
$$D_3 = (\$6667)(0.8333)^{3-1} = \$4629$$
$$BV_3 = \$23,148$$
$$D_4 = (\$6667)(0.8333)^{4-1} = \$3858$$
$$BV_4 = \$19,290$$
$$D_5 = (\$6667)(0.8333)^{5-1} = \$3215$$
$$BV_5 = \$16,075$$
$$D_6 = (\$6667)(0.8333)^{6-1} = \$2679$$
$$BV_6 = \$13,396$$
$$D_7 = (\$6667)(0.8333)^{7-1} = \$2232$$
$$BV_7 = \$11,164$$
$$D_8 = (\$6667)(0.8333)^{8-1} = \$1860$$
$$BV_8 = \$9304$$
$$D_9 = (\$6667)(0.8333)^{9-1} = \$1550$$
$$BV_9 = \$7754$$
$$D_{10} = (\$6667)(0.8333)^{10-1} = \$1292$$
$$BV_{10} = \$6462$$
$$D_{11} = \$463$$
$$BV_{11} = \$5999$$

Present worth of the depreciation recovery of the year $j = D_j(P/F, 6\%, j)$

$$P_1(\text{DR}) = D_1(P/F, 6\%, 1) = (\$6667)(0.9434)$$
$$= \$6290$$
$$P_2(\text{DR}) = D_2(P/F, 6\%, 2) = (\$5556)(0.8900)$$
$$= \$4945$$
$$P_3(\text{DR}) = D_3(P/F, 6\%, 3) = (\$4629)(0.8396)$$
$$= \$3886$$
$$P_4(\text{DR}) = D_4(P/F, 6\%, 4) = (\$3858)(0.7921)$$
$$= \$3056$$
$$P_5(\text{DR}) = D_5(P/F, 6\%, 5) = (\$3215)(0.7473)$$
$$= \$2402$$
$$P_6(\text{DR}) = D_6(P/F, 6\%, 6) = (\$2679)(0.7050)$$
$$= \$1889$$
$$P_7(\text{DR}) = D_7(P/F, 6\%, 7) = (\$2232)(0.6651)$$
$$= \$1484$$
$$P_8(\text{DR}) = D_8(P/F, 6\%, 8) = (\$1860)(0.6274)$$
$$= \$1167$$
$$P_9(\text{DR}) = D_9(P/F, 6\%, 9) = (\$1550)(0.5919)$$
$$= \$917$$
$$P_{10}(\text{DR}) = D_{10}(P/F, 6\%, 10) = (\$1292)(0.5584)$$
$$= \$721$$
$$P_{11}(\text{DR}) = D_{11}(P/F, 6\%, 11) = (\$463)(0.5268)$$
$$= \$244$$

$$\sum P_j(\text{DR}) = \$27,001$$

$$P = -\$40,000 + (\$30,000 - \$23,000)$$
$$\times (1 - 0.52)(P/A, 6\%, 12) + (\$27,001)(0.52)$$
$$+ (\$6000)(P/F, 6\%, 12)$$
$$= -\$40,000 + (\$7000)(0.48)(8.3838)$$
$$+ \$14,040 + (\$6000)(0.4970)$$
$$= \boxed{\$5192}$$

12. The available fund is \$7120. Unused funds will be invested at 5%.

$$\text{EUAC}(\$4500) = (\$4500)(A/P, 8\%, 10) + \$2700$$
$$= (\$4500)(0.1490) + \$2700$$
$$= \$3371$$

$$\text{EUAC}(\$5700) = (\$5700)(A/P, 8\%, 10) + \$2500$$
$$= (\$5700)(0.1490) + \$2500$$
$$= \$3350$$

ECONOMICS

$$\text{EUAC(\$6750)} = (\$6750)(0.1490) + \$2200$$
$$= \$3206$$

$$\text{EUAC(\$7120)} = (\$7120)(0.1490) + \$2030$$
$$= \$3090$$

The $7120 precipitator should be recommended.

PROFESSIONAL PUBLICATIONS, INC. ● Belmont, CA

FLUID PROPERTIES

1. $p_{ab} = p_{atm} - p_{vac} = \gamma h - p_{vac}$

$$= \left(0.491 \frac{lbf}{in^3}\right)(29\ in) - 9.5\frac{lbf}{in^2}$$

$$= \boxed{4.74\ lbf/in^2\ (psia)}$$

2. $\rho = \dfrac{mass}{volume} = \dfrac{m_{water} + m_{liquid}}{V_{water} + V_{liquid}}$

$$= \frac{m_{water} + m_{liquid}}{\dfrac{m_{water}}{\rho_{water}} + \dfrac{m_{liquid}}{\rho_{liquid}}}$$

$$= \frac{0.7\ lbm + 1.1\ lbm}{\dfrac{0.7\ lbm}{62.4\dfrac{lbm}{ft^3}} + \dfrac{1.1\ lbm}{56\dfrac{lbm}{ft^3}}}$$

$$= \boxed{58.33\ lbm/ft^3}$$

$$v = \frac{1}{\rho} = \boxed{0.01714\ ft^3/lbm}$$

3. (a) $\rho = \dfrac{p}{RT} = \dfrac{p(MW)}{R^*T}$

$$= \frac{\left(14.8\dfrac{lbf}{in^2}\right)\left(4.003\dfrac{lbm}{lbmole}\right)\left(\dfrac{144\ in^2}{ft^2}\right)}{\left(1545.33\dfrac{ft\text{-}lbf}{lbmole\text{-}°R}\right)(460 + 68)°R}$$

$$= 0.0105\ lbm/ft^3$$

$$\gamma = \frac{g}{g_c}\rho = \left(\frac{32.2\dfrac{ft}{sec^2}}{32.2\dfrac{lbm\text{-}ft}{lbf\text{-}sec^2}}\right)\left(0.0105\frac{lbm}{ft^3}\right)$$

$$= \boxed{0.0105\ lbf/ft^3}$$

(b) $SG = \dfrac{\rho_{He}}{\rho_{air}} = \dfrac{0.0105\dfrac{lbm}{ft^3}}{0.0807\dfrac{lbm}{ft^3}} = \boxed{0.130}$

4. For component A,

$$p_A = \rho_A\frac{R^*}{(MW)_A}T = \left(\frac{m_A}{(MW)_A}\right)\left(\frac{R^*T}{V}\right) = n_A\frac{R^*T}{V}$$

$p = p_A + p_B + \cdots$

$$= (n_A + n_B + \cdots)\left(\frac{R^*T}{V}\right) = n\frac{R^*T}{V}$$

$$\boxed{x_A = \frac{n_A}{n} = \frac{p_A\left(\dfrac{V}{R^*T}\right)}{p\left(\dfrac{V}{R^*T}\right)} = \frac{p_A}{p}}$$

5. $\mu = \dfrac{\nu}{g_c}\rho$

$$= \left(\frac{6.0\times10^{-3}\dfrac{ft^2}{sec}}{32.2\dfrac{lbm\text{-}ft}{lbf\text{-}sec^2}}\right)(0.92)\left(62.4\frac{lbm}{ft^2}\right)$$

$$= 0.0107\ lbf\text{-}sec/ft^2$$

$$F = A_{cylinder}\mu_{oil}\frac{dv}{dy} = A_{cylinder}\mu_{oil}\frac{\Delta v}{\Delta y}$$

$$= (1.5\ ft)(2)(\pi)(1.5\ in)\left(\frac{ft}{12\ in}\right)$$

$$\times\left(0.0107\frac{lbf\text{-}sec}{ft^2}\right)\left[\frac{3\dfrac{ft}{sec}}{\left(\dfrac{3.2\ in - 3\ in}{2}\right)\left(\dfrac{ft}{12\ in}\right)}\right]$$

$$= \boxed{4.54\ lbf}$$

6. (a) $\nu = (1000\ cs)\left(1.0764\times10^{-5}\dfrac{\dfrac{ft^2}{sec}}{cs}\right)$

$$= \boxed{1.0764\times10^{-2}\ ft^2/sec}$$

(b) $\rho_{oil} = (SG_{oil})(\rho_{water})$

$\mu = \rho_{oil}\nu$

$$= (0.92)\left(1.94\frac{slug}{ft^3}\right)\left(1.0764\times10^{-2}\frac{ft^2}{sec}\right)\left(1.0\frac{lbf\text{-}sec^2}{ft\text{-}slug}\right)$$

$$= \boxed{1.92\times10^{-2}\ lbf\text{-}sec/ft^2}$$

FLUIDS Properties

(c) $\mu = \left(1.92 \times 10^{-2} \dfrac{\text{lbf-sec}}{\text{ft}^2}\right)\left(47.88 \dfrac{\text{Pa·s}}{\dfrac{\text{lbf-sec}}{\text{ft}^2}}\right)$

$= \boxed{0.919 \text{ Pa·s}}$

7. From a table of vapor pressures,

(a) $\boxed{0.00362 \text{ lbf/ft}^2 \text{ (psf)}}$

(b) $\boxed{122.4 \text{ lbf/ft}^2 \text{ (psf)}}$

(c) $\boxed{48.9 \text{ lbf/ft}^2 \text{ (psf)}}$

From the saturated steam (temperatures) table,

(d) $\boxed{0.9503 \text{ lbf/in}^2 \text{ (psi)}}$

(e) $\boxed{14.70 \text{ lbf/in}^2 \text{ (psi)}}$

8. (a) $h = \dfrac{p_a - p_v}{\gamma} = \dfrac{13.9 \dfrac{\text{lbf}}{\text{in}^2} - 2.5 \times 10^{-5} \dfrac{\text{lbf}}{\text{in}^2}}{0.491 \dfrac{\text{lbf}}{\text{in}^3}}$

$= \boxed{28.3 \text{ in}}$

(b) $h = \dfrac{p_a - p_v}{\gamma} = \dfrac{13.9 \dfrac{\text{lbf}}{\text{in}^2} - 0.34 \dfrac{\text{lbf}}{\text{in}^2}}{0.036 \dfrac{\text{lbf}}{\text{in}^3}} = \boxed{376.7 \text{ in}}$

(c) $h = \dfrac{p_a - p_v}{\gamma} = \dfrac{13.9 \dfrac{\text{lbf}}{\text{in}^2} - 0.85 \dfrac{\text{lbf}}{\text{in}^2}}{0.0285 \dfrac{\text{lbf}}{\text{in}^3}} = \boxed{457.9 \text{ in}}$

9. For a sphere,

$$V = \frac{4}{3}\pi r^3 = \frac{1}{6}\pi d^3$$

For ideal gases,

$$p_1 V_1 = p_2 V_2 \qquad p = \gamma h$$

$$(\gamma h_1)\left(\frac{1}{6}\pi d_1^3\right) = (\gamma h_2)\left(\frac{1}{6}\pi d_2^3\right)$$

$$h_1 d_1^3 = h_2 d_2^3$$

(a) $d_2 = \sqrt[3]{\dfrac{h_1 d_1^3}{h_2}} = \sqrt[3]{\dfrac{(200 \text{ ft})(0.5 \text{ in})^3}{50 \text{ ft}}} = \boxed{0.794 \text{ in}}$

(b) $h_2 = h_1 \left(\dfrac{d_1}{d_2}\right)^3 = (200 \text{ ft})\left(\dfrac{0.5 \text{ in}}{0.25 \text{ in}}\right)^3$

$= \boxed{1600 \text{ ft}}$

(c) $p_{50} = \gamma h_{50} + p_{\text{atm}}$

$= \left(62.4 \dfrac{\text{lbf}}{\text{ft}^3}\right)(50 \text{ ft})\left(\dfrac{\text{ft}}{12 \text{ in}}\right)^2 + 14.7 \dfrac{\text{lbf}}{\text{in}^2}$

$= \boxed{36.4 \text{ lbf/in}^2 \text{ (psia)}}$

$p_{200} = \gamma h_{200} + p_{\text{atm}}$

$= \left(62.4 \dfrac{\text{lbf}}{\text{ft}^3}\right)(200 \text{ ft})\left(\dfrac{\text{ft}}{12 \text{ in}}\right)^2 + 14.7 \dfrac{\text{lbf}}{\text{in}^2}$

$= \boxed{101.4 \text{ lbf/in}^2 \text{ (psia)}}$

$p_{400} = \gamma h_{400} + p_{\text{atm}}$

$= \left(62.4 \dfrac{\text{lbf}}{\text{ft}^3}\right)(400 \text{ ft})\left(\dfrac{\text{ft}}{12 \text{ in}}\right)^2 + 14.7 \dfrac{\text{lbf}}{\text{in}^2}$

$= \boxed{188.0 \text{ lbf/in}^2 \text{ (psia)}}$

10. $p_a = p_{\text{vapor}} + \gamma h = p_{\text{vapor}} + \rho\left(\dfrac{g}{g_c}\right)h$

$= \left(48.9 \dfrac{\text{lbf}}{\text{ft}^2}\right)\left(\dfrac{\text{ft}^2}{144 \text{ in}^2}\right) + \left(62.4 \dfrac{\text{lbm}}{\text{ft}^3}\right)$

$\times \left(\dfrac{32.2 \dfrac{\text{ft}}{\text{sec}^2}}{32.2 \dfrac{\text{lbm-ft}}{\text{lbf-sec}^2}}\right)(33.20 \text{ ft})\left(\dfrac{\text{ft}^2}{144 \text{ in}^2}\right)$

$= \boxed{14.73 \text{ psi}}$

11. $\sigma_{\text{droplet}} = \dfrac{d\Delta p}{4}$

$d = \dfrac{4\sigma}{\Delta p}$

$= \dfrac{(4)\left(0.00499 \dfrac{\text{lbf}}{\text{ft}}\right)\left(\dfrac{\text{ft}}{12 \text{ in}}\right)}{0.1 \dfrac{\text{lbf}}{\text{in}^2}}$

$= \boxed{0.0166 \text{ in}}$

12. $h = \dfrac{4\sigma \cos\beta}{\gamma d_{\text{tube}}}$

$= \left[\dfrac{(4)\left(0.005\,\dfrac{\text{lbf}}{\text{ft}}\right)(\cos 0°)}{\left(62.4\,\dfrac{\text{lbf}}{\text{ft}^3}\right)(0.1\,\text{in})}\right]\left(\dfrac{12\,\text{in}}{\text{ft}}\right)^2 = \boxed{0.46\,\text{in}}$

13. $h = \left(\dfrac{4\sigma \cos\beta}{\rho d_{\text{tube}}}\right)\left(\dfrac{g_c}{g}\right)$

$= \left[\dfrac{(4)\left(0.0356\,\dfrac{\text{lbf}}{\text{ft}}\right)(\cos 140°)}{\left(848\,\dfrac{\text{lbm}}{\text{ft}^3}\right)(0.04\,\text{in})}\right]\left(\dfrac{144\,\text{in}^2}{\text{ft}^2}\right)$

$\times \left(\dfrac{32.2\,\dfrac{\text{lbm-ft}}{\text{lbf-sec}^2}}{32.2\,\dfrac{\text{ft}}{\text{sec}^2}}\right) = -0.463\,\text{in}$

$\boxed{\text{The mercury depression is } 0.463\,\text{in.}}$

14. $\Delta V = \dfrac{V\Delta p}{E} = \dfrac{(4.9\,\text{ft}^3)\left(5000\,\dfrac{\text{lbf}}{\text{in}^2}\right)}{3.11 \times 10^5\,\dfrac{\text{lbf}}{\text{in}^2}} = \boxed{0.079\,\text{ft}^3}$

15.

$\beta = \dfrac{-\dfrac{\Delta V}{V_o}}{\Delta p} = \dfrac{\dfrac{\Delta\rho}{\rho_o}}{\Delta p}$

$= \dfrac{\dfrac{56.6\,\dfrac{\text{lbm}}{\text{ft}^3} - 56.0\,\dfrac{\text{lbm}}{\text{ft}^3}}{56.0\,\dfrac{\text{lbm}}{\text{ft}^3}}}{5000\,\dfrac{\text{lbf}}{\text{in}^2} - 1000\,\dfrac{\text{lbf}}{\text{in}^2}}$

$= \boxed{2.68 \times 10^{-6}\,\text{in}^2/\text{lbf}}$

16. $a = \dfrac{1}{\sqrt{\beta\rho}}$

$= \dfrac{1}{\sqrt{\dfrac{\left(3.4 \times 10^{-6}\,\dfrac{\text{in}^2}{\text{lbf}}\right)\left(62.42\,\dfrac{\text{lbm}}{\text{ft}^3}\right)}{\left(32.2\,\dfrac{\text{lbm-ft}}{\text{sec}^2\text{-lbf}}\right)\left(144\,\dfrac{\text{in}^2}{\text{ft}^2}\right)}}}$

$= \boxed{4674\,\text{ft/sec}}$

17.(a) $a = \sqrt{\dfrac{k g_c R^* T}{(\text{MW})}}$

$= \sqrt{\dfrac{(1.4)\left(32.2\,\dfrac{\text{lbm-ft}}{\text{lbf-sec}^2}\right)\left(1545.3\,\dfrac{\text{ft-lbf}}{\text{lbmole-°R}}\right)(460 + 570)°\text{R}}{29.0\,\dfrac{\text{lbm}}{\text{lbmole}}}}$

$= \boxed{1573\,\text{ft/sec}}$

(b) $M = \dfrac{\text{v}}{a}$

$\text{v} = Ma = (0.8)\left(1573\,\dfrac{\text{ft}}{\text{sec}}\right)$

$= \boxed{1258\,\text{ft/sec}}$

18. $a = \sqrt{\dfrac{g_c}{\beta\rho}}$

$\rho = \dfrac{g_c}{a^2 \beta} = \dfrac{\left(32.2\,\dfrac{\text{lbm-ft}}{\text{lbf-sec}^2}\right)\left(144\,\dfrac{\text{in}^2}{\text{ft}^2}\right)}{\left(4100\,\dfrac{\text{ft}}{\text{sec}}\right)^2\left(2.68 \times 10^{-6}\,\dfrac{\text{in}^2}{\text{lbf}}\right)}$

$= \boxed{102.9\,\text{lbm/ft}^3}$

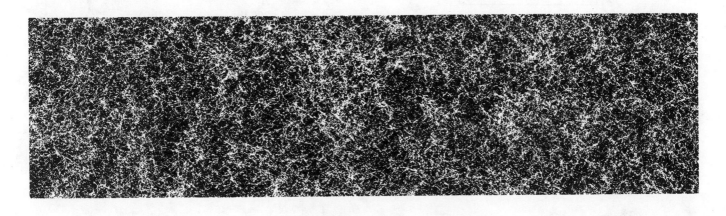

FLUID STATICS

1. $\Delta p = \gamma_m h_m - \gamma_w h_w$

$$= (6\,\text{in})\left(0.491\,\frac{\text{lbf}}{\text{in}^3}\right) - (8\,\text{in})\left(0.0361\,\frac{\text{lbf}}{\text{in}^3}\right)$$

$$= \boxed{2.66\,\text{lbf/in}^2\ (\text{psig})}$$

2. $\Delta p = \gamma_m h_m - \gamma_w h_w$

$$= (18\,\text{in})\left(0.491\,\frac{\text{lbf}}{\text{in}^3}\right) - (18\,\text{in})\left(0.0361\,\frac{\text{lbf}}{\text{in}^3}\right)$$

$$= \boxed{8.19\,\text{lbf/in}^2\ (\text{psig})}$$

3. Since the specific gravity of water is 1.0, the density of kerosene is

$$\gamma_k = (\text{SG}_k)(\gamma_w) = (0.82)\left(0.0361\,\frac{\text{lbf}}{\text{in}^3}\right)$$

$$= 0.0296\,\frac{\text{lbf}}{\text{in}^3}$$

$$\Delta p = \gamma_m h_m - \gamma_k(-h_k)$$

$$= (15\,\text{in})\left(0.491\,\frac{\text{lbf}}{\text{in}^3}\right) + (30\,\text{in})\left(0.0296\,\frac{\text{lbf}}{\text{in}^3}\right)$$

$$= \boxed{8.25\,\text{lbf/in}^2\ (\text{psig})}$$

4. The density of sea water is about 64.0 lbf/ft³.

$$p = \gamma h = \left(64.0\,\frac{\text{lbf}}{\text{ft}^3}\right)(8000\,\text{ft})$$

$$= \boxed{512{,}000\,\text{lbf/ft}^2\ (\text{psfg})}$$

5. If the two chambers did not communicate, the pressure under the left chamber would be

$$p = \Sigma\gamma h = (3\,\text{ft})(0.7)\left(62.4\,\frac{\text{lbf}}{\text{ft}^3}\right)$$

$$+ (6\,\text{ft})(1.3)\left(62.4\,\frac{\text{lbf}}{\text{ft}^3}\right)$$

$$= 617.76\,\text{lbf/ft}^2\ (\text{psfg})$$

The pressure under the right chamber would be

$$p = \Sigma\gamma h = (3\,\text{ft})\left(62.4\,\frac{\text{lbf}}{\text{ft}^3}\right) + (4\,\text{ft})(1.3)\left(62.4\,\frac{\text{lbf}}{\text{ft}^3}\right)$$

$$= 511.68\,\text{lbf/ft}^2\ (\text{psfg})$$

Since the two chambers do communicate, the pressure must be 617.76 psf everywhere on the bottom. Working upwards on the right-hand side,

$$p_A = 617.76\,\text{psfg} - 511.68\,\text{psfg} = 106.08\,\text{psfg}$$

$$= \boxed{0.737\,\text{psig}}$$

6. The pressures at various depths are

$$p_4 = \Sigma\gamma h = (2\,\text{ft})\left(55\,\frac{\text{lbf}}{\text{ft}^3}\right) + (2\,\text{ft})\left(62.4\,\frac{\text{lbf}}{\text{ft}^3}\right)$$

$$= 234.8\,\text{psfg}$$

$$p_6 = 234.8\,\frac{\text{lbf}}{\text{ft}^3} + (2\,\text{ft})\left(62.4\,\frac{\text{lbf}}{\text{ft}^3}\right)$$

$$= 359.6\,\text{psfg}$$

$$\bar{p} = \tfrac{1}{2}(p_4 + p_6) = 297.2\,\text{psfg}$$

$$R_x = \bar{p}A = (297.2\,\text{psf})(5\,\text{ft})(2\,\text{ft})$$

$$= \boxed{2972\,\text{lbf}}$$

$$R_y = R_x \tan\theta = (2972\,\text{lbf})(\tan 45°)$$

$$= \boxed{2972\,\text{lbf}}$$

7. The pressure at the bottom of the left chamber is

$$p = \frac{F}{A} + \gamma h$$

$$= \frac{200\,\text{lbf}}{\frac{\pi}{4}\,\text{in}^2} + \left(0.0361\,\frac{\text{lbf}}{\text{in}^2}\right)(0.7)(8\,\text{in})$$

$$= 254.85\,\text{lbf/in}^2\ (\text{psig})$$

Working upward, the pressure at B is

$$p_B = 254.85\,\frac{\text{lbf}}{\text{in}^2} - (10\,\text{ft})\left(12\,\frac{\text{in}}{\text{ft}}\right)(0.7)\left(0.0361\,\frac{\text{lbf}}{\text{in}^3}\right)$$

$$= 251.82\,\text{lbf/in}^2\ (\text{psig})$$

The force at B is

$$R = p_B A$$

$$= \left(251.82\ \frac{\text{lbf}}{\text{in}^2}\right)(6\text{ in})(6\text{ in})(\pi)$$

$$= \boxed{28{,}480\text{ lbf}}$$

8. The specific gas constant for air is 53.3 ft-lbf/lbm-°R.

$$\rho_{\text{air}} = \frac{p}{RT} = \frac{\left(26\ \dfrac{\text{lbf}}{\text{in}^2} + 14.7\ \dfrac{\text{lbf}}{\text{in}^2}\right)\left(\dfrac{144\text{ in}^2}{\text{ft}^2}\right)}{\left(53.3\ \dfrac{\text{ft-lbf}}{\text{lbm-°R}}\right)(460 + 70)°\text{R}}$$

$$= 0.2075\text{ lbm/ft}^3$$

$$p_{\text{absolute}} = p_a + p_g + \Sigma\gamma h$$

$$= 14.7\text{ psia} + 26\text{ psia}$$

$$+ \frac{(62.4)(9.3) + (62.4)(0.71)(6.5) + (0.2075)(5)}{144}$$

$$= \boxed{46.737\text{ psia}}$$

9. (a) Since $h_1 = 0$, $h_2 = 5$ ft.

$$\bar{p} = \tfrac{1}{2}(h_1 + h_2)\gamma$$

$$= \left(\tfrac{1}{2}\right)(5\text{ ft})\left(62.4\ \frac{\text{lbf}}{\text{ft}^3}\right)$$

$$= 156\text{ lbf/ft}^2\ (\text{psf})$$

The total force is

$$R = \bar{p}A = (156\text{ lbf})(5\text{ ft})(10\text{ ft})$$

$$= \boxed{7800\text{ lbf}}$$

(b) $$h_R = \tfrac{2}{3}\left(h_1 + h_2 - \frac{h_1 h_2}{h_1 + h_2}\right)$$

$$= \left(\tfrac{2}{3}\right)(5) = \boxed{3.33\text{ ft}}$$

The resultant acts 4.33 ft below the upper edge of the gate.

10. The pressure in the jar is the same as the pressure at the lower water surface.

$$p = \gamma h = \left(0.0361\ \frac{\text{lbf}}{\text{in}^3}\right)(16\text{ in})$$

$$= \boxed{0.5776\text{ lbf/in}^2\ (\text{psig})}$$

11. (a) $h_2 = 15$, $h_1 = (15) - (8)(\sin 20°) = 12.26$

$$\bar{p} = \tfrac{1}{2}\gamma(h_1 + h_2)$$

$$= 850.6\text{ psf}$$

The normal force acting on the plate is

$$R = \bar{p}A = \left(850.6\ \frac{\text{lbf}}{\text{ft}^2}\right)(4\text{ ft})(8\text{ ft})$$

$$= \boxed{27{,}219\text{ lbf}}$$

(b) $$h = \tfrac{2}{3}\left(h_1 + h_2 - \frac{h_1 h_2}{h_1 + h_2}\right)$$

$$= 13.68\text{ ft (vertical)}$$

$$h_R = \frac{h}{\sin\theta} = \frac{13.68\text{ ft}}{\sin 20°}$$

$$= \boxed{40.0\text{ ft (inclined)}}$$

12. (a) $h_1 = 0$, $h_2 = 2 + 3 = 5$ ft

$$\bar{p} = \tfrac{1}{2}\gamma(h_1 + h_2) = \left(\tfrac{1}{2}\right)\left(62.4\ \frac{\text{lbf}}{\text{ft}^2}\right)(5\text{ ft})$$

$$= 156\text{ psf}$$

$$R_x = \bar{p}A = \left(156\ \frac{\text{lbf}}{\text{ft}^2}\right)(5\text{ ft})(1\text{ ft})$$

$$= \boxed{780\text{ lbf}}$$

(b) $R_y = \gamma V$

$$= \left(62.4\ \frac{\text{lbf}}{\text{ft}^3}\right)$$

$$\times \left[(2\text{ ft})(3\text{ ft})(1\text{ ft}) + (3)(3)(\pi)(1)\left(\tfrac{1}{4}\right)\right]$$

$$= \boxed{815.5\text{ lbf}}$$

(c) $R = \sqrt{R_x^2 + R_y^2} = \sqrt{(780\text{ lbf})^2 + (815.5\text{ lbf})^2}$

$$= \boxed{1128.5\text{ lbf}}$$

$$\theta = \arctan\left(\frac{R_y}{R_x}\right) = \arctan\left(\frac{815.5\text{ lbf}}{780\text{ lbf}}\right)$$

$$= \boxed{46.27°\text{ from horizontal}}$$

13. Consider the left side of the gate.

$$h_1 = 6\text{ ft}$$

$$h_2 = 6 + 3 + 2 = 11\text{ ft}$$

$$\overline{p} = \tfrac{1}{2}\gamma(h_1 + h_2) = \left(62.4\,\frac{\text{lbf}}{\text{ft}^2}\right)(8.5\,\text{ft})$$

$$= 530.4\,\text{psf}$$

$$R = \overline{p}A = \left(530.4\,\frac{\text{lbf}}{\text{ft}^2}\right)(5\,\text{ft})(1\,\text{ft})$$

$$= 2652\,\text{lbf}$$

$$h_R = \tfrac{2}{3}\left(h_1 + h_2 - \frac{h_1 h_2}{h_1 + h_2}\right)$$

$$= 8.745\,\text{ft from the top}$$

$$= 2.255\,\text{ft from the bottom}$$

For the right side, $h_1 = 0$, $h_2 = 2$ ft.

$$\overline{p} = 62.4\,\text{psf}$$

$$\overline{R} = 124.8\,\text{lbf}$$

$$h_R = 0.67\,\text{ft from the bottom}$$

Draw a freebody diagram of the gate.

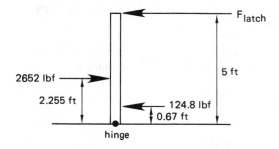

Equating moments about the hinge,

$$(2652\,\text{lbf})(2.255\,\text{ft}) = (F_{\text{latch}})(5\,\text{ft})$$
$$+ (124.8\,\text{lbf})(0.67\,\text{ft})$$

$$\boxed{F_{\text{latch}} = 1179.4\,\text{lbf}}$$

The latch force should be at least 1179.4 lbf at point B.

14. $h_1 = 12\,\text{ft} - (3.5\,\text{ft})(\sin 30°) = 10.25\,\text{ft}$

$h_2 = 12\,\text{ft} + (3.5\,\text{ft})(\sin 30°) = 13.75\,\text{ft}$

$$\overline{p} = \tfrac{1}{2}\gamma(h_1 + h_2) = \left(62.4\,\frac{\text{lbf}}{\text{ft}^3}\right)(12\,\text{ft})$$

$$= 748.8\,\text{psf}$$

$$\overline{R} = \overline{p}A = \left(748.8\,\frac{\text{lbf}}{\text{ft}^2}\right)(3.5\,\text{ft})(3.5\,\text{ft})(\pi)$$

$$= \boxed{28{,}817\,\text{lbf (60° from horizontal)}}$$

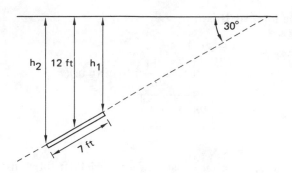

Since this is not a rectangle,

$$h_R = h_c + \frac{I_c}{Ah_c}$$

$$= \frac{12\,\text{ft}}{\sin 30°} + \frac{\tfrac{1}{4}(\pi)(3.5\,\text{ft})^4}{(3.5\,\text{ft})(3.5\,\text{ft})(\pi)\left(\dfrac{12\,\text{ft}}{\sin 30°}\right)}$$

$$= 24 + 0.13 = 24.13\,\text{ft}$$

Both h_R and h_c are parallel to the surface. The vertical depth corresponding to h_R is

$$h = h_R \sin\theta = (24.13\,\text{ft})(\sin 30°)$$

$$= \boxed{12.06\,\text{ft}}$$

15. $$R = pA = \left(900\,\frac{\text{lbf}}{\text{in}^2}\right)(2\,\text{in})^2\pi$$

$$= 11{,}310\,\text{lbf}$$

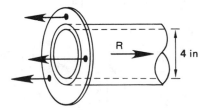

The force exerted by each bolt is

$$\frac{11{,}310\,\text{lbf}}{3} = \boxed{3770\,\text{lbf}}$$

16. At 60°F, the vapor pressure of water is 0.2563 psia.

$$h = \frac{p_a - p_v}{\gamma} = \frac{p_a - p_v}{\rho\,\dfrac{g}{g_c}}$$

$$= \frac{\left(14.6\,\dfrac{\text{lbf}}{\text{in}^2} - 0.2563\,\dfrac{\text{lbf}}{\text{in}^2}\right)\left(\dfrac{144\,\text{in}^2}{\text{ft}^2}\right)}{\left(62.4\,\dfrac{\text{lbm}}{\text{ft}^3}\right)\left(\dfrac{28.0\,\dfrac{\text{ft}}{\text{sec}^2}}{32.2\,\dfrac{\text{ft-lbf}}{\text{sec}^2\text{-lbm}}}\right)}$$

$$= \boxed{38.07\,\text{ft}}$$

17. Polytropic compression with $n = 1.235$ can be assumed in the troposphere.

The initial temperature is

$$T_1 = 59°\text{F} + 460 = 519°\text{R}$$

The pressure at $h_2 = 35,000$ ft is

$$p_2 = p_1 \left[1 - \left(\frac{n-1}{n}\right)\left(\frac{g}{g_c}\right)\left(\frac{h_2 - h_1}{RT_1}\right)\right]^{\frac{n}{n-1}}$$

$$= (14.7\,\text{psia})\left[1 - \left(\frac{1.235 - 1}{1.235}\right)\left(\frac{32.2\,\frac{\text{ft}}{\text{sec}^2}}{32.2\,\frac{\text{ft-lbm}}{\text{lbf-sec}^2}}\right)\right.$$

$$\left. \times \left(\frac{35,000\,\text{ft} - 0}{\left(53.3\,\frac{\text{ft-lbf}}{\text{lbm-°R}}\right)(519°\text{R})}\right)\right]^{\frac{1.235}{1.235-1}}$$

$$= (14.7\,\text{psia})(0.7592)^{5.255}$$

$$= \boxed{3.456\,\text{psia}}$$

18. (a) $\rho = \dfrac{p}{RT}$

$$= \frac{\left(14.7\,\frac{\text{lbf}}{\text{in}^2}\right)\left(144\,\frac{\text{in}^2}{\text{ft}^2}\right)}{\left(53.3\,\frac{\text{lbf-ft}}{\text{lbm-°R}}\right)(460 + 60)°\text{R}}$$

$$= 0.076\,\text{lbm/ft}^3$$

The pressure at 12,000 feet altitude is

$$p_2 = p_1 - \gamma h$$

$$= 14.7\,\frac{\text{lbf}}{\text{in}^2} - \left(0.076\,\frac{\text{lbf}}{\text{ft}^3}\right)(12,000\,\text{ft})\left(\frac{\text{ft}^2}{144\,\text{in}^2}\right)$$

$$= \boxed{8.34\,\text{psia}}$$

(b) $p_2 = (p_1)\exp\left[\dfrac{g(h_1 - h_2)}{g_c RT}\right]$

$$= (14.7\,\text{psia})\exp\left[\left(\frac{32.2\,\frac{\text{ft}}{\text{sec}^2}}{32.2\,\frac{\text{ft-lbf}}{\text{sec}^2\text{-lbm}}}\right)\right.$$

$$\left. \times \left(\frac{-12,000\,\text{ft}}{\left(53.3\,\frac{\text{ft-lbf}}{\text{lbm-°R}}\right)(460 + 60)°\text{R}}\right)\right]$$

$$= \boxed{9.53\,\text{psia}}$$

(c) The ratio of specific heats, k, for air is 1.4.

$$p_2 = p_1\left[1 - \left(\frac{k-1}{k}\right)\left(\frac{g}{g_c}\right)\left(\frac{h_2 - h_1}{RT_1}\right)\right]^{\frac{k}{k-1}}$$

$$= (14.7\,\text{psia})\left[1 - \left(\frac{1.4 - 1}{1.4}\right)\left(\frac{32.2\,\frac{\text{ft}}{\text{sec}^2}}{32.2\,\frac{\text{ft-lbf}}{\text{sec}^2\text{-lbm}}}\right)\right.$$

$$\left. \times \left(\frac{12,000 - 0\,\text{ft}}{\left(53.3\,\frac{\text{ft-lbf}}{\text{lbm-°R}}\right)(460 + 60)°\text{R}}\right)\right]^{\frac{1.4}{1.4-1}}$$

$$= \boxed{9.26\,\text{psia}}$$

19. Disregarding the weight of the air in the box, the weight of the box is

$$(12\,\text{in})^3 - (11.5\,\text{in})^3 = 207.125\,\text{in}^3$$

The weight of the box is

$$\gamma V = (7.7)\left(0.0361\,\frac{\text{lbf}}{\text{in}^3}\right)(207.125\,\text{in}^3)$$

$$= 57.6\,\text{lbf}$$

The volume of the box is 1 ft^3. When submerged, the buoyant force is

$$F_{\text{buoyant}} = \gamma V = \left(62.4\,\frac{\text{lbf}}{\text{ft}^3}\right)(1\,\text{ft}^3)$$

$$= 62.4\,\text{lbf}$$

$$\boxed{\text{Since the buoyant force is larger, the box will float.}}$$

20. The specific gravities of lead and cork are 11.34 and 0.25, respectively.

$$\rho_{\text{lead}} = (11.34)\left(0.0361\,\frac{\text{lbm}}{\text{in}^3}\right) = 0.409\,\text{lbm/in}^3$$

$$\rho_{\text{cork}} = (0.25)\left(0.0361\,\frac{\text{lbm}}{\text{in}^3}\right) = 0.009\,\text{lbm/in}^3$$

Since the combination floats in water, the buoyant force equals the weight.

$$\gamma_{\text{water}}(V_{\text{cork}} + V_{\text{lead}}) = \gamma_{\text{cork}}V_{\text{cork}} + \gamma_{\text{lead}}V_{\text{lead}}$$

Dividing through by the density of water,

$$8\,\text{in}^3 + V_{\text{lead}} = (0.25)(8) + 11.34\,V_{\text{lead}}$$

$$V_{\text{lead}} = 0.58\,\text{in}^3$$

FLUIDS
Statics

The mass of the lead is

$$\rho V = (0.58 \text{ in}^3)\left(0.409\,\frac{\text{lbm}}{\text{in}^3}\right)$$

$$= \boxed{0.237 \text{ lbm}}$$

21. Since W in air $= F_{\text{buoyant}} + W$ in water,

$$F_{\text{buoyant}} = 19.9\,\text{lbf} - 12.4\,\text{lbf}$$

$$= 7.5\,\text{lbf}$$

$$= \gamma_{\text{water}}V_{\text{stone}}$$

$$V_{\text{stone}} = \frac{7.5\,\text{lbf}}{62.4\,\dfrac{\text{lbf}}{\text{ft}^3}} = 0.12\,\text{ft}^3$$

$$\rho = \frac{m}{V} = \frac{19.9\,\text{lbm}}{0.12\,\text{ft}^3}$$

$$= \boxed{165.8\ \text{lbm/ft}^3}$$

22. The additional displaced volume is

$$(4700\,\text{ft}^2)\left(\frac{3\,\text{in}}{12\,\dfrac{\text{in}}{\text{ft}}}\right) = 1175\,\text{ft}^3$$

The increased cargo load equals the buoyant force.

$$F_{\text{buoyant}} = \gamma V = \left(64.0\,\frac{\text{lbf}}{\text{ft}^3}\right)(1175\,\text{ft}^3)$$

$$= \boxed{75{,}200\,\text{lbf}}$$

23. The densities of air, ρ_a, and hydrogen, ρ_h, are 0.0771 lbm/ft^3 and 0.00536 lbm/ft^3, respectively.

$$F_{\text{buoyant}} = \gamma_a V = \left(0.0771\,\frac{\text{lbf}}{\text{ft}^3}\right)(16{,}000\,\text{ft}^3)$$

$$= 1233.6\,\text{lbf}$$

The weight of sand to prevent lift-off should be at least

$$W_{\text{sand}} = F_{\text{buoyant}} - W_h - 650\,\text{lbf}$$

$$= 1233.6\,\text{lbf} - \left(0.00536\,\frac{\text{lbf}}{\text{ft}^3}\right)(16{,}000\,\text{ft}^3) - 650$$

$$= \boxed{497.84\,\text{lbf}}$$

24. Let x equal the percent of submerged iceberg and let V equal the volume, since weight equals buoyant force.

$$\gamma_{\text{ice}}V = \gamma_{\text{water}}xV$$

$$\left(57.1\,\frac{\text{lbf}}{\text{ft}^3}\right)V = \left(62.4\,\frac{\text{lbf}}{\text{ft}^3}\right)xV$$

$$x = 0.915 = 91.5\%$$

The visible part of the volume is

$$1 - 91.5\% = \boxed{8.5\%}$$

25. The volume of the satellite is

$$V = \left(\frac{m}{\rho}\right)\left(\frac{2000\,\text{lbm}}{27\,\dfrac{\text{lbm}}{\text{ft}^3}}\right) = 74.07\,\text{ft}^3$$

$$F_{\text{buoyant}} = \gamma V = \rho\frac{g}{g_c}V$$

$$= \left(0.011\,\frac{\text{lbm}}{\text{ft}^3}\right)\left(\frac{30.6\,\dfrac{\text{ft}}{\text{sec}^2}}{32.2\,\dfrac{\text{ft-lbm}}{\text{sec}^2\text{-lbf}}}\right)(74.07\,\text{ft}^3)$$

$$= \boxed{0.774\,\text{lbf}}$$

26. (a) $\bar{y}_{\text{CG}} = \dfrac{\Sigma A_i \bar{y}_i}{\Sigma A_i}$ (all dimensions are in feet)

$$= \frac{(3)(5)\left(\dfrac{3}{2}+1\right) + \left(\dfrac{1}{2}\right)(1)(5)\left(\dfrac{2}{3}\right)}{(3)(5) + \left(\dfrac{1}{2}\right)(1)(5)}$$

$$= 2.238\,\text{ft from the bottom}$$

$$\bar{y}_{\text{CB}} = \frac{(2.5)(5)\left(\dfrac{2.5}{2}+1\right) + \left(\dfrac{1}{2}\right)(1)(5)\left(\dfrac{2}{3}\right)}{(2.5)(5) + \left(\dfrac{1}{2}\right)(1)(5)}$$

$$= 1.986\,\text{ft from the bottom}$$

$$y_{\text{bg}} = \bar{y}_{\text{CG}} - \bar{y}_{\text{CB}} = 2.238 - 1.986$$

$$= 0.252\,\text{ft}$$

The moment of inertia of the top part of the submerged portion is

$$I_{x,1} = \tfrac{1}{3}bh^3 = \left(\tfrac{1}{3}\right)(5\,\text{ft})(2.5\,\text{ft})^3$$

$$= 26.042\,\text{ft}^4$$

The bottom part is

$$I_{x,2} = I_c + Ad^2$$

$$= \tfrac{1}{36}bh^3 + Ad^2$$

$$= \left(\tfrac{1}{36}\right)(5)(1)^3 + \left(\tfrac{1}{2}\right)(1)(5)\left(\tfrac{1}{3}+2.5\right)^2$$

$$= 20.208\,\text{ft}^4$$

FLUIDS
Statics

$$I = I_{x,1} + I_{x,2} = 26.042 + 20.208$$

$$= 46.25 \text{ ft}^4$$

$$h_m = \frac{I}{V} + y_{bg}$$

$$= \frac{46.25 \text{ ft}^4}{\left[(2.5)(5) + \left(\frac{1}{2}\right)(1)(5)\right](20)} + 0.252$$

$$= \boxed{0.4062 \text{ ft}}$$

(b) The displaced volume is

$$\left[(2.5)(5) + \left(\frac{1}{2}\right)(1)(5)\right](20) = 300 \text{ ft}^3$$

$$F_{\text{buoyant}} = \gamma V = \left(62.4 \frac{\text{lbf}}{\text{ft}^3}\right)(300 \text{ ft}^3)$$

$$= \boxed{18,720 \text{ lbf}}$$

27. Approximate the bottom part of the vessel by two symmetric parabolic spandrels.

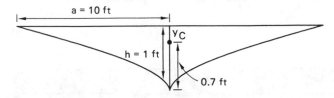

The area of this part is

$$(2)\left(\frac{1}{3}\right)ah = 6.67 \text{ ft}^2$$

$$\bar{y}_{CG} = \frac{\Sigma A_i \bar{y}_i}{\Sigma A_i}$$

$$= \frac{(5 \text{ ft})(20 \text{ ft})\left(\frac{5}{2} + 1 \text{ ft}\right) + (6.67 \text{ ft}^2)\left(\frac{7}{10} \text{ ft}\right)}{(5)(20) + 6.67}$$

$$= 3.325 \text{ ft from the bottom}$$

$$\bar{y}_{CB} = \frac{(4)(20)\left(\frac{4}{2} + 1\right) + (6.67)\left(\frac{7}{10}\right)}{(4)(20) + 6.67}$$

$$= 2.823 \text{ ft from the bottom}$$

$$y_{bg} = \bar{y}_{CG} - \bar{y}_{CB} = 3.325 - 2.823$$

$$= 0.502 \text{ ft}$$

$$I_{\text{top}} = \frac{1}{3}bh^3 = \left(\frac{1}{3}\right)(20 \text{ ft})(4 \text{ ft})^3$$

$$= 426.67 \text{ ft}^4$$

$$I_{\text{bottom}} = I_c + Ad^2$$

$$= (2)\left(\frac{ah^3}{21}\right) + (6.67)(4 + 0.3)^2$$

$$= \frac{20}{21} + 123.33 = 124.28 \text{ ft}^4$$

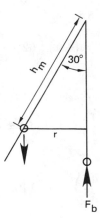

$$I_{\text{total}} = I_{\text{top}} + I_{\text{bottom}} = 426.67 \text{ ft}^4 + 124.28 \text{ ft}^4$$

$$= 550.95 \text{ ft}^4$$

$$h_m = \frac{I}{V} + \bar{y}_{bg} = \frac{550.95 \text{ ft}^4}{10,000 \text{ ft}^3} + 0.502 \text{ ft}$$

$$= 0.557 \text{ ft}$$

$$r = h_m \sin \phi = (0.557 \text{ ft})(\sin 30°)$$

$$= 0.279 \text{ ft}$$

$$F_b = \gamma V = \left(64.0 \frac{\text{lbf}}{\text{ft}^3}\right)(10,000 \text{ ft}^3)$$

$$= 640,000 \text{ lbf}$$

$$\text{couple} = F_b r$$

$$= (640,000 \text{ lbf})(0.279 \text{ ft})$$

$$= \boxed{178,560 \text{ ft-lbf}}$$

28.

Assume that the width of the box is 1 foot and the volume of water is $(3)(8)(1) = 24 \text{ ft}^3$. When the water reaches the rear top of the tank, the water level at the front of the tank is

$$\frac{1}{2}(h + 5 \text{ ft})(8 \text{ ft})(1 \text{ ft}) = 24 \text{ ft}^3$$

$$h = 1 \text{ ft}$$

(a) $\quad \phi = \arctan\left(\frac{a_x}{a_y + g}\right) = \arctan\left(\frac{5 - 1}{8}\right)$

$$\frac{a_x}{0 + 32.2} = \frac{4}{8}$$

$$a_x = \boxed{16.1 \text{ ft/sec}^2}$$

(b) $h_A = 0$, $h_B = 5$

$$\bar{p} = \tfrac{1}{2}\gamma(h_A + h_B) = \left(\tfrac{5}{2}\text{ ft}\right)\left(62.4\,\frac{\text{lbf}}{\text{ft}^3}\right)$$

$$= 156\text{ lbf/ft}^2$$

The force on a one-foot section of the rear wall is

$$F_{\text{rear}} = \bar{p}A = \left(156\,\frac{\text{lbf}}{\text{ft}^2}\right)(1\text{ ft})(5\text{ ft}) = \boxed{780\text{ lbf}}$$

$h_C = 0$, $h_D = 1$

$$\bar{p} = \left(\tfrac{1}{2}\text{ ft}\right)\left(62.4\,\frac{\text{lbf}}{\text{ft}^3}\right) = 31.2\text{ lbf/ft}^2$$

The force on a one-foot section of the front wall is

$$F_{\text{front}} = \left(31.2\,\frac{\text{lbf}}{\text{ft}^2}\right)(1\text{ ft})(1\text{ ft}) = \boxed{31.2\text{ lbf}}$$

29. Shape EIH is parabolic. The height, h, can be obtained from

$$\text{vol (ABCD)} = \text{vol (EFGH)} - \text{vol (EIH)}$$

$$\left(\frac{\pi}{4}\right)(h-1) = \left(\frac{\pi}{4}\right)(h) - \left(\frac{1}{2}\right)\left(\frac{\pi}{4}\right)h$$

$$h = 2\text{ ft}$$

$$\omega = \frac{\sqrt{2gh}}{r} = \frac{\sqrt{(2)\left(32.2\,\dfrac{\text{ft}}{\text{sec}^2}\right)(2\text{ ft})}}{\tfrac{1}{2}\text{ ft}}$$

$$= \boxed{22.7\text{ rad/sec}}$$

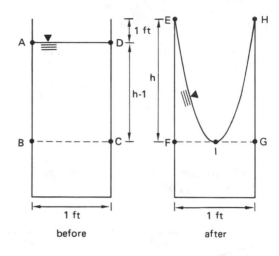

FLUID FLOW PARAMETERS

1.(a) $E_v = \dfrac{v^2}{2g_c} = \dfrac{\left(15\,\dfrac{ft}{sec}\right)^2}{(2)\left(32.2\,\dfrac{ft\text{-}lbm}{lbf\text{-}sec^2}\right)}$

$= \boxed{3.5 \text{ ft-lbf/lbm}}$

(b) $E_v = \dfrac{v^2}{2g_c} = \dfrac{\left(15\,\dfrac{ft}{sec}\right)^2}{(2)\left(32.2\,\dfrac{ft\text{-}lbm}{lbf\text{-}sec^2}\right)}$

$= \boxed{3.5 \text{ ft-lbf/lbm}}$

2. $v_A = \dfrac{\dot{V}}{A_A}$

$= \dfrac{2.5\,\dfrac{ft^3}{sec}}{(12.5\,in^2)\left(\dfrac{ft}{12\,in}\right)^2} = 28.8\,\dfrac{ft}{sec}$

$p_A = \left(10\,\dfrac{lbf}{in^2} + 14.7\,\dfrac{lbf}{in^2}\right) + \dfrac{(1.5\,ft)\rho_{H_2O}g}{g_c}$

$= 24.7\,lbf/in^2\ (psia)$

$+ \dfrac{(1.5\,ft)\left(62.32\,\dfrac{lbm}{ft^3}\right)\left(\dfrac{ft}{12\,in}\right)^2\left(32.2\,\dfrac{ft}{sec^2}\right)}{32.2\dfrac{lbm\text{-}ft}{lbf\text{-}sec^2}}$

$= 25.35\,lbf/in^2\ (psia)$

$p_B = 14.7\,\dfrac{lbf}{in^2} + \dfrac{(2.5\,in)\rho_{Hg}g}{g_c} - \dfrac{(12\,in)\rho_{H_2O}g}{g_c}$

$= 14.7\,\dfrac{lbf}{in^2} + \dfrac{(2.5\,in)\left(847\,\dfrac{lbm}{ft^3}\right)\left(\dfrac{ft}{12\,in}\right)^2\left(32.2\,\dfrac{ft}{sec^2}\right)}{32.2\dfrac{lbm\text{-}ft}{lbf\text{-}sec^2}}$

$- \dfrac{(12\,in)\left(62.32\,\dfrac{lbm}{ft^3}\right)\left(\dfrac{ft}{12\,in}\right)^2\left(32.2\,\dfrac{ft}{sec^2}\right)}{32.2\dfrac{lbm\text{-}ft}{lbf\text{-}sec^2}}$

$= 24.21\,lbf/in^2\ (psia)$

$E_{p,A} = \dfrac{p_A}{\rho_A} = \dfrac{\left(25.35\,\dfrac{lbf}{in^2}\right)\left(\dfrac{12\,in}{ft}\right)^2}{62.32\,\dfrac{lbm}{ft^3}}$

$= 58.58 \text{ ft-lbf/lbm}$

$E_{v,A} = \dfrac{v_A^2}{2g_c} = \dfrac{\left(28.8\,\dfrac{ft}{sec}\right)^2}{(2)\left(32.2\,\dfrac{lbm\text{-}ft}{lbf\text{-}sec^2}\right)}$

$= 12.88 \text{ ft-lbf/lbm}$

Arbitrarily choose $z_A = 0$, so $E_{z,A} = 0$

$E_{p,B} = \dfrac{p_B}{\rho_B}$

$= \dfrac{\left(24.21\,\dfrac{lbf}{in^2}\right)\left(\dfrac{12\,in}{ft}\right)^2}{62.32\,\dfrac{lbm}{ft^3}}$

$= 55.94 \text{ ft-lbf/lbm}$

$E_{z,B} = \dfrac{z_B g}{g_c} = \dfrac{(10\,ft)\left(32.2\,\dfrac{ft}{sec^2}\right)}{32.2\,\dfrac{lbm\text{-}ft}{lbf\text{-}sec^2}}$

$= 10 \text{ ft-lbf/lbm}$

The Bernoulli equation is:

$$E_{t,A} = E_{t,B}$$
$$E_{p,A} + E_{v,A} + E_{z,A} = E_{p,B} + E_{v,B} + E_{z,B}$$
$$E_{v,B} = E_{p,A} + E_{v,A} + E_{z,A} - E_{p,B} - E_{z,B}$$
$$\dfrac{v_B^2}{2g_c} = 58.58\,\dfrac{ft\text{-}lbf}{lbm} + 12.88\,\dfrac{ft\text{-}lbf}{lbm} + 0$$
$$- 55.94\,\dfrac{ft\text{-}lbf}{lbm} - 10\,\dfrac{ft\text{-}lbf}{lbm}$$
$$= 5.52 \text{ ft-lbf/lbm}$$

$v_B = \sqrt{(2)\left(32.2\,\dfrac{lbm\text{-}ft}{lbf\text{-}sec^2}\right)\left(5.52\,\dfrac{ft\text{-}lbf}{lbm}\right)}$

$= \boxed{18.85 \text{ ft/sec}}$

FLUIDS
Flow Params

3. Assume the manometer is open to the atmosphere; use gage pressures.

$$p_{NH_3} = \left(20\,\frac{lbf}{in^2} - 14.7\,\frac{lbf}{in^2}\right) = 5.3\ lbf/in^2\ (psig)$$

$$(E_p)_{NH_3} = \frac{p_{NH_3}}{\rho_{NH_3}}$$

$$= \frac{\left(5.3\,\dfrac{lbf}{in^2}\right)\left(\dfrac{12\ in}{ft}\right)^2}{0.0617\,\dfrac{lbm}{ft^3}}$$

$$= 12{,}370\ ft\text{-}lbf/lbm$$

Neglect the weight of the ammonia in the manometer.

$$(E_i)_{H_2O} = \frac{h_{manometer}\,g}{g_c}$$

$$= \frac{(160\ in)\left(\dfrac{ft}{12\ in}\right)\left(32.2\,\dfrac{ft}{sec^2}\right)}{32.2\,\dfrac{lbm\text{-}ft}{lbf\text{-}sec^2}}$$

$$= 13.3\ ft\text{-}lbf/lbm$$

$$(E_i)_{NH_3} = (E_i)_{H_2O}\,\frac{\rho_{H_2O}}{\rho_{NH_3}}$$

$$= \left(13.3\,\frac{ft\text{-}lbf}{lbm}\right)\left(\frac{62.4\,\dfrac{lbm}{ft^3}}{0.0617\,\dfrac{lbm}{ft^3}}\right)$$

$$= 13{,}451\ ft\text{-}lbf/lbm$$

$$(E_v)_{NH_3} = (E_i)_{NH_3} - (E_p)_{NH_3}$$

$$= 13{,}451\,\frac{ft\text{-}lbf}{lbm} - 12{,}370\,\frac{ft\text{-}lbf}{lbm}$$

$$= 1081\ ft\text{-}lbf/lbm$$

$$v_{NH_3} = \sqrt{2g_c(E_v)_{NH_3}}$$

$$= \sqrt{(2)\left(32.2\,\frac{lbm\text{-}ft}{lbf\text{-}sec^2}\right)\left(1081\,\frac{ft\text{-}lbf}{lbm}\right)}$$

$$= \boxed{263.8\ ft/sec}$$

4. $h = \dfrac{p}{\gamma} = \dfrac{p g_c}{\rho g}$

$$= \frac{\left(28\,\dfrac{lbf}{in^2}\right)\left(\dfrac{12\ in}{ft}\right)^2\left(32.2\,\dfrac{lbm\text{-}ft}{lbf\text{-}sec^2}\right)}{\left(56\,\dfrac{lbm}{ft^3}\right)\left(32.2\,\dfrac{ft}{sec^2}\right)}$$

$$= \boxed{72\ ft}$$

5. Neglecting the weight of the air in the manometer,

$$p_i = (2.0\ in\ Hg)\left(0.491\,\frac{\frac{lbf}{in^2}}{in\ Hg}\right)$$

$$= 0.982\ lbf/in^2\ (psig)$$

$$p_i = \left(0.982\,\frac{lbf}{in^2} + 14.7\,\frac{lbf}{in^2}\right) = 15.68\ lbf/in^2\ (psia)$$

$$p_v = p_i - p_s = 15.68\,\frac{lbf}{in^2} - 14.7\,\frac{lbf}{in^2}$$

$$= 0.982\ lbf/in^2\ (psig)$$

$$\rho_{air} = \frac{p}{RT} = \frac{\left(14.7\,\dfrac{lbf}{in^2}\right)\left(\dfrac{12\ in}{ft}\right)^2}{\left(53.35\,\dfrac{ft\text{-}lbf}{lbm\text{-}°R}\right)(460 + 70)°R}$$

$$= 0.0749\ lbm/ft^3$$

$$v = \sqrt{2g_c h_v} = \sqrt{\frac{2g_c p_v}{\rho_{air}}}$$

$$= \sqrt{\frac{(2)\left(32.2\,\dfrac{lbm\text{-}ft}{lbf\text{-}sec^2}\right)\left(0.982\,\dfrac{lbf}{in^2}\right)\left(\dfrac{12\ in}{ft}\right)^2}{0.0749\,\dfrac{lbm}{ft^3}}}$$

$$= \boxed{348.7\ ft/sec}$$

6. (a)

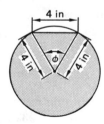

$$r_h = \frac{\pi r^2 - \dfrac{1}{2}r^2(\phi - \sin\phi)}{2\pi r - r\phi}$$

$$= \frac{\pi(4\ in) - \left(\dfrac{1}{2}\right)(4\ in)\left[\dfrac{2\pi}{6} - \sin\left(\dfrac{2\pi}{6}\right)\right]}{2\pi - \dfrac{2\pi}{6}}$$

$$= \boxed{2.33\ in}$$

(b)

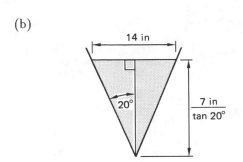

$$r_h = \frac{bh}{2\sqrt{b^2 + h^2}}$$

$$= \frac{(7 \text{ in})\left(\dfrac{7 \text{ in}}{\tan 20^\circ}\right)}{2\sqrt{(7 \text{ in})^2 + \left(\dfrac{7 \text{ in}}{\tan 20^\circ}\right)^2}} = \boxed{3.29 \text{ in}}$$

(c) $r_h = \dfrac{\dfrac{1}{2}\pi ab}{\pi\sqrt{\dfrac{1}{2}(a^2 + b^2)}}$

$$= \frac{\dfrac{1}{2}(8 \text{ in})(6 \text{ in})}{\sqrt{\dfrac{1}{2}\left[(8 \text{ in})^2 + (6 \text{ in})^2\right]}} = \boxed{3.39 \text{ in}}$$

7. $D_e = \dfrac{4hL}{L + 2h} = \dfrac{(4)(5.0 \text{ ft})(30 \text{ ft})}{30 \text{ ft} + (2)(5.0 \text{ ft})} = \boxed{15 \text{ ft}}$

8. $\text{Re} = \dfrac{D_e \mathrm{v}}{\nu} = \dfrac{\left(\dfrac{2}{12} \text{ ft}\right)\left(120 \dfrac{\text{ft}}{\text{sec}}\right)}{27.3 \times 10^{-5} \dfrac{\text{ft}^2}{\text{sec}}}$

$$= \boxed{7.33 \times 10^4}$$

9. $\text{Re} = \dfrac{D_e \mathrm{v}}{\nu} = \dfrac{\left(\dfrac{4}{12} \text{ ft}\right)\left(20 \dfrac{\text{ft}}{\text{sec}}\right)}{0.739 \times 10^{-5} \dfrac{\text{ft}^2}{\text{sec}}}$

$$= \boxed{9.02 \times 10^5}$$

10. $\text{Re} = \dfrac{D_e \mathrm{v}}{\nu} = \dfrac{\left(\dfrac{3}{12} \text{ ft}\right)\left(10 \dfrac{\text{ft}}{\text{sec}}\right)}{0.005 \dfrac{\text{ft}^2}{\text{sec}}}$

$$= 500$$

Since $500 < 2100$, $\boxed{\text{the flow is laminar.}}$

FLUID DYNAMICS

1. $v_B = \left(\dfrac{A_A}{A_B}\right) v_A$

$= \left(\dfrac{\pi\left[\left(\dfrac{2\ \text{in}}{2}\right)\left(\dfrac{\text{ft}}{12\ \text{in}}\right)\right]^2}{\pi\left[\left(\dfrac{5\ \text{in}}{2}\right)\left(\dfrac{\text{ft}}{12\ \text{in}}\right)\right]^2}\right)\left(12\ \dfrac{\text{ft}}{\text{sec}}\right)$

$= \boxed{1.92\ \text{ft/sec}}$

2. $\rho_B = \left(\dfrac{A_A v_A}{A_B v_B}\right)\rho_A$

$= \left(\dfrac{\pi\left[\left(\dfrac{4\ \text{in}}{2}\right)\left(\dfrac{\text{ft}}{12\ \text{in}}\right)\right]^2\left(50\ \dfrac{\text{ft}}{\text{sec}}\right)}{\pi\left[\left(\dfrac{7\ \text{in}}{2}\right)\left(\dfrac{\text{ft}}{12\ \text{in}}\right)\right]^2\left(20\ \dfrac{\text{ft}}{\text{sec}}\right)}\right)$

$\times\left(0.065\ \dfrac{\text{lbm}}{\text{ft}^3}\right) = \boxed{0.053\ \text{lbm/ft}^3}$

3. The kinematic viscosity, ν, of air at 100°F is 18×10^{-5} ft²/sec.

$\text{Re} = \dfrac{D_e v}{\nu} = \dfrac{\left[(12\ \text{in})\left(\dfrac{\text{ft}}{12\ \text{in}}\right)\right]\left(14\ \dfrac{\text{ft}}{\text{sec}}\right)}{18 \times 10^{-5}\ \dfrac{\text{ft}^2}{\text{sec}}}$

$= 7.78 \times 10^4$

The specific roughness of concrete is $\epsilon = 0.004$ ft. The relative roughness is

$\dfrac{\epsilon}{D_e} = \dfrac{0.004\ \text{ft}}{1\ \text{ft}} = 0.004$

From the Moody friction factor table or chart,

$$f \approx 0.298$$

$h_f = \dfrac{fLv^2}{2D_e g} = \dfrac{(0.0298)(100\ \text{ft})\left(14\ \dfrac{\text{ft}}{\text{sec}}\right)^2}{(2)(1\ \text{ft})\left(32.2\ \dfrac{\text{ft}}{\text{sec}^2}\right)}$

$= \boxed{9.07\ \text{ft}}$

4. The kinematic viscosity, ν, of water at 60°F is 1.217×10^{-5} ft²/sec.

$\text{Re} = \dfrac{D_e v}{\nu} = \dfrac{(2\ \text{in})\left(\dfrac{\text{ft}}{12\ \text{in}}\right)\left(4\ \dfrac{\text{ft}}{\text{sec}}\right)}{1.217 \times 10^{-5}\ \dfrac{\text{ft}^2}{\text{sec}}}$

$= 5.48 \times 10^4$

The specific roughness of galvanized iron is $\epsilon = 0.0005$ ft. The relative roughness is

$\dfrac{\epsilon}{D_e} = \dfrac{0.0005\ \text{ft}}{(2\ \text{in})\left(\dfrac{\text{ft}}{12\ \text{in}}\right)} = 0.003$

From the Moody friction factor table or chart,

$$f \approx 0.0284$$

$h_f = \dfrac{fLv^2}{2D_e g} = \dfrac{(0.0284)(1000\ \text{ft})\left(4\ \dfrac{\text{ft}}{\text{sec}}\right)^2}{(2)(2\ \text{in})\left(\dfrac{\text{ft}}{12\ \text{in}}\right)\left(32.2\ \dfrac{\text{ft}}{\text{sec}^2}\right)}$

$= \boxed{42.34\ \text{ft}}$

5. $v = \dfrac{\text{flow rate}}{\text{area}} = \dfrac{\left(750\ \dfrac{\text{gal}}{\text{min}}\right)\left(0.002228\ \dfrac{\text{ft}^3\text{-min}}{\text{sec-gal}}\right)}{\pi\left(\dfrac{1}{4}\ \text{ft}\right)^2}$

$= 8.51\ \text{ft/sec}$

At A: The velocity at A is the same as at B, so it may be neglected. The gage pressure head is $h_{p,A}$. The gravitational head is 0.

At B: The pressure head is

$h_{p,B} = \dfrac{p}{\gamma} = \dfrac{\left(50\ \dfrac{\text{lbf}}{\text{in}^2}\right)\left(\dfrac{12\ \text{in}}{\text{ft}}\right)^2}{(0.9)\left(62.4\ \dfrac{\text{lbf}}{\text{ft}^3}\right)} = 128.2\ \text{ft}$

The gravitational head is $h_{g,B} = 60$ ft.

The friction between points A and B is

$h_f = \dfrac{fLv^2}{2D_e g} = \dfrac{(0.03)(3000\ \text{ft})\left(8.51\ \dfrac{\text{ft}}{\text{sec}}\right)^2}{(2)\left[(6\ \text{in})\left(\dfrac{\text{ft}}{12\ \text{in}}\right)\right]\left(32.2\ \dfrac{\text{ft}}{\text{sec}^2}\right)}$

$= 202.4\ \text{ft}$

$h_{p,A} = h_{p,B} + h_{g,B} + h_f$

$= 128.2\ \text{ft} + 60\ \text{ft} + 202.4\ \text{ft}$

$= 390.6\ \text{ft}$

$p_A = \gamma h_{p,A}$

$= (0.9)\left(62.4\ \dfrac{\text{lbf}}{\text{ft}^3}\right)(390.6\ \text{ft})\left(\dfrac{\text{ft}}{12\ \text{in}}\right)^2$

$= \boxed{152.3\ \text{lbf/in}^2\ \text{(psig)}}$

6. $h_m = K h_v = K \left(\dfrac{v^2}{2g} \right)$

$$= (0.8) \left(\frac{\left(15 \, \frac{ft}{sec} \right)^2}{(2) \left(32.2 \, \frac{ft}{sec^2} \right)} \right) = \boxed{2.795 \text{ ft}}$$

7. When the fluid enters the second tank, all of the velocity head will be lost. The total loss is

$$h_t = h_f + h_v + h_m$$

For 70°F water, $\nu = 1.059 \times 10^{-5}$ ft²/sec.

$$\text{Re} = \frac{D_e v}{\nu} = \frac{(12 \text{ in}) \left(\frac{ft}{12 \text{ in}} \right) \left(10 \, \frac{ft}{sec} \right)}{1.059 \times 10^{-5} \, \frac{ft^2}{sec}} = 9.44 \times 10^5$$

The specific roughness, ϵ, is 0.0002 ft. The relative roughness is

$$\frac{\epsilon}{D_e} = \frac{0.0002}{1 \text{ ft}} = 0.0002$$

From the Moody friction factor table or chart,

$$f \approx 0.0147$$

$$h_f = \frac{fLv^2}{2D_e g} = \frac{(0.0147)(300 \text{ ft}) \left(10 \, \frac{ft}{sec} \right)^2}{(2)(1 \text{ ft}) \left(32.2 \, \frac{ft}{sec^2} \right)}$$

$$= 6.85 \text{ ft}$$

$$h_v = \frac{v^2}{2g} = \frac{\left(10 \, \frac{ft}{sec} \right)^2}{(2) \left(32.2 \, \frac{ft}{sec^2} \right)} = 1.55 \text{ ft}$$

$$h_m = K h_v = (0.5)(1.55 \text{ ft}) = 0.78 \text{ ft}$$

The total head loss is

$$h_t = h_f + h_v + h_m = 6.85 \text{ ft} + 1.55 \text{ ft} + 0.78 \text{ ft}$$

$$= \boxed{9.18 \text{ ft}}$$

8. The equivalent lengths are approximately

regular elbow	13 ft
globe valve	110 ft
straight pipe	150 ft
total	273 ft

For 70°F water, $\nu = 1.059 \times 10^{-5}$ ft²/sec.

$$\text{Re} = \frac{D_e v}{\nu} = \frac{(4 \text{ in}) \left(\frac{ft}{12 \text{ in}} \right) \left(12 \, \frac{ft}{sec} \right)}{1.059 \times 10^{-5} \, \frac{ft^2}{sec}}$$

$$= 3.78 \times 10^5$$

The specific roughness, ϵ, is 0.0002 ft.

$$\frac{\epsilon}{D_e} = \frac{0.0002 \text{ ft}}{\frac{4}{12} \text{ ft}} = 0.0006$$

From the Moody friction factor table or chart,

$$f \approx 0.0185$$

$$h_f = \frac{fLv^2}{2D_e g} = \frac{(0.0185)(273 \text{ ft}) \left(12 \, \frac{ft}{sec} \right)^2}{(2) \left(\frac{4}{12} \text{ ft} \right) \left(32.2 \, \frac{ft}{sec^2} \right)}$$

$$= \boxed{33.88 \text{ ft}}$$

9. At A: $h_{z,A} = 100$ ft
   $h_{v,A} = 0$
   $h_{p,A} = 0$

At B: $h_{z,B} = 0$

$$h_{v,B} = \frac{v^2}{2g} = \frac{v^2}{(2) \left(32.2 \, \frac{ft}{sec^2} \right)}$$

$$= \left(0.0155 \, \frac{sec^2}{ft} \right) v^2$$

$h_{p,B} = 0$

$$h_{f,B} = \frac{fLv^2}{2D_e g} = \frac{(0.02)(2000 \text{ ft})v^2}{(2)(3 \text{ in}) \left(\frac{ft}{12 \text{ in}} \right) \left(32.2 \, \frac{ft}{sec^2} \right)}$$

$$= \left(2.48 \, \frac{sec^2}{ft} \right) v^2$$

$$h_{z,A} = h_{v,B} + h_{f,B}$$

$$100 \text{ ft} = \left(0.0155 \, \frac{sec^2}{ft} \right) v^2 + \left(2.48 \, \frac{sec^2}{ft} \right) v^2$$

Solving, v = 6.33 ft/sec for gravity flow. If the flow is doubled by a pump, i.e.,

$$v_2 = 2v = (2) \left(6.33 \, \frac{ft}{sec} \right) = 12.66 \text{ ft/sec}$$

$\dot{m} = \rho A v$

$$= \left(62.4\,\frac{\text{lbm}}{\text{ft}^3}\right)\pi\left[\left(\frac{3\text{ in}}{2}\right)\left(\frac{\text{ft}}{12\text{ in}}\right)\right]^2\left(12.66\,\frac{\text{ft}}{\text{sec}}\right)$$

$$= 38.78\text{ lbm/sec}$$

$$h_v = \frac{v^2}{2g} = \frac{\left(12.66\,\frac{\text{ft}}{\text{sec}}\right)^2}{(2)\left(32.2\,\frac{\text{ft}}{\text{sec}^2}\right)} = 2.489\text{ ft}$$

$$h_f = \frac{fLv^2}{2D_e g}$$

$$= \frac{(0.02)(2000\text{ ft})\left(12.66\,\frac{\text{ft}}{\text{sec}}\right)^2}{(2)\left(32.2\,\frac{\text{ft}}{\text{sec}^2}\right)\left(\frac{3}{12}\text{ ft}\right)}$$

$$= 398.2\text{ ft}$$

$$h_{z,A} + h_A = h_v + h_f$$

$$100\text{ ft} + h_A = 2.489\text{ ft} + 398.2\text{ ft}$$

$$h_A = 300.69\text{ ft}$$

$$P = \dot{m}h_A \times \frac{g}{g_c}$$

$$= \left(38.78\,\frac{\text{lbm}}{\text{sec}}\right)(300.69\text{ ft})\left(\frac{1}{550\,\frac{\text{ft-lbf}}{\text{hp-sec}}}\right)\left(\frac{32.2\,\frac{\text{ft}}{\text{sec}^2}}{32.2\,\frac{\text{ft-lbm}}{\text{lbf-sec}^2}}\right)$$

$$= \boxed{21.20\text{ hp}}$$

10. $\dot{V} = 2000\,\dfrac{\text{gal}}{\text{min}}$

$$= \left(2000\,\frac{\text{gal}}{\text{min}}\right)\left(0.002228\,\frac{\text{ft}^3\text{-min}}{\text{gal-sec}}\right)$$

$$= 4.456\text{ ft}^3/\text{sec}$$

$$\dot{m} = \dot{V}\rho = \left(4.456\,\frac{\text{ft}^3}{\text{sec}}\right)(1.2)\left(62.4\,\frac{\text{lbm}}{\text{ft}^3}\right)$$

$$= 333.67\text{ lbm/sec}$$

At A: Since the velocity at A is the same as the velocity at B, velocity is neglected.

$$h_{p,A} = \frac{p_a}{\gamma} = \frac{p - p_v}{\gamma}$$

$$= \frac{\left[14.7\,\frac{\text{lbf}}{\text{in}^2} - (6\text{ in})\left(0.491\,\frac{\text{lbf}}{\text{in}^3}\right)\right]\left(\frac{12\text{ in}}{\text{ft}}\right)^2}{(1.2)\left(62.4\,\frac{\text{lbf}}{\text{ft}^3}\right)}$$

$$= 22.6\text{ ft}$$

$$h_{z,A} = 0$$

At B: $h_{p,B} = \dfrac{p_a}{\gamma} = \dfrac{p + p_g}{\gamma}$

$$= \frac{\left(14.7\,\frac{\text{lbf}}{\text{in}^2} + 20\,\frac{\text{lbf}}{\text{in}^2}\right)\left(\frac{12\text{ in}}{\text{ft}}\right)^2}{(1.2)\left(62.4\,\frac{\text{lbf}}{\text{ft}^3}\right)}$$

$$= 66.73\text{ ft}$$

$$h_{z,B} = 4\text{ ft}$$

If friction is neglected,

$$h_{p,A} + h_A = h_{p,B} + h_{z,B}$$

$$22.6\text{ ft} + h_A = 66.73\text{ ft} + 4\text{ ft}$$

$$h_A = 48.13\text{ ft}$$

$$\text{WHP} = \dot{m}h_A \times \frac{g}{g_c}$$

$$= \left(333.67\,\frac{\text{lbm}}{\text{sec}}\right)\left(48.13\,\frac{\text{ft-lbf}}{\text{lbm}}\right)$$

$$\times\left(\frac{1}{550\,\frac{\text{ft-lbf}}{\text{hp-sec}}}\right)\left(\frac{32.2\,\frac{\text{ft}}{\text{sec}^2}}{32.2\,\frac{\text{ft-lbm}}{\text{lbf-sec}^2}}\right)$$

$$= \boxed{29.2\text{ hp}}$$

11. $\dot{V} = 100\,\dfrac{\text{gal}}{\text{min}}$

$$= \left(100\,\frac{\text{gal}}{\text{min}}\right)\left(0.002228\,\frac{\text{ft}^3\text{-min}}{\text{gal-sec}}\right) = 0.2228\text{ ft}^3/\text{sec}$$

$$\dot{m} = \dot{V}\rho = \left(0.2228\,\frac{\text{ft}^3}{\text{sec}}\right)(0.85)\left(62.4\,\frac{\text{lbm}}{\text{ft}^3}\right)$$

$$= 11.82\text{ lbm/sec}$$

At A: $h_{p,A} = \dfrac{p_a}{\gamma} = \dfrac{p_g + p}{\gamma}$

$$= \frac{\left(5\,\frac{\text{lbf}}{\text{in}^2} + 14.7\,\frac{\text{lbf}}{\text{in}^2}\right)\left(12\,\frac{\text{in}}{\text{ft}}\right)^2}{(0.85)\left(62.4\,\frac{\text{lbf}}{\text{ft}^3}\right)} = 53.48\text{ ft}$$

$$v_A = \frac{\dot{V}}{A} = \frac{0.2228\,\frac{\text{ft}^3}{\text{sec}}}{\pi\left[\left(\frac{3\text{ in}}{2}\right)\left(\frac{\text{ft}}{12\text{ in}}\right)\right]^2} = 4.54\,\frac{\text{ft}}{\text{sec}}$$

$$h_{v,A} = \frac{v_A^2}{2g} = \frac{\left(4.54\,\frac{\text{ft}}{\text{sec}}\right)^2}{(2)\left(32.2\,\frac{\text{ft}}{\text{sec}^2}\right)} = 0.32\text{ ft}$$

FLUIDS
Dynamics

At B: $h_{p,B} = \dfrac{\left(100\,\dfrac{lbf}{in^2} + 14.7\,\dfrac{lbf}{in^2}\right)\left(\dfrac{12\ in}{ft}\right)^2}{(0.85)\left(62.4\,\dfrac{lbf}{ft^3}\right)}$

$= 311.4\ ft$

$v_B = \dfrac{\dot{V}}{A} = \dfrac{0.2228\,\dfrac{ft^3}{sec}}{\pi\left[\left(\dfrac{2\ in}{2}\right)\left(\dfrac{ft}{12\ in}\right)\right]^2} = 10.21\ ft/sec$

$h_{v,B} = \dfrac{v_B^2}{2g} = \dfrac{\left(10.21\,\dfrac{ft}{sec}\right)^2}{(2)\left(32.2\,\dfrac{ft}{sec^2}\right)} = 1.62\ ft$

$h_A + h_{p,A} + h_{v,A} = h_{p,B} + h_{v,B}$

$h_A + 53.48\ ft + 0.32\ ft = 311.4\ ft + 1.62\ ft$

$h_A = 259.22\ ft$

$WHP = \dot{m}h_A \times \dfrac{g}{g_c}$

$= \left(11.82\,\dfrac{lbm}{sec}\right)\left(259.22\,\dfrac{ft\text{-}lbf}{lbm}\right)$

$\times \left(\dfrac{1}{550\,\dfrac{ft\text{-}lbf}{hp\text{-}sec}}\right)\left(\dfrac{32.2\,\dfrac{ft}{sec^2}}{32.2\,\dfrac{ft\text{-}lbm}{lbf\text{-}sec^2}}\right)$

$= \boxed{5.57\ hp}$

12. From a table of steel pipe dimensions,

$A_{3.5''} = 0.06866\ ft^2$

$A_{8''} = 0.3474\ ft^2$

$v_{3.5''} = \dfrac{\dot{V}}{A_{3.5''}} = \dfrac{2\,\dfrac{ft^3}{sec}}{0.06866\ ft^2} = 29.13\ ft/sec$

$v_{8''} = \dfrac{\dot{V}}{A_{8''}} = \dfrac{2\,\dfrac{ft^3}{sec}}{0.3474\ ft^2} = 5.76\ ft/sec$

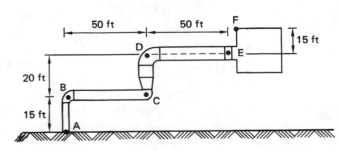

At F: $h_{p,F} = 0$

$h_{v,F} = 0$

$h_{z,F} = 15\ ft + 20\ ft + 15\ ft = 50\ ft$

At A: $h_{p,A}$ is unknown.

$h_{z,A} = 0$

$h_{v,A} = \dfrac{v_{3.5''}^2}{2g} = \dfrac{\left(29.13\,\dfrac{ft}{sec}\right)^2}{(2)\left(32.2\,\dfrac{ft}{sec^2}\right)} = 13.176\ ft$

$h_{t,A} = h_{t,F}$

$h_{p,A} + h_{v,A} = h_{z,F}$

$h_{p,A} + 13.176\ ft = 50\ ft$

$h_{p,A} = 36.824\ ft$

$HGL_A = h_{p,A} + h_{z,A} = 36.824\ ft + 0 = 36.824\ ft$

The velocity from A to C is constant; the elevation, however, rises steadily from A to B by 15 ft.

$HGL_B = HGL_A + 15\ ft = 36.824\ ft + 15\ ft = 51.824\ ft$

$HGL_C = HGL_B = 51.824\ ft$

At C: The velocity changes to 5.76 ft/sec.

$h_{v,C} = \dfrac{v^2}{2g} = \dfrac{\left(5.76\,\dfrac{ft}{sec}\right)^2}{(2)\left(32.2\,\dfrac{ft}{sec^2}\right)} = 0.515\ ft$

$h_{p,C} + h_{v,C} + h_{z,C} = h_{z,F}\ ft$

$h_{p,C} + 0.515\ ft + 15\ ft = 50\ ft$

$h_{p,C} = 34.485\ ft$

$HGL_C = 34.485\ ft + 15\ ft = 49.485\ ft$

$HGL_D = 49.485\ ft + 20\ ft = 69.485\ ft$

$HGL_E = HGL_D = 69.485\ ft$

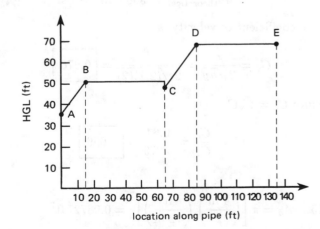

location along pipe (ft)

13. Let the nozzle be located y ft from the bottom. The discharge velocity is

$$v = \sqrt{2g(20 - y)}$$

Assume the stream flow strikes the ground after traveling horizontally x ft.

$$x = vt$$

$$y = \frac{1}{2}gt^2$$

$$x = v\sqrt{\frac{2y}{g}} = \sqrt{2g(20 - y)}\sqrt{\frac{2y}{g}}$$

$$= 2\sqrt{20y - y^2}$$

Maximizing y requires $\dfrac{dx}{dy} = 0$.

$$\frac{dx}{dy} = \frac{20 - 2y}{\sqrt{20y - y^2}}$$

If $\dfrac{dx}{dy} = 0$, $\boxed{y = 10 \text{ ft}}$

14. $A_o = \pi\left[\left(\dfrac{1 \text{ in}}{2}\right)\left(\dfrac{\text{ft}}{12 \text{ in}}\right)\right]^2 = 0.00545 \text{ ft}^2$

The theoretical velocity is

$$v = \sqrt{2gh} = \sqrt{(2)\left(32.2\,\frac{\text{ft}}{\text{sec}^2}\right)(2.1 \text{ ft})} = 11.63 \text{ ft/sec}$$

The theoretical discharge is

$$W_{\text{theoretical}} = vA_o\rho t$$

$$= \left(11.63\,\frac{\text{ft}}{\text{sec}}\right)(0.00545 \text{ ft}^2)\left(62.4\,\frac{\text{lbm}}{\text{ft}^3}\right)(90 \text{ sec})$$

$$= 356 \text{ lbm}$$

$$C_d = \frac{m_{\text{actual}}}{m_{\text{theoretical}}} = \frac{228 \text{ lbm}}{356 \text{ lbm}} = 0.64$$

The coefficient of velocity is

$$C_v = \frac{x}{2\sqrt{hy}} = \frac{4}{2\sqrt{(2.1)(2)}} = \boxed{0.976}$$

Since $C_d = C_vC_c$,

$$C_c = \frac{C_d}{C_v} = \frac{0.64}{0.976} = \boxed{0.656}$$

15. $A_o = \pi\left[\left(\dfrac{4 \text{ in}}{2}\right)\left(\dfrac{\text{ft}}{12 \text{ in}}\right)\right]^2 = 0.08727 \text{ ft}^2$

$$A_t = \pi\left(\frac{20 \text{ ft}}{2}\right)^2 = 314.16 \text{ ft}^2$$

The time required to lower the fluid elevation is

$$t = \frac{2A_t\left(\sqrt{z_1} - \sqrt{z_2}\right)}{C_dA_o\sqrt{2g}}$$

$$= \left[\frac{(2)(314.16 \text{ ft}^2)\left(\sqrt{40 \text{ ft}} - \sqrt{20 \text{ ft}}\right)}{(0.98)(0.08727 \text{ ft}^2)\sqrt{(2)\left(32.2\,\dfrac{\text{ft}}{\text{sec}^2}\right)}}\right]\left(\frac{\text{min}}{60 \text{ sec}}\right)$$

$$= \boxed{28.26 \text{ min}}$$

16. $C_f = C_dF_{\text{va}} = \dfrac{C_d}{\sqrt{1 - \left(\dfrac{D_2}{D_1}\right)^4}}$

$$= \frac{1}{\sqrt{1 - \left(\dfrac{8 \text{ in}}{12 \text{ in}}\right)^4}} = 1.116$$

$$A_2 = \pi\left[\left(\frac{8 \text{ in}}{2}\right)\left(\frac{\text{ft}}{12 \text{ in}}\right)\right]^2 = 0.349 \text{ ft}^2$$

$$\dot{V} = C_fA_2\sqrt{\frac{2g(\rho_m - \rho)h}{\rho}} = (1.116)(0.349 \text{ ft}^2)\times$$

$$\sqrt{\frac{(2)\left(32.2\,\dfrac{\text{ft}}{\text{sec}^2}\right)\left(848.5\,\dfrac{\text{lbm}}{\text{ft}^3} - 62.4\,\dfrac{\text{lbm}}{\text{ft}^3}\right)(4 \text{ in})\left(\dfrac{\text{ft}}{12 \text{ in}}\right)}{62.4\,\dfrac{\text{lbm}}{\text{ft}^3}}}$$

$$= \boxed{6.405 \text{ ft}^3/\text{sec}}$$

17. Neglect the compressibility of the air.

$$C_f = C_dF_{\text{va}} = \frac{C_d}{\sqrt{1 - \left(\dfrac{D_2}{D_1}\right)^4}}$$

$$= \frac{0.98}{\sqrt{1 - \left(\dfrac{1.5 \text{ in}}{3 \text{ in}}\right)^4}} = 1.012$$

$$A_2 = \pi\left[\left(\frac{1.5 \text{ in}}{2}\right)\left(\frac{\text{ft}}{12 \text{ in}}\right)\right]^2 = 0.0123 \text{ ft}^2$$

$$\rho_{air} = \frac{p_1}{RT} = \frac{\gamma h}{RT}$$

$$= \frac{(30.05 \text{ in} + 0.287 \text{ in})\left(\dfrac{\text{ft}}{12 \text{ in}}\right)\left(848.5 \dfrac{\text{lbf}}{\text{ft}^3}\right)}{\left(53.3 \dfrac{\text{ft-lbf}}{\text{lbm-°R}}\right)(460 + 114)\text{°R}}$$

$$= 0.0701 \text{ lbm/ft}^3$$

$$\Delta p = (0.838 \text{ in})\left(\frac{\text{ft}}{12 \text{ in}}\right)\left(848.5 \frac{\text{lbf}}{\text{ft}^3}\right)$$

$$= 59.25 \text{ lbf/ft}^2$$

$$\dot{V} = C_f A_2 v_{2,\text{ideal}} = C_f A_2 \sqrt{\frac{2g\Delta p}{\gamma_{air}}}$$

$$= (1.012)(0.0123 \text{ ft}^2)$$

$$\times \sqrt{\frac{(2)\left(32.2 \dfrac{\text{ft}}{\text{sec}^2}\right)\left(59.25 \dfrac{\text{lbf}}{\text{ft}^2}\right)}{0.0701 \dfrac{\text{lbf}}{\text{ft}^3}}}$$

$$= 2.904 \text{ ft}^3/\text{sec}$$

$$\dot{m} = \rho\dot{V} = \left(0.0701 \frac{\text{lbm}}{\text{ft}^3}\right)\left(2.904 \frac{\text{ft}^3}{\text{sec}}\right)$$

$$= \boxed{0.204 \text{ lbm/sec}}$$

18. $\Delta p = \dfrac{\gamma}{2g}\left(\dfrac{\dot{V}}{C_f A_2}\right)^2$

$$= \left[\frac{55 \dfrac{\text{lbf}}{\text{ft}^3}}{(2)\left(32.2 \dfrac{\text{ft}}{\text{sec}^2}\right)}\right]$$

$$\times \left(\frac{\left(1 \dfrac{\text{ft}}{\text{sec}}\right)\pi\left[\left(\dfrac{1 \text{ in}}{2}\right)\left(\dfrac{\text{ft}}{12 \text{ in}}\right)\right]^2}{(0.6)\pi\left[\left(\dfrac{0.2}{2}\right)\left(\dfrac{\text{ft}}{12 \text{ in}}\right)\right]^2}\right)^2$$

$$= \boxed{1482.7 \text{ lbf/ft}^2}$$

19. Assume that the water is at 70°F, and that the kinematic viscosity, ν, is 1.059×10^{-5} ft^2/sec.

$$\text{Re} = \frac{D_e v}{\nu} = \frac{(\text{ft})\left(2 \dfrac{\text{ft}}{\text{sec}}\right)}{1.059 \times 10^{-5} \dfrac{\text{ft}^2}{\text{sec}}} = 1.89 \times 10^5$$

$$A_o = \pi\left(\frac{0.2 \text{ ft}}{2}\right)^2 = 0.0314 \text{ ft}^2$$

$$A_2 = \pi\left(\frac{1 \text{ ft}}{2}\right)^2 = 0.7854 \text{ ft}^2$$

$$\frac{A_o}{A_2} = \frac{0.0314}{0.7854} = 0.04$$

The flow coefficient for this orifice plate is approximately 0.59.

$$\Delta p = \frac{\gamma}{2g} = \left(\frac{\dot{V}}{C_f A_2}\right)^2$$

$$= \left[\frac{62.4 \dfrac{\text{lbf}}{\text{ft}^3}}{(2)\left(32.2 \dfrac{\text{ft}}{\text{sec}^2}\right)}\right]\left[\frac{\left(2 \dfrac{\text{ft}}{\text{sec}}\right)\pi\left(\dfrac{1 \text{ ft}}{2}\right)^2}{(0.59)\pi\left(\dfrac{0.2 \text{ ft}}{2}\right)^2}\right]^2$$

$$= \boxed{6958.8 \text{ lbf/ft}^2}$$

20. $A_A = \pi\left[\left(\dfrac{24 \text{ in}}{2}\right)\left(\dfrac{\text{ft}}{12 \text{ in}}\right)\right]^2 = 3.142 \text{ ft}^2$

$$A_B = \pi\left[\left(\frac{12 \text{ in}}{2}\right)\left(\frac{\text{ft}}{12 \text{ in}}\right)\right]^2 = 0.7854 \text{ ft}^2$$

$$v_A = \frac{\dot{V}}{A_A} = \frac{8 \dfrac{\text{ft}^3}{\text{sec}}}{3.142 \text{ ft}^2} = 2.546 \text{ ft/sec}$$

$$v_B = \frac{\dot{V}}{A_B} = \frac{8 \dfrac{\text{ft}^3}{\text{sec}}}{0.7854 \text{ ft}^2} = 10.186 \text{ ft/sec}$$

$$p_A = (20 \text{ ft})\left(62.4 \frac{\text{lbf}}{\text{ft}^3}\right) = 1248 \text{ lbf/ft}^2$$

$$p_B = p_A + \frac{\gamma\left(v_A^2 - v_B^2\right)}{2g}$$

$$= 1248 \frac{\text{lbf}}{\text{ft}^2} + \frac{62.4 \dfrac{\text{lbf}}{\text{ft}^3}}{(2)\left(32.2 \dfrac{\text{ft}}{\text{sec}^2}\right)}$$

$$\times \left[\left(2.548 \frac{\text{ft}}{\text{sec}}\right)^2 - \left(10.186 \frac{\text{ft}}{\text{sec}}\right)^2\right]$$

$$= 1153.75 \text{ lbf/ft}^2$$

FLUIDS
Dynamics

$$F_x = p_B A_B \cos\theta - p_A A_A + \frac{\dot{m}(v_B \cos\theta - v_A)}{g_c}$$

$$= \left(1153.75\,\frac{\text{lbf}}{\text{ft}^2}\right)(0.7854\,\text{ft}^2)\cos\theta^\circ$$

$$- \left(1248\,\frac{\text{lbf}}{\text{ft}^2}\right)(3.142\,\text{ft}^2)$$

$$+ \frac{\left(8\,\frac{\text{ft}^3}{\text{sec}}\right)\left(62.4\,\frac{\text{lbm}}{\text{ft}^3}\right)\left[\left(10.186\,\frac{\text{ft}}{\text{sec}}\right)\cos\theta^\circ - 2.546\,\frac{\text{ft}}{\text{sec}}\right]}{32.2\,\frac{\text{lbm-ft}}{\text{lbf-sec}^2}}$$

$$= -2896.6\,\text{lbf}$$

$$F_y = 0$$

The resultant force is

$$\boxed{2896.6\ \text{lbf from B to A}}$$

21.
$$F = \frac{\dot{m}\Delta v}{g_c} = \frac{\rho\dot{V}(v_2 - v_1)}{g_c}$$

$$= \frac{\rho A v_1(0 - v_1)}{g_c} = \frac{-\rho A v_1^2}{g_c}$$

$$= \frac{-\left[0.075\,\frac{\text{lbm}}{\text{ft}^3}\right]\left[(0.75\,\text{in}^2)\left(\frac{\text{ft}}{12\,\text{in}}\right)^2\right]\left[250\,\frac{\text{ft}}{\text{sec}}\right]^2}{32.2\,\frac{\text{lbm-ft}}{\text{lbf-sec}^2}}$$

$$= -0.758\,\text{lbf}$$

$$\boxed{\text{The force required is } 0.758\ \text{lbf to the left.}}$$

22.
$$\dot{m} = \rho\dot{V} = \left(62.4\,\frac{\text{lbm}}{\text{ft}^3}\right)\left(100\,\frac{\text{ft}^3}{\text{sec}}\right)$$

$$= 6240\,\text{lbm/sec}$$

(a)
$$F = \frac{\dot{m}\Delta v}{g_c} = \frac{\left(6240\,\frac{\text{lbm}}{\text{sec}}\right)\left(-600\,\frac{\text{ft}}{\text{sec}}\right)}{32.2\,\frac{\text{lbm-ft}}{\text{lbf-sec}^2}}$$

$$= \boxed{-1.16 \times 10^5\ \text{lbf}}$$

(b)
$$F = \frac{\dot{m}\Delta v}{g_c} = \frac{\dot{m}(-v - v)}{g_c}$$

$$= \frac{\left(6240\,\frac{\text{lbm}}{\text{sec}}\right)\left(-1200\,\frac{\text{ft}}{\text{sec}}\right)}{32.2\,\frac{\text{lbm-ft}}{\text{lbf-sec}^2}}$$

$$= \boxed{-2.32 \times 10^5\ \text{lbf}}$$

23. The density of air is approximately 0.075 lbm/ft³.

$$F_D = \frac{C_D A \rho v^2}{2g_c}$$

$$= \frac{(0.2)\pi\left(\frac{1\,\text{ft}}{2}\right)^2\left(0.075\,\frac{\text{lbm}}{\text{ft}^3}\right)\left(100\,\frac{\text{ft}}{\text{sec}}\right)^2}{(2)\left(32.2\,\frac{\text{lbm-ft}}{\text{lbf-sec}^2}\right)}$$

$$= \boxed{1.83\ \text{lbf}}$$

24.
$$v = \left(90\,\frac{\text{mile}}{\text{hr}}\right)\left(\frac{\text{hr}}{3600\,\text{sec}}\right)\left(5280\,\frac{\text{ft}}{\text{mile}}\right)$$

$$= 132\,\text{ft/sec}$$

$$A = \frac{2g_c F_L}{C_L \rho v^2} = \frac{(2)\left(32.2\,\frac{\text{lbm-ft}}{\text{lbf-sec}^2}\right)(5000\,\text{lbf})}{(0.5)\left(0.075\,\frac{\text{lbm}}{\text{ft}^3}\right)\left(132\,\frac{\text{ft}}{\text{sec}}\right)^2}$$

$$= \boxed{492.8\ \text{ft}^2}$$

25. Since $F \propto v^2$, the force will increase by a factor of 4 when the velocity doubles.

$$F = (4)(800\,\text{lbf}) = \boxed{3200\ \text{lbf}}$$

26.
$$\text{Re}_m = \text{Re}_p$$

$$\frac{L_m v_m}{\nu_m} = \frac{L_p v_p}{\nu_p}$$

For air at the same temperature and pressure,

$$\nu_m = \nu_p$$

$$v_m = \left(\frac{L_p}{L_m}\right)v_p = \left(\frac{L_p}{\frac{1}{10}L_p}\right)v_p = 10v_p$$

$$= (10)(60\,\text{mph}) = \boxed{600\ \text{mph}}$$

27.
$$\text{Re}_m = \text{Re}_p$$

$$\frac{L_m v_m}{\nu_m} = \frac{L_p v_p}{\nu_p}$$

Assume $\nu_m = \nu_p$.

$$v_m = \left(\frac{L_p}{L_m}\right)v_p = \left(\frac{600\,\text{ft}}{20\,\text{ft}}\right)\left(10\,\frac{\text{ft}}{\text{sec}}\right)$$

$$= \boxed{300\ \text{ft/sec}}$$

FLUIDS
Dynamics

HYDRAULIC MACHINES

1. (a) $\text{WHP} = \dfrac{h_A \dot{V}(\text{SG})}{8.814}$

$$= \dfrac{(150\text{ ft})\left(10\,\dfrac{\text{ft}^3}{\text{sec}}\right)(0.82)}{8.814\,\dfrac{\text{ft}^4}{\text{hp-sec}}}$$

$$= 139.6\text{ hp}$$

$\text{motor power} = \dfrac{\text{WHP}}{\eta_p} = \dfrac{139.6\text{ hp}}{0.8}$

$$= 174.5\text{ hp}$$

A 200 hp motor should be chosen.

(b) $\text{NPSHA} = h_{\text{atm}} + h_{z(s)} - h_{f(s)} - h_{\text{vp}}$

$$= 0\text{ ft} + 150\text{ ft} - 8\text{ ft} - 0.8\text{ ft}$$

$$= 141.2\text{ ft}$$

(c) At cavitation, NPSHR = NPSHA. Using $\gamma = 62.4\text{ lbf/ft}^3$,

$100\text{ ft} = h_{p(i)} + h_{v(i)} - h_{\text{vp}}$

$$= \dfrac{\left(10\,\dfrac{\text{lbf}}{\text{in}^2}\right)\left(144\,\dfrac{\text{in}^2}{\text{ft}^2}\right)}{62.4\,\dfrac{\text{lbf}}{\text{ft}^3}} + \dfrac{v_{(i)}^2}{(2)\left(32.2\,\dfrac{\text{ft}}{\text{sec}^2}\right)} - 0.8\text{ ft}$$

$v_{(i)} = \boxed{70.75\text{ ft/sec}}$

(d) $A = \dfrac{\dot{V}}{v_{(i)}} = \dfrac{\pi d^2}{4}$

$$d = \sqrt{\dfrac{4\dot{V}}{v_{(i)}\pi}} = \sqrt{\dfrac{(4)\left(10\,\dfrac{\text{ft}^3}{\text{sec}}\right)}{\left(70.75\,\dfrac{\text{ft}}{\text{sec}}\right)\pi}}$$

$$= 0.424\text{ ft}$$

$d = (0.424\text{ ft})\left(\dfrac{12\text{ in}}{\text{ft}}\right) = \boxed{5.09\text{ in}}$

2. (a) $\text{BHP} = \dfrac{\text{WHP}}{\eta_p}$

The friction horsepower is

$\text{FHP} = \text{BHP} - \text{WHP} = \dfrac{\text{WHP}}{\eta_p} - \text{WHP}$

$\text{WHP} = \dfrac{\eta_p(\text{FHP})}{1 - \eta_p} = \dfrac{(0.8)(2\text{ hp})}{1 - 0.8} = 8\text{ hp}$

For SAE-30 oil, SG = 0.89.

$\text{WHP} = \dfrac{\Delta p\dot{m}}{(34,320)(\text{SG})}$

$\dot{m} = \dfrac{(\text{WHP})(34,320)(\text{SG})}{\Delta p}$

$$= \dfrac{(8\text{ hp})\left(34,320\,\dfrac{\text{lbf-lbm}}{\text{ft}^2\text{-sec-hp}}\right)(0.89)}{\left(100\,\dfrac{\text{lbf}}{\text{in}^2}\right)\left(144\,\dfrac{\text{in}^2}{\text{ft}^2}\right)}$$

$$= \boxed{17\text{ lbm/sec}}$$

(b) $\eta = \dfrac{\text{WHP}}{\text{motor power}} = \dfrac{8\text{ hp}}{12\text{ hp}} = \boxed{0.67}$

3. $h_A = z_d - z_s = 100\text{ ft} - 3\text{ ft} = 97\text{ ft}$

$\text{WHP} = \dfrac{h_A Q(\text{SG})}{3956}$

$$= \dfrac{(97\text{ ft})\left(800\,\dfrac{\text{gal}}{\text{hr}}\right)(0.85)}{\left(3956\,\dfrac{\text{gal-ft}}{\text{hp-min}}\right)\left(60\,\dfrac{\text{min}}{\text{hr}}\right)} = \boxed{0.2779\text{ hp}}$$

4. The total head added by the pump is

$h_A = h_{\text{pressure}} + h_{\text{kinetic}} + h_{\text{potential}}$

$$= h_p + \dfrac{v_d^2 - v_s^2}{2g} + 0$$

$$= 330\text{ ft} + \dfrac{\left(16\,\dfrac{\text{ft}}{\text{sec}}\right)^2 - \left(6.5\,\dfrac{\text{ft}}{\text{sec}}\right)^2}{(2)\left(32.2\,\dfrac{\text{ft}}{\text{sec}^2}\right)}$$

$$= 333.3\text{ ft}$$

$$\dot{m} = \left[\dfrac{(\text{WHP})}{h_A}\right]\left(\dfrac{g_c}{g}\right)$$

$$= \left[\dfrac{(67\text{ hp})\left(550\,\dfrac{\text{ft-lbf}}{\text{hp-sec}}\right)}{333.3\text{ ft}}\right]\left(\dfrac{32.2\,\dfrac{\text{lbm-ft}}{\text{lbf-sec}^2}}{32.2\,\dfrac{\text{ft}}{\text{sec}^2}}\right)$$

$$= \boxed{110.6\text{ lbm/sec}}$$

5. (a) $\text{WHP} = \dfrac{\Delta p Q}{1714}$

$$= \dfrac{\left(75 \dfrac{\text{lbf}}{\text{in}^2} - 14.7 \dfrac{\text{lbf}}{\text{in}^2}\right)\left(2.5 \dfrac{\text{gal}}{\text{sec}}\right)\left(60 \dfrac{\text{sec}}{\text{min}}\right)}{1714 \dfrac{\text{lbf-gal}}{\text{in}^2\text{-min-hp}}}$$

$$= \boxed{5.28 \text{ hp}}$$

(b) $h_A = \dfrac{\Delta p g_c}{\rho g} = \dfrac{\Delta p g_c}{(\text{SG})\rho_{\text{water}} g}$

$$= \dfrac{\left(75 \dfrac{\text{lbf}}{\text{in}^2} - 14.7 \dfrac{\text{lbf}}{\text{in}^2}\right)\left(\dfrac{144 \text{ in}^2}{\text{ft}^2}\right)\left(32.2 \dfrac{\text{lbm-ft}}{\text{lbf-sec}^2}\right)}{(0.9)\left(62.4 \dfrac{\text{lbm}}{\text{ft}^3}\right)\left(32.2 \dfrac{\text{ft}}{\text{sec}^2}\right)}$$

$$= \boxed{154.6 \text{ ft}}$$

(c) $W = \dfrac{(3956)(\text{WHP})}{Q(\text{SG})}$

$$= \dfrac{\left(3956 \dfrac{\text{gal-ft}}{\text{hp-min}}\right)(5.28 \text{ hp})}{\left(60 \dfrac{\text{sec}}{\text{min}}\right)\left(2.5 \dfrac{\text{gal}}{\text{sec}}\right)(0.9)}$$

$$= \boxed{154.7 \text{ ft-lbf/lbm}}$$

6. (a) $p_s = p_{\text{atm}} - \rho_{\text{Hg}}\left(\dfrac{g}{g_c}\right) h_{\text{Hg}}$

$$= p_{\text{atm}} - \left(848 \dfrac{\text{lbm}}{\text{ft}^3}\right)\left(\dfrac{32.2 \dfrac{\text{ft}}{\text{sec}^2}}{32.2 \dfrac{\text{lbm-ft}}{\text{lbf-sec}^2}}\right)$$

$$\times (4 \text{ in})\left(\dfrac{\text{ft}^3}{1728 \text{ in}^3}\right)$$

$$= p_{\text{atm}} - 1.963 \text{ psi}$$

$p_d = p_{\text{atm}} + 22 \text{ psi}$

$\Delta p = p_d - p_s = 22 \text{ psi} + 1.963 \text{ psi}$

$$= 23.96 \text{ psi}$$

$\dot{V} = \dfrac{17.5 \dfrac{\text{ft}^3}{\text{sec}}}{60 \dfrac{\text{sec}}{\text{min}}} = 0.2917 \text{ ft}^3/\text{sec}$

$\text{WHP} = \dfrac{\Delta p \dot{V}}{3.819} = \dfrac{\left(23.96 \dfrac{\text{lbf}}{\text{in}^2}\right)\left(0.2917 \dfrac{\text{ft}^3}{\text{sec}}\right)}{3.819 \dfrac{\text{lbf-sec}}{\text{in}^2\text{-ft}^3\text{-hp}}}$

$$= 1.83 \text{ hp}$$

motor output power = brake pump power

$$= \dfrac{\text{WHP}}{\eta_p} = \dfrac{1.83 \text{ hp}}{0.82}$$

$$= \boxed{2.23 \text{ hp}}$$

(b) friction power = BHP − WHP

$$= 2.23 \text{ hp} - 1.83 \text{ hp}$$

$$= \boxed{0.40 \text{ hp}}$$

7. (a) $h_A = z_d - z_s = 65 \text{ ft} - (-10 \text{ ft}) = 75 \text{ ft}$

$\text{WHP} = \dfrac{h_A \dot{V}(\text{SG})}{8.814} = \dfrac{(75 \text{ ft})\left(\dfrac{25 \dfrac{\text{ft}^3}{\text{min}}}{60 \dfrac{\text{sec}}{\text{min}}}\right)(1)}{8.814 \dfrac{\text{ft}^4}{\text{sec-hp}}}$

$$= 3.55 \text{ hp}$$

$\eta_p = \dfrac{\text{WHP}}{\text{BHP}} = \dfrac{3.55 \text{ hp}}{5.4 \text{ hp}}$

$$= \boxed{0.657 \ (65.7\%)}$$

(b) $\eta_m = \dfrac{\text{BHP}}{\text{motor input power}} = \dfrac{5.4 \text{ hp}}{6 \text{ hp}}$

$$= 0.90$$

$\eta = \eta_p \eta_m = (0.657)(0.90)$

$$= \boxed{0.591 \ (59.1\%)}$$

8. $t = \dfrac{V}{\dot{V}} = \dfrac{530 \text{ ft}^3}{\left(25 \dfrac{\text{ft}^3}{\text{min}}\right)\left(60 \dfrac{\text{min}}{\text{hr}}\right)}$

$$= 0.353 \text{ hr}$$

$W_{\text{kW-hr}} = \dfrac{(\text{BHP})t}{1.341 \dfrac{\text{hp}}{\text{kW}}}$

$$= \dfrac{(5.4 \text{ hp})(0.353 \text{ hr})}{1.341 \dfrac{\text{hp}}{\text{kW}}} = 1.42 \text{ kW-hr}$$

$\text{cost} = \dfrac{(W_{\text{kW-hr}})(\text{cost per kW-hr})}{\eta_m}$

$$= \dfrac{(1.42 \text{ kW-hr})\left(\dfrac{\$1.00}{\text{kW-hr}}\right)}{0.90}$$

$$= \boxed{\$1.58}$$

9. $$t = \frac{V}{\dot{V}} = \frac{700 \text{ ft}^3}{\left(800 \frac{\text{gal}}{\text{hr}}\right)\left(\frac{0.13368 \text{ ft}^3}{\text{gal}}\right)}$$

$$= 6.55 \text{ hr}$$

The amount of electrical power consumed by the motor is

$$\text{BHP} = \frac{\text{WHP}}{\eta} = \frac{0.2779 \text{ hp}}{0.75} = 0.371 \text{ hp}$$

The amount of electrical energy consumed is

$$W = Pt = \left(\frac{0.371 \text{ hp}}{1.341 \frac{\text{hp}}{\text{kW}}}\right)(6.55 \text{ hr}) = 1.81 \text{ kW-hr}$$

The cost per kW-hr is

$$\frac{\$2.00}{1.81 \text{ kW-hr}} = \boxed{\$1.10/\text{kW-hr}}$$

10. $$W_{\text{kW-hr}} = W = \frac{h_A Q (\text{SG})}{3956}$$

$$= \frac{\left(0.7457 \frac{\text{kW}}{\text{hp}}\right)(200 \text{ ft})(1000 \text{ gal})(1)}{\left(60 \frac{\text{min}}{\text{hr}}\right)\left(3956 \frac{\text{gal-ft}}{\text{hp-min}}\right)}$$

$$= 0.628 \text{ kW-hr}$$

$$\text{cost} = \left(\frac{\$1.00}{\text{kW-hr}}\right)(0.628 \text{ kW-hr}) = \boxed{\$0.628}$$

11. (a) $Q = (5 \text{ cfs})\left(448.83 \frac{\text{gal-sec}}{\text{ft}^3\text{-min}}\right) = 2244 \text{ gpm}$

$$\text{WHP} = \eta_m \eta_p (\text{motor power})$$

$$= (0.90)(50 \text{ hp}) = 45 \text{ hp}$$

$$= \frac{h_A \dot{V}(\text{SG})}{8.814}$$

$$h_A = \frac{(8.814)(\text{WHP})}{\dot{V}(\text{SG})}$$

$$= \frac{\left(8.814 \frac{\text{ft}^4}{\text{hp-sec}}\right)(45 \text{ hp})}{\left(5 \frac{\text{ft}^3}{\text{sec}}\right)(0.728)} = 109 \text{ ft}$$

$$n_s = \frac{n\sqrt{Q} \text{ in gpm}}{(h_A \text{ in ft})^{0.75}} = \frac{(900)\sqrt{2244}}{(109)^{0.75}}$$

$$= \boxed{1264}$$

(b) At the same efficiency,

$$\frac{P_2}{P_1} = \left(\frac{Q_2}{Q_1}\right)^3$$

$$P_2 = P_1 \left(\frac{Q_2}{Q_1}\right)^3 = P_1 \left(\frac{\dot{V}_2}{\dot{V}_1}\right)^3$$

$$= (50 \text{ hp})\left(\frac{8 \text{ cfs}}{5 \text{ cfs}}\right)^3 = \boxed{204.8 \text{ hp}}$$

12. $$Q = \frac{320 \frac{\text{ft}^3}{\text{min}}}{0.13368 \frac{\text{ft}^3}{\text{gal}}} = 2394 \text{ gpm}$$

$$n_s = \frac{n\sqrt{Q}}{(h_A)^{0.75}}$$

$$= \frac{(1800 \text{ rpm})\sqrt{2394 \text{ gpm}}}{(200 \text{ ft})^{0.75}} = \boxed{1656}$$

13. $$Q = \frac{300 \frac{\text{ft}^3}{\text{min}}}{0.13368 \frac{\text{ft}^3}{\text{gal}}} = 2244 \text{ gpm}$$

$$n_s = \frac{n\sqrt{Q}}{(h_A)^{0.75}} = \frac{(2600 \text{ rpm})\sqrt{2244 \text{ gpm}}}{(160 \text{ ft})^{0.75}}$$

$$= 2738$$

$$\boxed{\text{A Francis vane pump should be chosen.}}$$

14. The following conditions can cause cavitation in a pump.

- high suction lift
- low suction head
- high friction in suction line
- high liquid temperature (high vapor pressure)
- high pump speed
- low volumetric output compared to flow at high pump efficiency

15. $\text{NPSHA} = h_{t(s)} - h_{\text{vp}}$

$$h_{\text{vp}(86°\text{F})} = \left(\frac{p_{\text{vp}(86°\text{F})}}{\rho}\right)\left(\frac{g_c}{g}\right)$$

$$= \left[\frac{\left(0.6158 \frac{\text{lbf}}{\text{in}^2}\right)\left(\frac{144 \text{ in}^2}{\text{ft}^2}\right)}{62.4 \frac{\text{lbm}}{\text{ft}^3}}\right]\left(\frac{32.2 \frac{\text{lbm-ft}}{\text{lbf-sec}^2}}{32.2 \frac{\text{ft}}{\text{sec}^2}}\right)$$

$$= 1.42 \text{ ft}$$

FLUIDS
Hyd Machines

$$h_{t(s)} = h_{\text{atm}} + h_{z(s)}$$

$$= \left[\frac{\left(14.7 \frac{\text{lbf}}{\text{in}^2}\right)\left(144 \frac{\text{in}^2}{\text{ft}^2}\right)}{\left(62.4 \frac{\text{lbm}}{\text{ft}^3}\right)\left(32.2 \frac{\text{ft}}{\text{sec}^2}\right)} \right]$$

$$\times \left(32.2 \frac{\text{lbm-ft}}{\text{lbf-sec}^2}\right) - 16\,\text{ft}$$

$$= 17.92\,\text{ft}$$

$$\text{NPSHA} = 17.92\,\text{ft} - 1.42\,\text{ft} = \boxed{16.50\,\text{ft}}$$

16. $\text{NPSHA} = h_{t(s)} - h_{\text{vp}}$

$$= h_{\text{atm}} + h_{z(s)} - h_{\text{vp}}$$

$$= \left[\frac{\left(14.6 \frac{\text{lbf}}{\text{in}^2}\right)\left(144 \frac{\text{in}^2}{\text{ft}^2}\right)}{62.4 \frac{\text{lbm}}{\text{ft}^3}}\right]\left(\frac{32.2 \frac{\text{lbm-ft}}{\text{lbf-sec}^2}}{32.2 \frac{\text{ft}}{\text{sec}^2}}\right)$$

$$-10\,\text{ft} - 1.6\,\text{ft}$$

$$= 22.1\,\text{ft}$$

$$22.1\,\text{ft} < 30\,\text{ft}\ (\text{NPSHA} < \text{NPSHR})$$

$$\boxed{\text{Cavitation will occur.}}$$

17. $$\sigma = \frac{\text{NPSHA}}{h_A} = \frac{16.50\,\text{ft}}{65\,\text{ft}}$$

$$= 0.254 > 0.228$$

$$\boxed{\text{Cavitation will not occur.}}$$

18. $$Q = \frac{26 \frac{\text{ft}^3}{\text{min}}}{0.13368 \frac{\text{ft}^3}{\text{gal}}} = 194.5\,\text{gpm}$$

$$n_{ss} = \frac{n\sqrt{Q}}{\text{NPSHR}^{0.75}} = \frac{(1800\,\text{rpm})\sqrt{194.5\,\text{gpm}}}{(100\,\text{ft})^{0.75}}$$

$$= 793.8 \approx 800$$

$$\boxed{\text{The pump is a single-suction pump.}}$$

19. Since $h_z = 0$, the system curve is governed by the equation

$$h_A = h_f = 9.26 \times 10^{-3} Q^2$$

Plot the pump curve and the system curve.

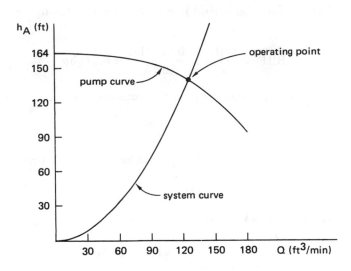

$$\boxed{Q = 123\,\text{ft}^3/\text{min and } h_A = 141\,\text{ft}}$$

20. Find the solution graphically. First, plot the two pump curves and construct the pump curve for the parallel combination. Plot the system curve and find its intersection with the combination pump curve.

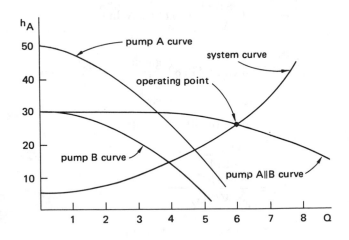

The operating point is

$$\boxed{Q \approx 6.0 \text{ and } h_A \approx 26}$$

21. Find the solution graphically. First, plot the two pump curves and construct the pump curve for the series combination. Plot the system curve and find its intersection with the combination pump curve.

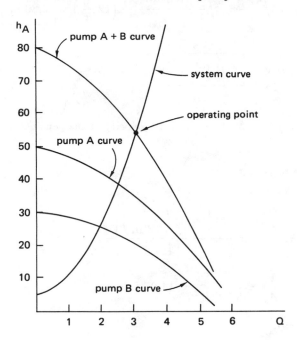

The operating point is

$$\boxed{Q \approx 3.1 \text{ and } h_A \approx 54}$$

22. $D_2 = D_1\sqrt{\dfrac{h_2}{h_1}} = (1\text{ ft})\sqrt{\dfrac{330\text{ ft}}{200\text{ ft}}} = \boxed{1.28\text{ ft}}$

23. $\qquad Q_2 = Q_1\left(\dfrac{n_2}{n_1}\right)$

$$= \left(70\,\frac{\text{ft}^3}{\text{min}}\right)\left(\frac{2500\text{ rpm}}{1850\text{ rpm}}\right)$$

$$= \boxed{94.6\text{ ft}^3/\text{min}}$$

24. Use subscript 1 for the prototype and subscript 2 for the model.

$$\text{WHP}_2 = \eta_p\,\text{BHP} = (0.75)(14.8\text{ hp}) = 11.1\text{ hp}$$

$$= \frac{h_2\dot{V}}{8.814}$$

$$h_2 = \frac{(8.814)(\text{WHP}_2)}{\dot{V}}$$

$$= \frac{\left(8.814\,\dfrac{\text{ft}^4}{\text{hp-sec}}\right)(11.1\text{ hp})\left(60\,\dfrac{\text{sec}}{\text{min}}\right)}{26\,\dfrac{\text{ft}^3}{\text{min}}}$$

$$= 225.8\text{ ft}$$

$$\frac{n_1\sqrt{Q_1}}{(h_1)^{0.75}} = \frac{n_2\sqrt{Q_2}}{(h_2)^{0.75}}$$

$$n_2 = n_1\sqrt{\frac{Q_1}{Q_2}}\left(\frac{h_2}{h_1}\right)^{0.75}$$

$$= (1200\text{ rpm})\sqrt{\frac{130\,\dfrac{\text{ft}^3}{\text{min}}}{26\,\dfrac{\text{ft}^3}{\text{min}}}\left(\frac{225.8\text{ ft}}{65\text{ ft}}\right)^{0.75}}$$

$$= \boxed{6828\text{ rpm}}$$

25. $\dfrac{n_1 D_1}{\sqrt{h_1}} = \dfrac{n_2 D_2}{\sqrt{h_2}}$

$$n_2 = n_1\left(\frac{D_1}{D_2}\right)\sqrt{\frac{h_2}{h_1}}$$

$$= (1500\text{ rpm})\left[\frac{16\text{ in}}{(6.2\text{ ft})\left(12\,\dfrac{\text{in}}{\text{ft}}\right)}\right]\sqrt{\frac{50\text{ ft}}{25\text{ ft}}}$$

$$= \boxed{456\text{ rpm}}$$

$$\frac{P_1}{D_1^2 h_1^{1.5}} = \frac{P_2}{D_2^2 h_2^{1.5}}$$

$$P_2 = P_1\left(\frac{D_2}{D_1}\right)^2\left(\frac{h_2}{h_1}\right)^{1.5}$$

$$= (12\text{ hp})\left[\frac{(6.2\text{ ft})\left(12\,\dfrac{\text{in}}{\text{ft}}\right)}{16\text{ in}}\right]^2\left(\frac{50\text{ ft}}{25\text{ ft}}\right)^{1.5}$$

$$= \boxed{734\text{ hp}}$$

26. $n_s = \dfrac{n\sqrt{P}}{(h_t)^{1.25}} = \dfrac{(600\text{ rpm})\sqrt{550\text{ hp}}}{(140\text{ ft})^{1.25}} = \boxed{29.2}$

27. (a) $v_b = 2\pi f\left(\dfrac{d}{2}\right)$

$$= (2\pi)\left(500\,\frac{\text{rev}}{\text{min}}\right)\left(\frac{10\text{ in}}{2}\right)\left(\frac{\text{ft}}{12\text{ in}}\right)\left(\frac{\text{min}}{60\text{ sec}}\right)$$

$$= 21.82\text{ ft/sec}$$

For a round-edged orifice, $C_v = 0.98$.

Neglecting h_f, $h' = h_t = 45$ ft.

$$v_j = C_v \sqrt{2gh'}$$

$$= (0.98)\sqrt{(2)\left(32.2 \frac{\text{ft}}{\text{sec}^2}\right)(45 \text{ ft})}$$

$$= 52.76 \text{ ft/sec}$$

$$E = \frac{v_b(v_j - v_b)}{g_c}(1 - \cos\theta)$$

$$= \left[\frac{\left(21.82 \dfrac{\text{ft}}{\text{sec}}\right)\left(52.76 \dfrac{\text{ft}}{\text{sec}} - 21.82 \dfrac{\text{ft}}{\text{sec}}\right)}{32.2 \dfrac{\text{ft-lbm}}{\text{lbf-sec}^2}}\right]$$

$$\times (1 - \cos 120°)$$

$$= \boxed{31.45 \text{ ft-lbf/lbm}}$$

(b) $P = \dot{m}E$

$$= \frac{\left(400 \dfrac{\text{lbm}}{\text{sec}}\right)\left(31.45 \dfrac{\text{ft-lbf}}{\text{lbm}}\right)}{550 \dfrac{\text{ft-lbf}}{\text{hp-sec}}}$$

$$= \boxed{22.87 \text{ hp}}$$

28. (a) Neglecting penstock friction loss, the jet velocity is

$$v_j = C_v \sqrt{2gh'}$$

$$= (0.97)\sqrt{(2)\left(32.2 \frac{\text{ft}}{\text{sec}^2}\right)(200 \text{ ft})}$$

$$= \boxed{110.1 \text{ ft/sec}}$$

Compute the flow rate.

$$\dot{V} = \frac{(\text{WHP})(8.814)}{h_A} = \frac{(\text{BHP})(8.814)}{\eta h_A}$$

$$h_A = h' - h_n = h' - h'(1 - C_v^2)$$

$$= h'C_v^2 = (200 \text{ ft})(0.97)^2 = 188.2 \text{ ft}$$

$$\dot{V} = \frac{(87 \text{ hp})\left(8.814 \dfrac{\text{ft}^4}{\text{hp-sec}}\right)}{(188.2 \text{ ft})(0.83)} = 4.91 \text{ ft}^3/\text{sec}$$

The jet area is

$$A_j = \frac{\dot{V}}{v_j} = \frac{4.91 \dfrac{\text{ft}^3}{\text{sec}}}{110.1 \dfrac{\text{ft}}{\text{sec}}} = 0.0446 \text{ ft}^2$$

The jet diameter is

$$D_j = \sqrt{\frac{4A_j}{\pi}} = \sqrt{\frac{(4)(0.0446 \text{ ft}^2)}{\pi}}\left(12 \frac{\text{in}}{\text{ft}}\right) = \boxed{2.86 \text{ in}}$$

(b) $\dot{V} = \boxed{4.91 \text{ ft}^3/\text{sec}}$

(c) $h' = h_p + \dfrac{v^2}{2g}$ [v is the velocity in the penstock]

The penstock area is

$$A_p = \frac{\pi D_p^2}{4} = \frac{(\pi)\left[(8 \text{ in})\left(\dfrac{\text{ft}}{12 \text{ in}}\right)\right]^2}{4} = 0.349 \text{ ft}^2$$

$$v_p = \frac{\dot{V}}{A_p} = \frac{4.91 \dfrac{\text{ft}^3}{\text{sec}}}{0.349 \text{ ft}^2} = 14.07 \text{ ft/sec}$$

$$h_p = h' - \frac{v^2}{2g}$$

$$= 200 \text{ ft} - \frac{\left(14.07 \dfrac{\text{ft}}{\text{sec}}\right)^2}{(2)\left(32.2 \dfrac{\text{ft}}{\text{sec}^2}\right)}$$

$$= \boxed{197 \text{ ft}}$$

29. $v_j = C_v\sqrt{2gh'} = (0.98)\sqrt{(2)\left(32.2 \dfrac{\text{ft}}{\text{sec}^2}\right)(900 \text{ ft})}$

$$= 236 \text{ ft/sec}$$

$$\dot{V} = A_j v_j$$

$$= \left(\frac{(\pi)\left[(5 \text{ in})\left(\dfrac{\text{ft}}{12 \text{ in}}\right)\right]^2}{4}\right)\left(236 \frac{\text{ft}}{\text{sec}}\right)$$

$$= 32.18 \text{ ft}^3/\text{sec}$$

$$h_A = h' - h_n = h'C_v^2$$

$$= (900 \text{ ft})(0.98)^2 = 864 \text{ ft}$$

$$\eta = \frac{\text{BHP}(8.814)}{h_A \dot{V}} = \frac{(2500 \text{ hp})\left(8.814 \dfrac{\text{ft}^4}{\text{hp-sec}}\right)}{(864 \text{ ft})\left(32.18 \dfrac{\text{ft}^3}{\text{sec}}\right)}$$

$$= \boxed{0.793 \ (79.3\%)}$$

30. $h_A = \Delta y - h_f = 625 \text{ ft} - 58 \text{ ft} = 567 \text{ ft}$

$$\text{WHP} = \frac{h_A Q(\text{SG})}{8.814} = \frac{(567 \text{ ft})\left(1000 \dfrac{\text{ft}^3}{\text{sec}}\right)(1.0)}{8.814 \dfrac{\text{ft}^4}{\text{hp-sec}}}$$

$$= 64,329 \text{ hp}$$

$$\text{BHP} = \eta_{\text{turbine}} \text{WHP}$$

$$= (0.89)(64,329 \text{ hp})\left(\frac{0.7457 \text{ kW}}{\text{hp}}\right)$$

$$= \boxed{4.27 \times 10^4 \text{ kW}}$$

31. (a) $\text{SG}_{25\% \text{ salt water}} = 1.19$

Neglecting h_f, $h_A = h_t = 40 \text{ ft}$.

$$\text{WHP} = \frac{h_A Q(\text{SG})}{3956}$$

$$= \frac{(40 \text{ ft})(100 \text{ gpm})(1.19)}{3956 \dfrac{\text{ft-gpm}}{\text{hp}}}$$

$$= 1.2 \text{ hp}$$

electrical output $= \eta_m \eta_p (\text{WHP})$

$$= (0.9)(0.9)(1.2 \text{ hp})\left(0.7457 \frac{\text{kW}}{\text{hp}}\right)$$

$$= \boxed{0.7248 \text{ kW}}$$

(b) The turbine power or brake horsepower is

$$P = (\text{WHP})\eta_p = (1.2 \text{ hp})(0.9) = 1.08 \text{ hp}$$

$$n_s = \frac{n\sqrt{P \text{ in hp}}}{(h_t \text{ in ft})^{1.25}} = \frac{(750)\sqrt{1.08}}{(40)^{1.25}} = \boxed{7.75}$$

(c) $n_1 = n_2$

$$\frac{P_2}{P_1} = \left(\frac{d_2}{d_1}\right)^3$$

$$d_2 = d_1 \left(\frac{P_2}{P_1}\right)^{1/3} = (6 \text{ in})\left(\frac{5}{1.08}\right)^{1/3}$$

$$= \boxed{10.0 \text{ in}}$$

(d) electrical output$_2 = P_2 \eta_m$

$$= (5 \text{ hp})\left(0.7457 \frac{\text{kW}}{\text{hp}}\right)(0.9)$$

$$= \boxed{3.36 \text{ kW}}$$

(e) $n_2 = n_1$

$$h_2 = h_1 \left(\frac{d_2}{d_1}\right)^2 = (40 \text{ ft})\left(\frac{10.0}{6}\right)^2$$

$$= \boxed{111.1 \text{ ft}}$$

32. (a) $h_t = h_p + h_v - z_{\text{tailwater}}$

$$= 92 \text{ ft} + \frac{\left(1.1 \dfrac{\text{ft}}{\text{sec}}\right)^2}{(2)\left(32.2 \dfrac{\text{ft}}{\text{sec}^2}\right)} - (-6.2 \text{ ft})$$

$$= \boxed{98.2 \text{ ft}}$$

(b) $$\text{WHP} = \frac{h_A \dot{V}}{8.814} = \frac{(98.2 \text{ ft})\left(26 \dfrac{\text{ft}^3}{\text{sec}}\right)}{8.814 \dfrac{\text{ft}^4}{\text{hp-sec}}}$$

$$= 290 \text{ hp}$$

$$\eta = \frac{\text{BHP}}{\text{WHP}} = \frac{260 \text{ hp}}{290 \text{ hp}} = 0.90$$

$$= \boxed{90\%}$$

(c) $$\frac{h_2}{h_1} = \left(\frac{n_2}{n_1}\right)^2$$

$$n_2 = n_1 \sqrt{\frac{h_2}{h_1}} = (600 \text{ rpm})\sqrt{\frac{226 \text{ ft}}{98.2 \text{ ft}}}$$

$$= \boxed{910 \text{ rpm}}$$

(d) $$\frac{P_2}{P_1} = \left(\frac{h_2}{h_1}\right)^{3/2}$$

$$\text{BHP}_2 = \text{BHP}_1 \left(\frac{h_2}{h_1}\right)^{3/2}$$

$$= (260 \text{ hp})\left(\frac{226 \text{ ft}}{98.2 \text{ ft}}\right)^{3/2} = \boxed{908 \text{ hp}}$$

(e) $$\left(\frac{Q_2}{Q_1}\right)^2 = \frac{h_2}{h_1} = \left(\frac{\dot{V}_2}{\dot{V}_1}\right)^2$$

$$\dot{V}_2 = \dot{V}_1 \sqrt{\frac{h_2}{h_1}}$$

$$= \left(26 \frac{\text{ft}^3}{\text{sec}}\right)\sqrt{\frac{226 \text{ ft}}{98.2 \text{ ft}}}$$

$$= \boxed{39.4 \text{ ft}^3/\text{sec}}$$

FLUIDS
Hyd Machines

OPEN CHANNEL FLOW

1.

$$A = wd$$

$$P = w + 2d$$

$$r_h = \frac{A}{P} = \frac{wd}{w + 2d}$$

$$Q = AC\sqrt{r_h S}$$

$$= A \left[\left(\frac{1.486}{n} \right) (r_h)^{\frac{1}{6}} \right] \sqrt{r_h S}$$

$$= \frac{(1.486)(wd) \left(\dfrac{wd}{w + 2d} \right)^{\frac{2}{3}} \sqrt{S}}{n}$$

$$3 = \frac{(1.486)(2d)^{\frac{5}{3}} (\sqrt{0.01})}{(0.012)(2 + 2d)^{\frac{2}{3}}}$$

$$\frac{d^{\frac{5}{3}}}{(1 + d)^{\frac{2}{3}}} = 0.1211$$

$$\left(\frac{d^{\frac{5}{2}}}{1 + d} \right)^{\frac{2}{3}} = 0.1211$$

By trial and error: $\boxed{d \cong 0.314 \text{ ft}}$

2. Area in flow is

$$A = \left(\frac{1}{2} \right) \left(\frac{\pi}{4} \right) D^2$$

$$= \left(\frac{1}{2} \right) \left(\frac{\pi}{4} \right) \left(\frac{18 \text{ in}}{12 \dfrac{\text{in}}{\text{ft}}} \right)^2 = 0.88357 \text{ ft}^2$$

The hydraulic radius is

$$r_h = \frac{D}{4} = \frac{18 \text{ in}}{(4) \left(12 \dfrac{\text{in}}{\text{ft}} \right)} = 0.375 \text{ ft}$$

The Chezy-Manning equation is

$$Q = \frac{1.486}{n} (A) r_h^{\frac{2}{3}} \sqrt{S}$$

$$= \left(\frac{1.486}{0.013} \right) (0.88357)(0.375)^{\frac{2}{3}} (\sqrt{0.001})$$

$$= \boxed{1.66 \text{ ft}^3/\text{sec}}$$

3. Assume full pipe flow: $d_o = D$.

$$Q = \left(100 \frac{\text{gal}}{\text{min}} \right) \left(0.1337 \frac{\text{ft}^3}{\text{gal}} \right) \left(\frac{\text{min}}{60 \text{ sec}} \right)$$

$$= 0.2228 \text{ ft}^3/\text{sec}$$

$$D = d_o = 1.335 \left(\frac{nQ}{\sqrt{S}} \right)^{\frac{3}{8}}$$

$$= (1.335) \left[\frac{(0.017)(0.2228)}{\sqrt{0.02}} \right]^{\frac{3}{8}} = 0.343 \text{ ft}$$

$$D = (0.343 \text{ ft}) \left(\frac{12 \text{ in}}{\text{ft}} \right) = 4.12 \text{ in}$$

$$\boxed{5 \text{ in nominal diameter pipe}}$$

4. (a)

$$A = wh = (3 \text{ ft})(1 \text{ ft}) = 3 \text{ ft}^2$$

$$P = w + 2h = 3 \text{ ft} + (2)(1 \text{ ft}) = 5 \text{ ft}$$

$$r_h = \frac{A}{P} = \frac{3 \text{ ft}^2}{5 \text{ ft}} = 0.6 \text{ ft}$$

$$v = \frac{Q}{A} = \frac{3 \dfrac{\text{ft}^3}{\text{sec}}}{3 \text{ ft}^2} = 1 \text{ ft/sec}$$

Using the Bazin formula,

$$C = \frac{157.6}{1 + \dfrac{m}{\sqrt{r_h}}}$$

$$= \frac{157.6}{1 + \dfrac{2.360}{\sqrt{0.6}}} = 38.94$$

$$v = C\sqrt{r_h S}$$

$$S = \left(\frac{v}{C} \right)^2 \left(\frac{1}{r_h} \right)$$

$$= \left(\frac{1}{38.94} \right)^2 \left(\frac{1}{0.6} \right) = \boxed{0.001099}$$

Using the Chezy-Manning formula, $n = 0.035$ for natural channel with stones and weeds.

$$v = \left(\frac{1.486}{n}\right)(r_h)^{\frac{2}{3}}\sqrt{S}$$

$$S = \left(\frac{vn}{1.486}\right)^2\left(\frac{1}{r_h}\right)^{\frac{4}{3}}$$

$$= \left[\frac{(1)(0.035)}{1.486}\right]^2\left(\frac{1}{0.6}\right)^{\frac{4}{3}} = \boxed{0.001096}$$

(b) Assuming uniform flow,

$$\frac{dE}{dL} = S = S_o\left(\frac{g}{g_c}\right)$$

$$E_{\text{lost}} = S_o\left(\frac{g}{g_c}\right)L$$

$$= (0.00110)\left(\frac{32.2\ \frac{\text{ft}}{\text{sec}^2}}{32.2\ \frac{\text{lbm-ft}}{\text{lbf-sec}^2}}\right)(300\ \text{ft})$$

$$= \boxed{0.33\ \text{ft-lbf/lbm}}$$

5. (a) Assume a rectangular channel for best efficiency.

$$d = \frac{w}{2}$$

$$A = \frac{w^2}{2}$$

$$P = 2w$$

$$r_h = \frac{w}{4}$$

Using the Chezy-Manning velocity formula, $n = 0.0165$.

$$Q = vA = \left[\left(\frac{1.486}{n}\right)\left(\frac{w}{4}\right)^{\frac{2}{3}}\sqrt{S}\right]\left(\frac{w^2}{2}\right)$$

$$Q = \left(\frac{1.486}{2}\right)\left(\frac{1}{4}\right)^{\frac{2}{3}}\left(\frac{\sqrt{S}}{n}\right)w^{\frac{8}{3}}$$

$$w = \left[\frac{2Q(4)^{\frac{2}{3}}n}{(1.486)\sqrt{S}}\right]^{\frac{3}{8}}$$

$$= \left[\frac{(2)(300)(4)^{\frac{2}{3}}(0.0165)}{(1.486)(\sqrt{0.0100})}\right]^{\frac{3}{8}} = 6.83\ \text{ft}$$

$$d = \frac{w}{2} = \frac{6.83\ \text{ft}}{2} = 3.415\ \text{ft}$$

$$\boxed{\text{A 3.415-ft deep, 6.83-ft wide channel will work best}}$$

(b) $$v = \left(\frac{1.486}{0.0165}\right)\left(\frac{6.83}{4}\right)^{\frac{2}{3}}(\sqrt{0.0100})$$

$$= \boxed{12.86\ \text{ft/sec}}$$

6. (a) Assuming uniform flow and using normal diameter equation with $n = 0.011$,

$$\frac{D}{2} = d_o = (0.8654)\left(\frac{nQ}{\sqrt{S}}\right)^{\frac{3}{8}}$$

$$D = (2)(0.8654)\left[\frac{(0.011)(500)}{\sqrt{0.0001}}\right]^{\frac{3}{8}}$$

$$= \boxed{18.45\ \text{ft (actual inner diameter)}}$$

(b) $$D = (2)(0.8654)\left[\frac{(0.014)(500)}{\sqrt{0.0001}}\right]^{\frac{3}{8}}$$

$$= \boxed{20.19\ \text{ft (actual inner diameter)}}$$

7. (a) $$d_c^3 = \frac{Q^2}{gw^2}$$

$$d_c = \left(\frac{Q^2}{gw^2}\right)^{\frac{1}{3}}$$

$$= \left[\frac{\left(375\ \frac{\text{ft}^3}{\text{sec}}\right)^2}{\left(32.2\ \frac{\text{ft}}{\text{sec}^2}\right)(5\ \text{ft})^2}\right]^{\frac{1}{3}}$$

$$= \boxed{5.59\ \text{ft}}$$

(b) $$v_c = \sqrt{gd_c} = \sqrt{\left(32.2\ \frac{\text{ft}}{\text{sec}^2}\right)(5.59\ \text{ft})}$$

$$= \boxed{13.42\ \text{ft/sec}}$$

(c) $$d_c = \frac{2}{3}E_c\left(\frac{g}{g_c}\right)$$

$$E_c = \frac{3}{2}d_c = \left(\frac{3}{2}\right)(5.59) = 8.39\ \text{ft-lbf/lbm}$$

$$P = \dot{m}E_c = \rho v_c A E_c = \rho v_c w d_c E_c$$

$$= \left(62.4 \, \frac{lbm}{ft^3}\right)\left(13.42 \, \frac{ft}{sec}\right)(5 \, ft)$$

$$\times \, (5.59 \, ft)\left(8.39 \, \frac{ft\text{-}lbf}{lbm}\right)$$

$$= 1.964 \times 10^5 \, ft\text{-}lbf/sec$$

$$P = \left(1.964 \times 10^5 \, \frac{ft\text{-}lbf}{sec}\right)\left(\frac{1.3558 \, W}{\frac{ft\text{-}lbf}{sec}}\right)$$

$$= 266.2 \times 10^3 \, W = \boxed{266.2 \, kW}$$

8. $$Q = Av = (14 \, ft)(2.5 \, ft)\left(2.9 \, \frac{ft}{sec}\right)$$

$$= 101.5 \, \frac{ft^3}{sec}$$

$$Q = \frac{2}{3} b \sqrt{2g}\left[\left(h + \frac{v_1^2}{2g}\right)^{\frac{3}{2}} - \left(\frac{v_1^2}{2g}\right)^{\frac{3}{2}}\right]$$

$$h = \left[\left(\frac{3}{2}\right)\left(\frac{Q}{b\sqrt{2g}}\right) + \left(\frac{v_1^2}{2g}\right)^{\frac{3}{2}}\right]^{\frac{2}{3}} - \frac{v_1^2}{2g}$$

$$= \left[\left(\frac{3}{2}\right)\left(\frac{101.5 \, \frac{ft^3}{sec}}{(14 \, ft)\sqrt{(2)\left(32.2 \, \frac{ft}{sec^2}\right)}}\right)\right.$$

$$+ \left.\left(\frac{\left(2.9 \, \frac{ft}{sec}\right)^2}{(2)\left(32.2 \, \frac{ft}{sec^2}\right)}\right)^{\frac{3}{2}}\right]^{\frac{2}{3}}$$

$$- \frac{\left(2.9 \, \frac{ft}{sec}\right)^2}{(2)\left(32.2 \, \frac{ft}{sec^2}\right)}$$

$$= 1.12 \, ft$$

depth just upstream $= y + h = 3.0 \, ft + 1.1 \, ft$

$$= \boxed{4.12 \, ft}$$

9. (a) $b_{eff} = b - 0.1Nh$

$$h = \frac{b - b_{eff}}{0.1N} = \frac{5 \, ft - 4.98 \, ft}{(0.1)(1)} = \boxed{0.2 \, ft}$$

(b) $$\frac{h}{b} = \frac{0.2}{5} = 0.04 < 5$$

$$C_d = 0.622 \text{ is close}$$

For greater accuracy, the Rehbock coefficient is

$$C_d = \left[0.6035 + (0.0813)\left(\frac{h}{y}\right) + \frac{0.000295}{y}\right]$$

$$\times \left[1 + \frac{0.00361}{h}\right]^{\frac{3}{2}}$$

$$= \left[0.6035 + (0.0813)\left(\frac{0.2 \, ft}{2 \, ft}\right) + \frac{0.000295}{2 \, ft}\right]$$

$$\times \left(1 + \frac{0.00361}{0.2}\right)^{\frac{3}{2}}$$

$$= \boxed{0.6284}$$

(c) Assume $v_1 \cong 0$,

$$Q = \frac{2}{3} C_d b_{eff} \sqrt{2g} \, h^{\frac{3}{2}}$$

$$= \left(\frac{2}{3}\right)(0.6284)(4.98 \, ft)\sqrt{(2)\left(32.2 \, \frac{ft}{sec^2}\right)}(0.2 \, ft)^{\frac{3}{2}}$$

$$= \boxed{1.50 \, ft^3/sec}$$

10. (a) $$d_2 = -\frac{1}{2}d_1 + \sqrt{\frac{2v_1^2 d_1}{g} + \frac{d_1^2}{4}}$$

$$= -\left(\frac{1}{2}\right)(1.5 \, ft)$$

$$+ \sqrt{\frac{(2)\left(9 \, \frac{ft}{sec}\right)^2(1.5 \, ft)}{32.2 \, \frac{ft}{sec^2}} + \frac{(1.5 \, ft)^2}{4}}$$

$$= \boxed{2.10 \, ft}$$

(b) Assuming mass conservation,

$$v_1 d_1 = v_2 d_2$$

$$v_2 = v_1 \frac{d_1}{d_2}$$

$$= \left(9 \, \frac{ft}{sec}\right)\left(\frac{1.5 \, ft}{2.10 \, ft}\right) = 6.43 \, \frac{ft}{sec}$$

$$E_{\text{lost}} = \left(d_1 - d_2 + \frac{v_1^2}{2g} - \frac{v_2^2}{2g}\right)\left(\frac{g}{g_c}\right)$$

$$= \left[1.5\text{ ft} - 2.10\text{ ft} + \frac{\left(9\,\dfrac{\text{ft}}{\text{sec}}\right)^2}{(2)\left(32.2\,\dfrac{\text{ft}}{\text{sec}^2}\right)}\right.$$

$$\left. - \frac{\left(6.43\,\dfrac{\text{ft}}{\text{sec}}\right)^2}{(2)\left(32.2\,\dfrac{\text{ft}}{\text{sec}^2}\right)}\right]\left(\frac{32.2\,\dfrac{\text{ft}}{\text{sec}^2}}{32.2\,\dfrac{\text{lbm-ft}}{\text{lbf-sec}^2}}\right)$$

$$= 0.0158\,\frac{\text{ft-lbf}}{\text{lbm}}$$

$$P_{\text{dissipated}} = \dot{m}\,E_{\text{lost}} = v_1 d_1 b \rho\, E_{\text{lost}}$$

$$= \left(9\,\frac{\text{ft}}{\text{sec}}\right)(1.5\text{ ft})(40\text{ ft})$$

$$\times \left(62.4\,\frac{\text{lbm}}{\text{ft}^3}\right)\left(0.0158\,\frac{\text{ft-lbf}}{\text{lbm}}\right)$$

$$= 532.4\,\frac{\text{ft-lbf}}{\text{sec}}$$

$$P_{\text{dissipated}} = \left(532.4\,\frac{\text{ft-lbf}}{\text{sec}}\right)\left(\frac{1.3558\text{ W}}{\dfrac{\text{ft-lbf}}{\text{sec}}}\right)$$

$$= \boxed{721.83\text{ W}}$$

(c) $d_c = \left(\dfrac{Q^2}{gw^2}\right)^{\frac{1}{3}}$

$$= \left(\frac{(v_1 d_1 b)^2}{gb^2}\right)^{\frac{1}{3}} = \left(\frac{v_1^2 d_1^2}{g}\right)^{\frac{1}{3}}$$

$$= \left[\frac{\left(9\,\dfrac{\text{ft}}{\text{sec}}\right)^2 (1.5\text{ ft})^2}{\left(32.2\,\dfrac{\text{ft}}{\text{sec}^2}\right)}\right]^{\frac{1}{3}} = \boxed{1.782\text{ ft}}$$

(d) Assuming a natural channel in good condition, $n = 0.025$.

$$r_h = \frac{d_c w}{2d_c + w} = \frac{(1.78\text{ ft})(40\text{ ft})}{(2)(1.78\text{ ft}) + 40\text{ ft}} = 1.635\text{ ft}$$

$$v_c = \sqrt{gd_c} = \sqrt{\left(32.2\,\frac{\text{ft}}{\text{sec}^2}\right)(1.78\text{ ft})} = 7.57\,\frac{\text{ft}}{\text{sec}}$$

Using the Chezy-Manning equation,

$$v_c = \frac{1.486}{n}(r_h)^{\frac{2}{3}}\sqrt{S}$$

$$S = \left(\frac{v_c n}{1.486}\right)^2 \left(\frac{1}{r_h}\right)^{\frac{4}{3}}$$

$$= \left[\frac{\left(7.57\,\dfrac{\text{ft}}{\text{sec}}\right)(0.025)}{1.486}\right]^2 \left(\frac{1}{1.635\text{ ft}}\right)^{\frac{4}{3}}$$

$$= \boxed{0.008421}$$

ENERGY, WORK, AND POWER

1. $E_{\text{kinetic}} = \frac{1}{2}mv^2 = \frac{1}{2}(V\rho)v^2$

$= \left(\frac{1}{2}\right)\left(\frac{4}{3}\pi r^3\right)\rho v^2 = \left(\frac{2}{3}\pi\right)\left(\frac{10}{2}\text{ in}\right)^3 \rho v^2$

$= \left(\frac{2}{3}\pi\right)\left[\left(\frac{10}{2}\text{ in}\right)\left(\frac{\text{ft}}{12\text{ in}}\right)\right]^3$

$\times \left(0.256\,\frac{\text{lbm}}{\text{in}^3}\right)\left(1728\,\frac{\text{in}^3}{\text{ft}^3}\right)$

$\times \left(30\,\frac{\text{ft}}{\text{sec}}\right)^2\left(\frac{1}{32.2}\,\frac{\text{lbf-sec}^2}{\text{lbm-ft}}\right)$

$= \boxed{1873\text{ ft-lbf}}$

2. $W = \Delta E_{\text{potential}} = m\dfrac{g}{g_c}\Delta h$

$= (12\text{ lbm})\left(\dfrac{32.2\,\dfrac{\text{ft}}{\text{sec}^2}}{32.2\,\dfrac{\text{lbm-ft}}{\text{sec}^2\text{-lbf}}}\right)(40{,}000\text{ ft})$

$= \boxed{4.8 \times 10^5\text{ ft-lbf}}$

3. $\Delta E_{\text{potential}} = \Delta E_{\text{spring}}$

$W(\Delta h + \Delta x) = \frac{1}{2}k(\Delta x)^2$

Rearranging,

$\frac{1}{2}k(\Delta x)^2 - W\Delta x - W\Delta h = 0$

$\left(\frac{1}{2}\right)\left(33.33\,\frac{\text{lbf}}{\text{in}}\right)(\Delta x)^2 - (100\text{ lbf})\Delta x - (100\text{ lbf})(8\text{ ft})\left(12\,\frac{\text{in}}{\text{ft}}\right) = 0$

$16.665\Delta x^2 - 100\Delta x = 9600$

$\Delta x^2 - 6\Delta x = 576$

$(\Delta x - 3)^2 = 576 + 9$

$\Delta x - 3 = \sqrt{585} = \pm 24.2$

$\Delta x = \boxed{27.2\text{ in}}$

4. $W_{\text{done by wheel}} = \Delta E_{\text{rotational}}$

$= \frac{1}{2}I\omega_{\text{initial}}^2 - \frac{1}{2}I\omega_{\text{final}}^2$

$\omega_{\text{final}} = \sqrt{(\omega_{\text{initial}})^2 - \dfrac{2W}{I}} = 2\pi f$

$f_{\text{final}} = \left(\dfrac{1}{2\pi}\right)\left(\dfrac{60\text{ rpm}}{\text{rps}}\right)$

$\times \sqrt{\left[(2\pi)(300\text{ rpm})\left(\dfrac{\text{rps}}{60\text{ rpm}}\right)\right]^2 - \dfrac{(2)(45\times 10^2\text{ ft-lbf})}{15\text{ slug-ft}^2}}$

$= \boxed{187.8\text{ rpm}}$

5. $W_{\text{done on box}} = F_x\Delta x = (F)(\cos\theta)\Delta x$

$= (550\text{ lbf})(\cos 40°)(20\text{ ft})$

$= \boxed{8430\text{ ft-lbf}}$

6. $W_{\text{to retrieve cable}} = \displaystyle\int_0^l F\,dh$

$= \displaystyle\int_0^l [(l-h)w]\,dh$

$= \frac{1}{2}wl^2 = \left(\frac{1}{2}\right)\left(2\,\frac{\text{lbf}}{\text{ft}}\right)(1000\text{ ft})^2$

$= \boxed{10^6\text{ ft-lbf}}$

7. $P_{\text{actual}}\Delta t = W_{\text{done by pump}}$

$\eta P_{\text{ideal}}\Delta t = \Delta E_{\text{potential}}$

$= m\dfrac{g}{g_c}\Delta h$

$= (\rho V)\dfrac{g}{g_c}\Delta h$

$V = \dfrac{\eta P_{\text{ideal}}\Delta t}{\rho\dfrac{g}{g_c}\Delta h}$

$= \dfrac{(0.85)(7\text{ hp})\left(550\,\dfrac{\text{ft-lbf}}{\text{hp-sec}}\right)(3600\text{ sec})}{\left(62.4\,\dfrac{\text{lbm}}{\text{ft}^3}\right)\left(\dfrac{32.2\,\dfrac{\text{ft}}{\text{sec}^2}}{32.2\,\dfrac{\text{lbm-ft}}{\text{sec}^2\text{-lbf}}}\right)(130\text{ ft})}$

$= \boxed{1450\text{ ft}^3}$

8. $P\Delta t = W = m\dfrac{g}{g_c}\Delta h$

$P = \dfrac{mg\Delta h}{g_c\Delta t}$

$= \dfrac{(3300\text{ lbm})\left(32.2\,\dfrac{\text{ft}}{\text{sec}^2}\right)(250\text{ ft})}{\left(32.2\,\dfrac{\text{lbm-ft}}{\text{sec}^2\text{-lbf}}\right)(14\text{ sec})\left(550\,\dfrac{\text{ft-lbf}}{\text{hp-sec}}\right)}$

$= \boxed{107\text{ hp}}$

THERMO
Work/Energy

THERMODYNAMIC PROPERTIES OF SUBSTANCES

1. (a) $T^\circ_R = T^\circ_F + 459.67$

$$= 70 + 459.67 = \boxed{529.67^\circ R}$$

   (b) $T^\circ_F = 32 + \left(\dfrac{9}{5}\right) T^\circ_C$

$$= 32 + \left(\dfrac{9}{5}\right)(20) = \boxed{68^\circ F}$$

   (c) $T^\circ_R = T^\circ_F + 459.67$

$$= 70 + 459.67 = \boxed{529.67^\circ R}$$

2. (a) Neglecting the slight effect of compressibility, the enthalpy of a subcooled liquid is a function of only temperature. From the saturated water table,

$$h \approx h_{f,80^\circ} = \boxed{48.09 \text{ BTU/lbm}}$$

   (b) From the superheated steam table,

$$h = \boxed{1534.3 \text{ BTU/lbm}}$$

   (c) From the saturated water table,

$$h_g = \boxed{1180.2 \text{ BTU/lbm}}$$

   (d) From the saturated water table,

$$h_g = \boxed{1164.3 \text{ BTU/lbm}}$$

   (e) Assume low pressure and use low pressure air table.

$$T^\circ_R = T^\circ_F + 460 = 400 + 460 = 860^\circ R$$

$$h = \boxed{206.46 \text{ BTU/lbm}}$$

3. As in Prob. 2(a), $h_i \approx 48.09$.

$$h_f = h_i + \frac{Q}{m} = 48.09\,\frac{\text{BTU}}{\text{lbm}} + \frac{240 \text{ BTU}}{4 \text{ lbm}}$$

$$= \boxed{108.09 \text{ BTU/lbm}}$$

4. (a) From the saturated water table,

$$s_f = \boxed{0.3682 \text{ BTU/lbm-}^\circ R}$$

   (b) From the superheated steam table,

$$s = \boxed{2.0458 \text{ BTU/lbm-}^\circ R}$$

5. Copper melts at $1981.4^\circ F$, so it is solid throughout heating.

$$Q = mc_p\Delta T$$

$$= (20 \text{ lbm})\left(0.094\,\frac{\text{BTU}}{\text{lbm-}^\circ F}\right)(500^\circ F - 30^\circ F)$$

$$= \boxed{883.6 \text{ BTU}}$$

6. Pure aluminum melts at $1220.4^\circ F$, so it remains solid throughout the heating.

$$Q = mc_p\Delta T$$

$$= (40 \text{ lbm})\left(0.208\,\frac{\text{BTU}}{\text{lbm-}^\circ F}\right)(1000^\circ F - 200^\circ F)$$

$$= \boxed{6656 \text{ BTU}}$$

7. $Q_t = Q_s + Q_l = mc_p\Delta T + q_l m$

$$= (60 \text{ lbm})\left(1.0\,\frac{\text{BTU}}{\text{lbm-}^\circ F}\right)(212^\circ F - 70^\circ F)$$

$$+ \left(970.3\,\frac{\text{BTU}}{\text{lbm}}\right)(60 \text{ lbm})$$

$$= \boxed{66,738 \text{ BTU}}$$

8. Approximate answers may be obtained by using c_v and c_p, but more accurate answers may be obtained by using air tables.

$$T_1 = 460 + 70 = 530^\circ R$$
$$T_2 = 460 + 180 = 640^\circ R$$

   (a) $Q|_v = (u_2 - u_1)m$

$$= \left(109.21\,\frac{\text{BTU}}{\text{lbm}} - 90.33\,\frac{\text{BTU}}{\text{lbm}}\right)(4 \text{ lbm})$$

$$= \boxed{75.52 \text{ BTU}}$$

   (b) $Q|_p = (h_2 - h_1)m$

$$= \left(153.09\,\frac{\text{BTU}}{\text{lbm}} - 126.66\,\frac{\text{BTU}}{\text{lbm}}\right)(4 \text{ lbm})$$

$$= \boxed{105.72 \text{ BTU}}$$

THERMO
Properties

9. Heat will flow from the hotter body (Fe) to the cooler body (H_2O) until equilibrium is reached. Assume an adiabatic process, so there is no heat loss to environment.

$$Q_{\text{lost by Fe}} = Q_{\text{gained by } H_2O}$$
$$m_{\text{Fe}}c_{\text{p,Fe}}(T_{\text{Fe}} - T_f) = m_{H_2O}c_{\text{p},H_2O}(T_f - T_{H_2O})$$
$$(2\text{ lbm})\left(0.1\frac{\text{BTU}}{\text{lbm-}^\circ\text{F}}\right)(200^\circ\text{F} - T_f)$$
$$= (1\text{ gal})\left(8.345\frac{\text{lbm}}{\text{gal}}\right)\left(1.0\frac{\text{BTU}}{\text{lbm-}^\circ\text{F}}\right)(T_f - 40^\circ\text{F})$$

$$\boxed{T_f = 43.74^\circ\text{F}}$$

10. Assume water at 1 atm and adiabatic process.

$$Q_t = Q_{s(\text{heat ice})} + Q_{l(\text{melt ice})} + Q_{s(\text{heat water})}$$
$$\quad + Q_{l(\text{vaporize water})} + Q_{s(\text{heat steam})}$$
$$= m\big[c_{p(\text{ice})}(T_m - T_i) + q_{l(\text{fusion})}$$
$$\quad + c_{p(\text{water})}(T_b - T_m)$$
$$\quad + q_{l(\text{vaporization})} + c_{p(\text{steam})}(T_f - T_b)\big]$$
$$= (20\text{ lbm})\bigg[\left(0.5\frac{\text{BTU}}{\text{lbm-}^\circ\text{F}}\right)(32^\circ\text{F} - 0^\circ\text{F})$$
$$+ \left(143.4\frac{\text{BTU}}{\text{lbm}}\right) + \left(1.0\frac{\text{BTU}}{\text{lbm-}^\circ\text{F}}\right)(212^\circ\text{F} - 32^\circ\text{F})$$
$$+ \left(970.3\frac{\text{BTU}}{\text{lbm}}\right) + \left(0.5\frac{\text{BTU}}{\text{lbm-}^\circ\text{F}}\right)(213^\circ\text{F} - 212^\circ\text{F})\bigg]$$
$$= 26{,}204\text{ BTU}$$

$$t = \frac{Q_t}{\dot{Q}} = \frac{26{,}204\text{ BTU}}{160\dfrac{\text{BTU}}{\text{sec}}} = \boxed{163.78\text{ sec}}$$

11. Neglecting the small effect of pressure, the enthalpy of the incoming water is a function of its temperature alone.

$$h_{iw} = 48.09\text{ BTU/lbm}$$

Similarly, the enthalpy of the outgoing water is

$$h_o = 147.99\text{ BTU/lbm}$$

$$h_{is} = h_f + x h_{fg}$$
$$= 298.6\frac{\text{BTU}}{\text{lbm}} + (0.80)\left(889.2\frac{\text{BTU}}{\text{lbm}}\right)$$
$$= 1009.96\text{ BTU/lbm}$$

$$\dot{H}_o = \dot{H}_{iw} + \dot{H}_{is}$$
$$\dot{m}_o h_o = \dot{m}_{iw}h_{iw} + \dot{m}_{is}h_{is} \qquad \text{[Eq. 1]}$$
$$\dot{m}_o = \dot{m}_{iw} + \dot{m}_{is} \qquad \text{[Eq. 2]}$$

Substitute Eq. 2 into Eq. 1.

$$\dot{m}_{is} = \frac{\dot{m}_o(h_o - h_{iw})}{h_{is} - h_{iw}}$$
$$= \frac{\left(4000\dfrac{\text{lbm}}{\text{hr}}\right)\left(147.99\dfrac{\text{BTU}}{\text{lbm}} - 48.09\dfrac{\text{BTU}}{\text{lbm}}\right)}{1009.96\dfrac{\text{BTU}}{\text{lbm}} - 48.09\dfrac{\text{BTU}}{\text{lbm}}}$$
$$= \boxed{415.44\text{ lbm/hr}}$$

$$\dot{m}_{iw} = \dot{m}_o - \dot{m}_{is} = 4000\frac{\text{lbm}}{\text{hr}} - 415.44\frac{\text{lbm}}{\text{hr}}$$
$$= \boxed{3584.6\text{ lbm/hr}}$$

12.
$$\dot{Q}_{\text{lost by A}} = \dot{Q}_{\text{gained by B}}$$
$$\dot{V}_A c_{p,A}(T_A - T_{A+B}) = \dot{V}_B c_{p,B}(T_{A+B} - T_B)$$
$$= (\dot{V}_{A+B} - \dot{V}_A)c_{p,B}(T_{A+B} - T_B)$$
$$\dot{V}_A = \frac{\dot{V}_{A+B}c_{p,B}(T_{A+B} - T_B)}{c_{p,A}(T_A - T_{A+B}) + c_{p,B}(T_{A+B} - T_B)}$$
$$= \frac{\left(50\dfrac{\text{gal}}{\text{min}}\right)\left(8.33\dfrac{\text{BTU}}{\text{gal-}^\circ\text{F}}\right)(80^\circ\text{F}-65^\circ\text{F})}{\left(10\dfrac{\text{BTU}}{\text{gal-}^\circ\text{F}}\right)(140^\circ\text{F}-80^\circ\text{F})+\left(8.33\dfrac{\text{BTU}}{\text{gal-}^\circ\text{F}}\right)(80^\circ\text{F}-65^\circ\text{F})}$$
$$= \boxed{8.62\text{ gal/min (gpm)}}$$

$$\dot{V}_B = \dot{V}_{A+B} - \dot{V}_A = 50\frac{\text{gal}}{\text{min}} - 8.62\frac{\text{gal}}{\text{min}}$$
$$= \boxed{41.38\text{ gal/min (gpm)}}$$

$$\dot{H}_{A+B} = \dot{H}_A + \dot{H}_B$$
$$\dot{V}_{A+B}c_{p,A+B}T_{A+B} = \dot{V}_A c_{p,A}T_A + \dot{V}_B c_{p,B}T_B$$
$$c_{p,A+B} = \frac{\dot{V}_A c_{p,A}T_A + \dot{V}_B c_{p,B}T_B}{\dot{V}_{A+B}T_{A+B}}$$
$$= \frac{\left(8.62\dfrac{\text{gal}}{\text{min}}\right)\left(10\dfrac{\text{BTU}}{\text{gal-}^\circ\text{F}}\right)(140^\circ\text{F})}{\left(50\dfrac{\text{gal}}{\text{min}}\right)(80^\circ\text{F})}$$
$$+ \frac{\left(41.38\dfrac{\text{gal}}{\text{min}}\right)\left(8.33\dfrac{\text{BTU}}{\text{gal-}^\circ\text{F}}\right)(65^\circ\text{F})}{\left(50\dfrac{\text{gal}}{\text{min}}\right)(80^\circ\text{F})}$$
$$= \boxed{8.62\text{ BTU/gal-}^\circ\text{F}}$$

THERMO
Properties

13. (a) From the saturated water tables at $p = 80$ psia, $T_{sat} = 312.07°F$.

$T = T_{sat} + T_{superheat} = 312.07°F + 388°F = 700.07°F$

Use the superheated steam tables and interpolation.

At $T = 700.07°F$ and $p = 80$ psia,

$$\boxed{h \approx 1380.3 \text{ BTU/lbm}}$$

(b) Use the saturated steam tables.

$h = h_f + xh_{fg} = 298.6 \frac{\text{BTU}}{\text{lbm}} + (0.45)\left(889.2 \frac{\text{BTU}}{\text{lbm}}\right)$

$$= \boxed{698.74 \text{ BTU/lbm}}$$

(c) Use the superheated steam tables and interpolation.

$$\boxed{h \approx 1431.6 \text{ BTU/lbm}}$$

14. $h = h_f + xh_{fg}$

$x = \dfrac{h - h_f}{h_{fg}} = \dfrac{600 \frac{\text{BTU}}{\text{lbm}} - 424.2 \frac{\text{BTU}}{\text{lbm}}}{781.2 \frac{\text{BTU}}{\text{lbm}}}$

$$= \boxed{0.225 \ (22.5\%)}$$

15. Interpolating between 60 psia and 100 psia,

$v \approx \left(\dfrac{1}{2}\right)\left(10.425 \frac{\text{ft}^3}{\text{lbm}} + 6.216 \frac{\text{ft}^3}{\text{lbm}}\right)$

$$= \boxed{8.32 \text{ ft}^3/\text{lbm}}$$

Using a more detailed superheat table,

$$v = \boxed{7.80 \text{ ft}^3/\text{lbm}}$$

16. $pV = \dfrac{m}{(MW)}R^*T$

$(MW) = \dfrac{mR^*T}{pV}$

$= \dfrac{(0.0016 \text{ lbm})\left(1545.33 \frac{\text{ft-lbf}}{\text{lbmole-}°\text{R}}\right)(212 + 460)°\text{R}}{\left(15 \frac{\text{lbf}}{\text{in}^2}\right)(17.54 \text{ in}^3)\left(\frac{\text{ft}}{12 \text{ in}}\right)}$

$$= \boxed{75.78 \text{ lbm/lbmole}}$$

17. $\dfrac{p_1V_1}{T_1} = \dfrac{p_2V_2}{T_2}$

$V_2 = \dfrac{p_1V_1T_2}{p_2T_1}$

$= \dfrac{(13.5 \text{ psia})(20 \text{ in}^3)(40 + 460)°\text{R}}{(14.2 \text{ psia})(70 + 460)°\text{R}}$

$$= \boxed{17.94 \text{ in}^3}$$

18. At STP,

$$T = 32°F = 492°R$$
$$p = 1 \text{ atm} = 14.7 \text{ psia}$$
$$\frac{p_1v_1}{T_1} = \frac{p_2v_2}{T_2}$$
$$\frac{p_1}{\rho_1T_1} = \frac{p_2}{\rho_2T_2}$$

$\rho_2 = \dfrac{p_2\rho_1T_1}{p_1T_2} = \dfrac{\left(14.2 \frac{\text{lbf}}{\text{in}^2}\right)\left(0.0111 \frac{\text{lbm}}{\text{ft}^3}\right)(492°\text{R})}{\left(14.7 \frac{\text{lbf}}{\text{in}^2}\right)(460 + 75)°\text{R}}$

$$= \boxed{0.00986 \text{ lbm/ft}^3}$$

19. $pV = \dfrac{m}{(MW)}R^*T$

$T = \dfrac{pV(MW)}{mR^*}$

$= \dfrac{\left(14.7 \frac{\text{lbf}}{\text{in}^2}\right)(60 \text{ in}^3)\left(32 \frac{\text{lbm}}{\text{lbmole}}\right)\left(\frac{\text{ft}}{12 \text{ in}}\right)}{(0.002 \text{ lbm})\left(1545.33 \frac{\text{ft-lbf}}{\text{lbmole-}°\text{R}}\right)}$

$$= \boxed{761°\text{R}}$$

20. At STP, $T = 32°F = 492°R$

$$p = 1 \text{ atm} = 14.7 \text{ lbf/in}^2$$
$$\frac{p_1v_1}{T_1} = \frac{p_2v_2}{T_2}$$
$$\frac{p_1}{\rho_1T_1} = \frac{p_2}{\rho_2T_2}$$

$\rho_2 = \dfrac{p_2\rho_1T_1}{p_1T_2}$

$= \dfrac{\left(14.5 \frac{\text{lbf}}{\text{in}^2}\right)\left(0.201 \frac{\text{lbm}}{\text{ft}^3}\right)(492°\text{R})}{\left(14.7 \frac{\text{lbf}}{\text{in}^2}\right)(460 + 75)°\text{R}}$

$= 0.182 \text{ lbm/ft}^3$

$m_2 = \rho_2V_2 = \left(0.182 \frac{\text{lbm}}{\text{ft}^3}\right)(6 \text{ in}^3)\left(\frac{\text{ft}^3}{1728 \text{ in}^3}\right)$

$$= \boxed{0.000632 \text{ lbm}}$$

21. $\dfrac{p_1 V_1}{T_1} = \dfrac{p_2 V_2}{T_2}$

$V_1 = V_2$

$p_2 = p_1 \dfrac{T_2}{T_1} = (14.7 \text{ psia}) \left[\dfrac{(460 + 212)°\text{R}}{(460 + 77)°\text{R}} \right]$

$= \boxed{18.40 \text{ psia}}$

22. At STP, $T = 32 + 460 = 492°\text{R}$

$p = 1 \text{ atm} = 14.7 \text{ lbf/in}^2$

$\dfrac{p_1 V_1}{T_1} = \dfrac{p_2 V_2}{T_2}$

$T_2 = \dfrac{p_2 V_2 T_1}{p_1 V_1} = \dfrac{\left(15 \dfrac{\text{lbf}}{\text{in}^2} \right)(1.5 \text{ ft}^3)(492°\text{R})}{\left(14.7 \dfrac{\text{lbf}}{\text{in}^2} \right)(1.2 \text{ ft}^3)}$

$= \boxed{627.6°\text{R}}$

23. $\dfrac{p_1 v_1}{T_1} = \dfrac{p_2 v_2}{T_2}$

$\dfrac{p_1}{\rho_1 T_1} = \dfrac{p_2}{\rho_2 T_2}$

$p_2 = \dfrac{p_1 \rho_2 T_2}{\rho_1 T_1} = \dfrac{(2 \text{ atm}) \left(0.270 \dfrac{\text{lbm}}{\text{ft}^3} \right)(250 + 460)°\text{R}}{\left(0.094 \dfrac{\text{lbm}}{\text{ft}^3} \right)(100 + 460)°\text{R}}$

$= \boxed{7.28 \text{ atm}}$

24. $\dfrac{p_1 V_1}{T_1} = \dfrac{p_2 V_2}{T_2}$

$p_2 = \dfrac{p_1 V_1 T_2}{V_2 T_1}$

$= \dfrac{\left(24 \dfrac{\text{lbf}}{\text{in}^2} + 14.7 \dfrac{\text{lbf}}{\text{in}^2} \right)(1000 \text{ in}^3)(35 + 460)°\text{R}}{(1020 \text{ in}^3)(32 + 460)°\text{R}}$

$= \boxed{38.17 \text{ lbf/in}^2 \text{ (psia)}}$

25. $pV = \dfrac{m}{(\text{MW})} R^* T$

$p = \dfrac{m R^* T}{V (\text{MW})}$

$= \dfrac{(0.0107 \text{ lbm}) \left(1545.33 \dfrac{\text{ft-lbf}}{\text{lbmole-}°\text{R}} \right)(50 + 460)°\text{R}}{(0.24 \text{ ft}^3) \left(2 \dfrac{\text{lbm}}{\text{lbmole}} \right) \left(12 \dfrac{\text{in}}{\text{ft}} \right)^2}$

$= \boxed{122 \text{ lbf/in}^2 \text{ (psia)}}$

26. STP is $T = 492°\text{R}$.

$p = 1 \text{ atm} = 14.7 \text{ lbf/ft}^2$

$V = \dfrac{\pi d^3}{6} = \left(\dfrac{\pi}{6} \right)(60 \text{ ft})^3 = 1.131 \times 10^5 \text{ ft}^3$

The lifting power (buoyant force), LP, is

$(\text{LP}) = [m_{\text{air}} - m_{\text{He}}] \times \dfrac{g}{g_c}$

$pV = mRT$, so

$(\text{LP}) = \dfrac{pV}{T} \left(\dfrac{1}{R_{\text{air}}} - \dfrac{1}{R_{\text{He}}} \right) \times \dfrac{g}{g_c}$

$= \dfrac{\left(14.7 \dfrac{\text{lbf}}{\text{in}^2} \right) \left(144 \dfrac{\text{in}^2}{\text{ft}^2} \right)(1.131 \times 10^5 \text{ ft}^3)}{492°\text{R}}$

$\times \left(\dfrac{1}{53.35 \dfrac{\text{ft-lbf}}{\text{lbm-}°\text{R}}} - \dfrac{1}{386.04 \dfrac{\text{ft-lbf}}{\text{lbm-}°\text{R}}} \right) \left(\dfrac{32.2 \dfrac{\text{ft}}{\text{sec}^2}}{32.2 \dfrac{\text{ft-lbm}}{\text{lbf-sec}^2}} \right)$

$= \boxed{7860 \text{ lbf}}$

27. Assume that the can is at $1 \text{ atm} = 14.696 \text{ psia}$. Assume no change in temperature during the process.

$pV = nR^* T$

$\dfrac{p}{n} = \dfrac{R^* T}{V} = \text{constant}$

$\dfrac{p_1}{n_1} = \dfrac{p_2}{n_2}$

$p_2 = \dfrac{n_2 p_1}{n_1} = \dfrac{\left(\dfrac{1}{3} n_1 \right) p_1}{n_1} = \dfrac{p_1}{3} = \dfrac{14.696 \dfrac{\text{lbf}}{\text{in}^2}}{3}$

$= 4.90 \text{ lbf/in}^2$

$F = pA = p \left(\dfrac{\pi d^2}{4} \right) = \left(4.90 \dfrac{\text{lbf}}{\text{in}^2} \right) \left(\dfrac{\pi}{4} \right)(4 \text{ in})^2$

$= 61.58 \text{ lbf}$

$F_{\text{net}} = F_{\text{outside}} - F_{\text{inside}} = p_{\text{out}} A - p_{\text{in}} A$

$= (p_{\text{out}} - p_{\text{in}}) \left(\dfrac{\pi d^2}{4} \right)$

$= \left(14.696 \dfrac{\text{lbf}}{\text{in}^2} - 4.9 \dfrac{\text{lbf}}{\text{in}^2} \right) \dfrac{\pi}{4} (4 \text{ in})^2$

$= \boxed{123.10 \text{ lbf}}$

THERMO Properties

28. When no more tires can be filled, the tank will be at tire pressure.

Tank alone:

$$pV = mRT$$

$$m_i = \frac{p_i V_{\text{tank}}}{RT} = \frac{\left[(120 + 14.7)\,\dfrac{\text{lbf}}{\text{in}^2}\right]\left(\dfrac{12\,\text{in}}{\text{ft}}\right)^2 (3\,\text{ft}^3)}{\left(53.35\,\dfrac{\text{ft-lbf}}{\text{lbm-}^\circ\text{R}}\right)(460 + 80)^\circ\text{R}}$$

$$= 2.020\,\text{lbm}$$

$$m_f = \frac{p_f V_{\text{tank}}}{RT}$$

$$= \frac{\left[(28 + 14.7)\,\dfrac{\text{lbf}}{\text{in}^2}\right]\left(\dfrac{12\,\text{in}}{\text{ft}}\right)^2 (3\,\text{ft}^3)}{\left(53.35\,\dfrac{\text{ft-lbf}}{\text{lbm-}^\circ\text{R}}\right)(460 + 80)^\circ\text{R}}$$

$$= 0.640\,\text{lbm}$$

$$m_{\text{used in tires}} = m_i - m_f = 2.020\,\text{lbm} - 0.640\,\text{lbm}$$

$$= 1.380\,\text{lbm}$$

The tires start out at atmospheric pressure, so the tank only increases the pressure by 28 psi.

$$m_{\text{air per tire}} = \frac{\Delta p_{\text{tire}} V_{\text{tire}}}{RT}$$

$$= \frac{\left(28\,\dfrac{\text{lbf}}{\text{in}^2}\right)\left(\dfrac{12\,\text{in}}{\text{ft}}\right)^2 (1.2\,\text{ft}^3)}{\left(53.35\,\dfrac{\text{ft-lbf}}{\text{lbm-}^\circ\text{R}}\right)(460 + 80)^\circ\text{R}}$$

$$= 0.168\,\text{lbm}$$

$$\text{no. tires} = \frac{m_{\text{used in tires}}}{m_{\text{air per tire}}} = \frac{1.380\,\text{lbm}}{0.168\,\text{lbm}} = 8.21$$

$$= \boxed{8\ \text{complete tires}}$$

29. $k = \dfrac{c_p}{c_v} = 1.4$

$$c_v = \frac{c_p}{1.4} = \frac{0.245\,\dfrac{\text{BTU}}{\text{lbm-}^\circ\text{R}}}{1.4} = 0.175\,\frac{\text{BTU}}{\text{lbm-}^\circ\text{R}}$$

$$Q = mc_v \Delta T$$

$$= (10\,\text{lbm})\left(0.175\,\frac{\text{BTU}}{\text{lbm-}^\circ\text{R}}\right)\left(\frac{^\circ\text{R}}{^\circ\text{F}}\right)(360^\circ\text{F} - 90^\circ\text{F})$$

$$= \boxed{472.5\,\text{BTU}}$$

30. $Q = nC_v \Delta T$

$$C_v = \frac{Q}{n\Delta T} = \frac{200\,\text{BTU}}{\left(\dfrac{1}{2}\,\text{lbmole}\right)(100^\circ\text{F})\left(\dfrac{^\circ\text{R}}{^\circ\text{F}}\right)}$$

$$= 4\,\text{BTU/lbmole-}^\circ\text{R}$$

$$C_p - C_v = R^*$$

$$C_p = R^* + C_v$$

$$= 1.986\,\frac{\text{BTU}}{\text{lbmole-}^\circ\text{R}} + 4\,\frac{\text{BTU}}{\text{lbmole-}^\circ\text{R}}$$

$$= \boxed{5.986\,\text{BTU/lbmole-}^\circ\text{R}}$$

31. $T_K = (460 + T_{^\circ F})^\circ\text{R}\left(\dfrac{5\text{K}}{9^\circ\text{R}}\right)$

$$= (460 + 70)^\circ\text{R}\left(\frac{5\text{K}}{9^\circ\text{R}}\right) = 294.4\text{K}$$

$$v_{\text{rms}} = \sqrt{\frac{3\kappa T}{m}} = \sqrt{\frac{3\kappa T}{\dfrac{(\text{MW})}{N_o}}}$$

$$= \sqrt{\frac{(3)\left(1.3803\times10^{-23}\,\dfrac{\text{J}}{\text{molecule}\cdot\text{K}}\right)(294.4\text{K})}{\dfrac{0.032\,\dfrac{\text{kg}}{\text{mol}}}{6.023\times10^{23}\,\dfrac{\text{molecule}}{\text{mol}}}}}$$

$$= \boxed{479.01\,\text{m/s}}$$

32. $T_K = (460 + T_{^\circ F})^\circ\text{R}\left(\dfrac{5\text{K}}{9^\circ\text{R}}\right)$

$$= (460 + 0)^\circ\text{R}\left(\frac{5\text{K}}{9^\circ\text{R}}\right) = 255.56\text{K}$$

$$E_K = \frac{3}{2}\kappa T$$

$$= \left(\frac{3}{2}\right)\left(1.3803\times10^{-23}\,\frac{\text{J}}{\text{molecule}\cdot\text{K}}\right)(255.56\text{K})$$

$$= \boxed{5.291\times10^{-21}\,\text{J}}$$

33. $v_m = (2)\sqrt{\dfrac{2\kappa T}{\pi m}} = (2)\sqrt{\dfrac{2\kappa T}{\pi\left(\dfrac{(\text{MW})}{N_o}\right)}}$

$$= (2)\sqrt{\frac{(2)\left(1.3803\times10^{-23}\,\dfrac{\text{J}}{\text{molecule}\cdot\text{K}}\right)(273.15 + 20)\text{K}}{(\pi)\left(\dfrac{0.028\,\dfrac{\text{kg}}{\text{mol}}}{6.023\times10^{23}\,\dfrac{\text{molecule}}{\text{mol}}}\right)}}$$

$$= \boxed{470.8\,\text{m/s}}$$

THERMO
Properties

34. $\kappa = \left(1.38 \times 10^{-23} \dfrac{J}{\text{molecule} \cdot K}\right)\left(\dfrac{\frac{kg \cdot m^2}{s^2}}{J}\right)\left(\dfrac{0.06852 \, \text{slug}}{kg}\right)$

$\times \left(\dfrac{3.281 \, \text{ft}}{m}\right)^2 \left(\dfrac{5K}{9°R}\right)\left(\dfrac{1 \, \text{lbf}}{\frac{\text{slug-ft}}{\text{sec}^2}}\right)$

$= \boxed{5.66 \times 10^{-24} \text{ ft-lbf/°R-molecule}}$

35. Assume ideal gas behavior. The molecular weight of the mixture is 16.114 lbm/lbmole. (See part (b).)

(a) $G_{CO} = \dfrac{B_{CO}(MW)_{CO}}{(MW)_{mix}}$

$= \dfrac{(0.30)\left(28.011 \dfrac{\text{lbm}}{\text{lbmole}}\right)}{16.114 \dfrac{\text{lbm}}{\text{lbmole}}} = \boxed{0.5215}$

$G_{CO_2} = \dfrac{B_{CO_2}(MW)_{CO_2}}{(MW)_{mix}}$

$= \dfrac{(0.15)\left(44.011 \dfrac{\text{lbm}}{\text{lbmole}}\right)}{16.114 \dfrac{\text{lbm}}{\text{lbmole}}} = \boxed{0.4097}$

$G_{H_2} = \dfrac{B_{H_2}(MW)_{H_2}}{(MW)_{mix}}$

$= \dfrac{(0.55)\left(2.0158 \dfrac{\text{lbm}}{\text{lbmole}}\right)}{16.114 \dfrac{\text{lbm}}{\text{lbmole}}} = \boxed{0.0688}$

(b) $(MW)_{mix} = B_{CO}(MW)_{CO} + B_{CO_2}(MW)_{CO_2}$
$+ B_{H_2}(MW)_{H_2}$

$= (0.30)\left(28.011 \dfrac{\text{lbm}}{\text{lbmole}}\right)$

$+ (0.15)\left(44.011 \dfrac{\text{lbm}}{\text{lbmole}}\right)$

$+ (0.55)\left(2.0158 \dfrac{\text{lbm}}{\text{lbmole}}\right)$

$= \boxed{16.114 \text{ lbm/lbmole}}$

(c) $R_{mix} = G_{CO}R_{CO} + G_{CO_2}R_{CO_2} + G_{H_2}R_{H_2}$

$= (0.5215)\left(55.17 \dfrac{\text{ft-lbf}}{\text{lbm-°R}}\right)$

$+ (0.4097)\left(35.11 \dfrac{\text{ft-lbf}}{\text{lbm-°R}}\right)$

$+ (0.0688)\left(766.53 \dfrac{\text{ft-lbf}}{\text{lbm-°R}}\right)$

$= \boxed{95.89 \text{ ft-lbf/lbm-°R}}$

(d) $c_{p,mix} = G_{CO}c_{p,CO} + G_{CO_2}c_{p,CO_2} + G_{H_2}c_{p,H_2}$

$= (0.5215)\left(0.249 \dfrac{\text{BTU}}{\text{lbm-°R}}\right)$

$+ (0.4097)\left(0.207 \dfrac{\text{BTU}}{\text{lbm-°R}}\right)$

$+ (0.0688)\left(3.420 \dfrac{\text{BTU}}{\text{lbm-°R}}\right)$

$= \boxed{0.44996 \ (0.4500) \text{ BTU/lbm-°R}}$

(e) $c_{v,mix} = G_{CO}c_{v,CO} + G_{CO_2}c_{v,CO_2} + G_{H_2}c_{v,H_2}$

$= (0.5215)\left(0.178 \dfrac{\text{BTU}}{\text{lbm-°R}}\right)$

$+ (0.4097)\left(0.162 \dfrac{\text{BTU}}{\text{lbm-°R}}\right)$

$+ (0.0688)\left(2.435 \dfrac{\text{BTU}}{\text{lbm-°R}}\right)$

$= \boxed{0.3267 \text{ BTU/lbm-°R}}$

36. $V = \dfrac{G_{He}}{(MW)_{He}} + \dfrac{G_{air}}{(MW)_{air}} + \dfrac{G_{CO_2}}{(MW)_{CO_2}}$

$= \dfrac{0.20}{4.003 \dfrac{\text{lbm}}{\text{lbmole}}} + \dfrac{0.40}{28.967 \dfrac{\text{lbm}}{\text{lbmole}}}$

$+ \dfrac{0.40}{44.011 \dfrac{\text{lbm}}{\text{lbmole}}} = 0.07286 \text{ lbmole/lbm}$

(a) $B_{He} = \dfrac{\frac{G_{He}}{(MW)_{He}}}{V} = \dfrac{\dfrac{0.20}{4.003 \dfrac{\text{lbm}}{\text{lbmole}}}}{0.07286 \dfrac{\text{lbmole}}{\text{lbm}}} = \boxed{0.6857}$

$$B_{air} = \frac{\dfrac{G_{air}}{(MW)_{air}}}{V} = \frac{\dfrac{0.40}{28.967 \,\dfrac{lbm}{lbmole}}}{0.07286 \,\dfrac{lbm}{lbmole}} = \boxed{0.1895}$$

$$B_{CO_2} = \frac{\dfrac{G_{CO_2}}{(MW)_{CO_2}}}{V} = \frac{\dfrac{0.40}{44.011 \,\dfrac{lbm}{lbmole}}}{0.07286 \,\dfrac{lbmole}{lbm}} = \boxed{0.1247}$$

(b) $(MW)_{avg} = B_{He}(MW)_{He} + B_{air}(MW)_{air}$
$$\qquad + B_{CO_2}(MW)_{CO_2}$$

$$= (0.6857)\left(4.003\,\frac{lbm}{lbmole}\right)$$

$$\quad + (0.1895)\left(28.967\,\frac{lbm}{lbmole}\right)$$

$$\quad + (0.1247)\left(44.011\,\frac{lbm}{lbmole}\right)$$

$$= \boxed{13.72 \text{ lbm/lbmole}}$$

(c) $R_{composite} = G_{He}R_{He} + G_{air}R_{air} + G_{CO_2}R_{CO_2}$

$$= (0.20)\left(386.04\,\frac{ft\text{-}lbf}{lbm\text{-}°R}\right)$$

$$\quad + (0.40)\left(53.35\,\frac{ft\text{-}lbf}{lbm\text{-}°R}\right)$$

$$\quad + (0.40)\left(35.11\,\frac{ft\text{-}lbf}{lbm\text{-}°R}\right)$$

$$= \boxed{112.59 \text{ ft-lbf/lbm-°R}}$$

37. (a) $(MW)_{avg} = B_{N_2}(MW)_{N_2} + B_{CO_2}(MW)_{CO_2}$
$$\qquad\qquad + B_{H_2}(MW)_{H_2}$$

$$= (0.60)\left(28.016\,\frac{g}{mol}\right) + (0.10)\left(44.011\,\frac{g}{mol}\right)$$

$$\quad + (0.30)\left(2.016\,\frac{g}{mol}\right) = \left(21.8155\,\frac{g}{mol}\right)$$

$$= \boxed{21.8155 \text{ g/mol or lbm/lbmole}}$$

Assume ideal gas behavior.

$$pV = \frac{m}{(MW)}R^*T$$

$$m_{mix} = \frac{pV_{initial}(MW)_{avg}}{R^*T_{initial}}$$

$$= \frac{(1 \text{ atm})(5 \text{ ft}^3)\left(21.8155\,\dfrac{lbm}{lbmole}\right)}{\left(0.7302\,\dfrac{atm\text{-}ft^3}{lbmole\text{-}°R}\right)(460+40)°R}$$

$$= \boxed{0.299 \text{ lbm}}$$

(b) $G_{N_2} = \dfrac{B_{N_2}(MW)_{N_2}}{(MW)_{avg}}$

$$= \frac{(0.60)\left(28.016\,\dfrac{g}{mol}\right)}{21.8155\,\dfrac{g}{mol}} = 0.771$$

$$G_{CO_2} = \frac{B_{CO_2}(MW)_{CO_2}}{(MW)_{avg}}$$

$$= \frac{(0.10)\left(44.011\,\dfrac{g}{mol}\right)}{21.8155\,\dfrac{g}{mol}} = 0.202$$

$$G_{H_2} = \frac{B_{H_2}(MW)_{H_2}}{(MW)_{avg}}$$

$$= \frac{(0.30)\left(2.016\,\dfrac{g}{mol}\right)}{21.8155\,\dfrac{g}{mol}} = 0.028$$

$$c_{p,mix} = G_{N_2}c_{p,N_2} + G_{CO_2}c_{p,CO_2} + G_{H_2}c_{p,H_2}$$

$$= (0.771)\left(0.249\,\frac{BTU}{lbm\text{-}°R}\right) + (0.202)\left(0.207\,\frac{BTU}{lbm\text{-}°R}\right)$$

$$\quad + (0.028)\left(3.420\,\frac{BTU}{lbm\text{-}°R}\right)$$

$$= \boxed{0.3296 \text{ BTU/lbm-°R}}$$

(c) $Q = m_{mix}c_{p,mix}\Delta T$

$$= (0.2988 \text{ lbm})\left(0.3296\,\frac{BTU}{lbm\text{-}°R}\right)$$

$$\quad \times \left(\frac{°R}{°F}\right)(250°F - 40°F)$$

$$= \boxed{20.68 \text{ BTU}}$$

THERMO
Properties

CHANGES IN THERMODYNAMIC PROPERTIES

1. $Q = \Delta U + W$

$\Delta U = Q - W$

$\quad = 10 \text{ BTU} - (1500 \text{ ft-lbf}) \left(\dfrac{1 \text{ BTU}}{778.2 \text{ ft-lbf}} \right)$

$\quad = \boxed{8.07 \text{ BTU}}$

2. $q = h_2 - h_1 + \dfrac{v_2^2 - v_1^2}{2 g_c J}$

$\quad + \dfrac{(z_2 - z_1)g}{g_c J} + \dfrac{W_{\text{shaft}}}{J}$

$Q = H_2 - H_1 + \dfrac{m}{2 g_c J}(v_2^2 - v_1^2)$

$\quad + \dfrac{(z_2 - z_1)mg}{g_c J} + \dfrac{W_{\text{shaft}}}{J}$

$\quad = \left(\dfrac{m}{2} \right) v_2^2$

$v_2 = \sqrt{\dfrac{2 g_c J Q}{m}}$

$\quad = \sqrt{\dfrac{(2)\left(32.2 \dfrac{\text{ft-lbm}}{\text{lbf-sec}^2} \right)\left(778.2 \dfrac{\text{ft-lbf}}{\text{BTU}} \right)(2 \text{ BTU})}{0.011 \text{ lbm}}}$

$\quad = \boxed{3019 \text{ ft/sec}}$

3. $\qquad Q = \Delta U + \dfrac{W}{J} = 0$

$Q = 0$ since the process is adiabatic.

$$\Delta U = \dfrac{-W}{J}$$

$$m c_p \Delta T = \dfrac{-mgh}{J g_c}$$

$\Delta T = \dfrac{-gh}{c_p g_c J}$

$\quad = \dfrac{-\left(32.2 \dfrac{\text{ft}}{\text{sec}^2} \right)(200 \text{ ft})}{\left(32.2 \dfrac{\text{lbm-ft}}{\text{lbf-sec}^2} \right)\left(1.0 \dfrac{\text{BTU}}{\text{lbm-}^\circ\text{F}} \right)\left(778.2 \dfrac{\text{ft-lbf}}{\text{BTU}} \right)}$

$\quad = \boxed{-0.257\,^\circ\text{F}}$

4. $\qquad Q = \Delta U + W$

Since there is no change in pressure or volume, $W = 0$ and $\Delta H = \Delta U$.

$$P \Delta t = \Delta H = (m_G c_{pG} + m_W c_{pW})(T_{\text{final}} - T_{\text{initial}})$$

$T_{\text{final}} = \dfrac{P \Delta t}{m_G c_{pG} + m_W c_{pW}} + T_{\text{initial}}$

$\quad = \dfrac{\left(\dfrac{1}{10} \text{ hp} \right)\left(\dfrac{42.42 \text{ BTU}}{\text{hp-min}} \right)(15 \text{ min})}{(0.5 \text{ lbm})\left(0.2 \dfrac{\text{BTU}}{\text{lbm-}^\circ\text{F}} \right) + (5 \text{ lbm})\left(1.0 \dfrac{\text{BTU}}{\text{lbm-}^\circ\text{F}} \right)}$

$\quad + 70^\circ\text{F} = \boxed{82.48^\circ\text{F}}$

5. $q = \Delta h + \dfrac{1}{2}\left(\dfrac{v_2^2 - v_1^2}{g_c J} \right) + \dfrac{(z_2 - z_1)g}{g_c J}$

$\quad = \dfrac{\left(\dfrac{1}{2} \right)\left[\left(30 \dfrac{\text{ft}}{\text{sec}} \right)^2 - \left(10 \dfrac{\text{ft}}{\text{sec}} \right)^2 \right] + (-40 \text{ ft})\left(32.2 \dfrac{\text{ft}}{\text{sec}^2} \right)}{\left(32.2 \dfrac{\text{lbm-ft}}{\text{lbf-sec}^2} \right)\left(778.2 \dfrac{\text{ft-lbf}}{\text{BTU}} \right)}$

$\quad = \boxed{-0.03544 \text{ BTU/lbm}}$

6. To a good approximation, a jet engine operates on an air standard cycle (i.e., the turbine input and exit gases may be assumed to be air). Because a jet engine uses 200% to 400% excess air for cooling purposes, neglecting products of combustion in the exit gases is justifiable.

The pressures $p_1 = 144$ psia and $p_2 = 30.6$ psia are low enough that the low pressure air tables may be used to determine the enthalpies.

$$T_1 = (1340 + 460)^\circ\text{R} = 1800^\circ\text{R}$$

$$h_1 = 449.71 \text{ BTU/lbm}$$

$$T_2 = (820 + 460)^\circ\text{R} = 1280^\circ\text{R}$$

$$\quad = 311.79 \text{ BTU/lbm}$$

$$q = h_2 - h_1 + \dfrac{v_1^2 - v_1^2}{2 g_c J}$$

$$\quad + \dfrac{g(z_2 - z_1)}{g_c J} + w_{\text{shaft}}$$

$$\Delta z = 0$$

$$q = 0$$

THERMO
Prop Changes

$$w_{\text{shaft}} = h_1 - h_2 + \frac{v_1^2 - v_2^2}{2g_c J}$$

$$= 449.71 \frac{\text{BTU}}{\text{lbm}} - 311.79 \frac{\text{BTU}}{\text{lbm}}$$

$$+ \frac{\left(540 \frac{\text{ft}}{\text{sec}}\right)^2 - \left(1000 \frac{\text{ft}}{\text{sec}}\right)^2}{(2)\left(32.2 \frac{\text{lbm-ft}}{\text{lbf-sec}^2}\right)\left(778.2 \frac{\text{ft-lbf}}{\text{BTU}}\right)}$$

$$= \boxed{123.78 \text{ BTU/lbm}}$$

7. $q = h_2 - h_1 + \dfrac{v_2^2 - v_1^2}{2g_c J} + \dfrac{g(z_2 - z_1)}{g_c J} + w_{\text{shaft}}$

$$q = 0 \quad \text{(adiabatic expansion)}$$

$$\Delta z \approx 0 \text{ in a turbine}$$

$$0 = h_2 - h_1 + \frac{v_2^2 - v_1^2}{2g_c J} + \frac{P_{\text{shaft}}}{\dot{m}}$$

$$h_2 = h_1 + \frac{v_1^2 - v_2^2}{2g_c J} - \frac{P_{\text{shaft}}}{\dot{m}}$$

$$= 1400 \frac{\text{BTU}}{\text{lbm}}$$

$$+ \frac{0 - \left(500 \frac{\text{ft}}{\text{sec}}\right)^2}{(2)\left(32.2 \frac{\text{lbm-ft}}{\text{lbf-sec}^2}\right)\left(778.2 \frac{\text{ft-lbf}}{\text{BTU}}\right)}$$

$$- \frac{(10{,}000 \text{ hp})\left(2545 \frac{\frac{\text{BTU}}{\text{hr}}}{\text{hp}}\right)}{\left(100{,}000 \frac{\text{lbm}}{\text{hr}}\right)}$$

$$= \boxed{1140.5 \text{ BTU/lbm}}$$

8. $q = h_2 - h_1 + \dfrac{v_2^2 - v_1^2}{2g_c J} = \dfrac{g(z_2 - z_1)}{g_c J} + w_{\text{shaft}}$

Multiply through by

$$\dot{m} = \left(275 \frac{\text{lbm}}{\text{min}}\right)\left(60 \frac{\text{min}}{\text{hr}}\right) = 16{,}500 \text{ lbm/hr}$$

$$\dot{Q} = \dot{m}(h_2 - h_1) + \frac{\dot{m}(v_2^2 - v_1^2)}{2g_c J}$$

$$+ \frac{\dot{m}g(z_2 - z_1)}{g_c J} + P_{\text{shaft}}$$

$$= \left(16{,}500 \frac{\text{lbm}}{\text{hr}}\right)\left[\left(1012.5 \frac{\text{BTU}}{\text{lbm}} - 1217.6 \frac{\text{BTU}}{\text{lbm}}\right)\right.$$

$$+ \frac{\left[\left(400 \frac{\text{ft}}{\text{sec}}\right)^2 - \left(70 \frac{\text{ft}}{\text{sec}}\right)^2\right]}{(2)\left(32.2 \frac{\text{lbm-ft}}{\text{lbf-sec}^2}\right)\left(778.2 \frac{\text{ft-lbf}}{\text{BTU}}\right)}$$

$$+ \left.\frac{\left(32.2 \frac{\text{ft}}{\text{sec}^2}\right)(1.8 \text{ ft} - 10.2 \text{ ft})}{\left(32.2 \frac{\text{lbm-ft}}{\text{lbf-sec}^2}\right)\left(778.2 \frac{\text{ft-lbf}}{\text{BTU}}\right)}\right]$$

$$+ (1000 \text{ hp})\left(2545 \frac{\frac{\text{BTU}}{\text{hr}}}{\text{hp}}\right)$$

$$= \boxed{-7.88 \times 10^5 \text{ BTU/hr (a loss)}}$$

9. P_{shaft}

$$= \dot{Q} + \dot{m}\left(h_1 - h_2 + \frac{v_1^2 - v_1^2}{2g_c J}\right) + \dot{m}g\left[\frac{z_1 - z_2}{g_c J}\right]$$

$$= -50 \frac{\text{BTU}}{\text{sec}} + \left(5 \frac{\text{lbm}}{\text{sec}}\right)\left[1000 \frac{\text{BTU}}{\text{lbm}} - 1020 \frac{\text{BTU}}{\text{lbm}}\right.$$

$$+ \frac{\left(100 \frac{\text{ft}}{\text{sec}}\right)^2 - \left(50 \frac{\text{ft}}{\text{sec}}\right)^2}{(2)\left(32.2 \frac{\text{lbm-ft}}{\text{lbf-sec}^2}\right)\left(778.2 \frac{\text{ft-lbf}}{\text{BTU}}\right)}$$

$$+ \left(5 \frac{\text{lbm}}{\text{sec}}\right)\left(32.2 \frac{\text{ft}}{\text{sec}^2}\right)$$

$$\times \left[\frac{100 \text{ ft} - 0}{\left(32.2 \frac{\text{lbm-ft}}{\text{lbf-sec}^2}\right)\left(778.2 \frac{\text{ft-lbf}}{\text{BTU}}\right)}\right]$$

$$= -148.61 \frac{\text{BTU}}{\text{sec}} \quad \text{(negative because work input)}$$

$$P_{\text{shaft}} = \left(148.61 \frac{\text{BTU}}{\text{sec}}\right)\left(\frac{1}{1.285 \times 10^{-3}} \frac{\text{ft-lbf}}{\text{BTU}}\right)$$

$$\times \left(\frac{\text{hp}}{550 \frac{\text{ft-lbf}}{\text{sec}}}\right)$$

$$= \boxed{210.3 \text{ hp}}$$

10. $q = RT \ln\left(\dfrac{v_2}{v_1}\right)$

$= \dfrac{R^*}{(MW)} T \ln\left(\dfrac{V_2}{V_1}\right)$

$Q = m\dfrac{R^*}{(MW)} T \ln\left(\dfrac{V_2}{V_1}\right)$

$= (3\ \text{lbm}) \left(\dfrac{1.986\ \dfrac{\text{BTU}}{\text{lbmole-}^\circ\text{R}}}{28.014\ \dfrac{\text{lbm}}{\text{lbmole}}}\right)$

$\times (300 + 460)^\circ\text{R} \ln\left(\dfrac{22.5\ \text{ft}^3}{40\ \text{ft}^3}\right)$

$= \boxed{-93\ \text{BTU (flows out of system)}}$

11. $p_2 = p_1 \left(\dfrac{v_1}{v_2}\right)^n = p_1 \left(\dfrac{V_1}{V_2}\right)^n$

$W = \dfrac{p_1 V_1 - p_2 V_2}{n - 1}$

$= \dfrac{p_1 V_1 - p_1 \left(\dfrac{V_1}{V_2}\right)^n V_2}{n - 1}$

$= \dfrac{\left(200\ \dfrac{\text{lbf}}{\text{in}^2}\right)\left(\dfrac{12\ \text{in}}{\text{ft}}\right)^2 (0.02\ \text{ft}^3)}{-1.2 - 1}$

$- \dfrac{\left(200\ \dfrac{\text{lbf}}{\text{in}^2}\right)\left(\dfrac{12\ \text{in}}{\text{ft}}\right)^2 \left(\dfrac{0.02\ \text{ft}^3}{0.028\ \text{ft}^3}\right)^{-1.2} (0.028\ \text{ft}^3)}{-1.2 - 1}$

$= \boxed{287.07\ \text{ft-lbf}}$

12. $q = RT \ln\left(\dfrac{p_1}{p_2}\right)$

$Q = mRT \ln\left(\dfrac{p_1}{p_2}\right)$

$= (4\ \text{lbm}) \left(53.3\ \dfrac{\text{ft-lbf}}{\text{lbm-}^\circ\text{R}}\right)$

$\times (240 + 460)^\circ\text{R} \ln\left(\dfrac{p_1}{2p_1}\right)$

$= (-103{,}445\ \text{ft-lbf}) \left(\dfrac{1\ \text{BTU}}{778.2\ \text{ft-lbf}}\right)$

$= \boxed{-132.93\ \text{BTU (exothermic process)}}$

13. $\boxed{Q = 0\ \text{(adiabatic)}}$

14. (a) If a steady flow system is not isentropic or a throttling process, it may be assumed to be polytropic. Treating air as an ideal gas,

$p_1 v_1 = RT_1$

$v_1 = \dfrac{RT_1}{p_1} = \dfrac{\left(53.35\ \dfrac{\text{ft-lbf}}{\text{lbm-}^\circ\text{R}}\right)(460 + 70)^\circ\text{R}}{\left(14.7\ \dfrac{\text{lbf}}{\text{in}^2}\right)\left(\dfrac{144\ \text{in}^2}{\text{ft}^2}\right)}$

$= \boxed{13.36\ \text{ft}^3/\text{lbm}}$

(b) $v_2 = \left(\dfrac{v_2}{v_1}\right) v_1 = \left(\dfrac{1}{10}\right)\left(13.36\ \dfrac{\text{ft}^3}{\text{lbm}}\right)$

$= \boxed{1.336\ \text{ft}^3/\text{lbm}}$

(c) The actual process is polytropic:

$w = h_1 - h_2$

$h_2 = h_1 - w$

$= 126.66\ \dfrac{\text{BTU}}{\text{lbm}} - \left(-209.13\ \dfrac{\text{BTU}}{\text{lbm}}\right)$

$= 335.79\ \text{BTU/lbm}$

This value of h corresponds in the air tables to

$\boxed{T = 1372.5^\circ\text{R}}$

(d) Using the low pressure air tables, at $T_1 = 530^\circ\text{R}$, by interpolation,

$v_{r1} = 151.38$

$h_1 = 126.66\ \dfrac{\text{BTU}}{\text{lbm}}$

For an isentropic process,

$\dfrac{v_{r2}}{v_{r1}} = \dfrac{V_2}{V_1}$

$v_{r2} = v_{r1}\left(\dfrac{V_2}{V_1}\right) = (151.38)\left(\dfrac{1\ \text{ft}^3}{10\ \text{ft}^3}\right)$

$\phantom{v_{r2}}= 15.14$

Locating this value for v_{r2} in the table yields, by interpolation, $h_{2,\Delta s=0} = 314.88\ \text{BTU/lbm}$.

$w|_s = h_1 - h_{2,\Delta s=0}$

$= 126.66\ \dfrac{\text{BTU}}{\text{lbm}} - 314.88\ \dfrac{\text{BTU}}{\text{lbm}}$

$= -188.22\ \text{BTU/lbm}$

Since the actual process was, in fact, not isentropic: for compressions, $\eta = \dfrac{w|_s}{w}$

$$w = \frac{w|_s}{\eta} = \frac{-188.22 \, \frac{\text{BTU}}{\text{lbm}}}{0.9}$$

$$= -209.13 \, \text{BTU/lbm}$$

$w_{\text{compression}} = -w$

$$\boxed{= 209.13 \, \text{BTU/lbm}}$$

15. $T_1 = (60 + 460)^\circ\text{R} = 520^\circ\text{R}$

$T_2 = (-100 + 460)^\circ\text{R} = 360^\circ\text{R}$

Use low pressure air tables:

$$s_2 - s_1 = \phi_2 - \phi_1 - R \ln\left(\frac{p_2}{p_1}\right)$$

$$= 0.50369 \, \frac{\text{BTU}}{\text{lbm-}^\circ\text{R}} - 0.59172 \, \frac{\text{BTU}}{\text{lbm-}^\circ\text{R}}$$

$$- \left(53.3 \, \frac{\text{ft-lbf}}{\text{lbm-}^\circ\text{R}}\right)\left(\frac{1 \, \text{BTU}}{778.2 \, \text{ft-lbf}}\right) \ln\left(\frac{14.7 \, \text{psia}}{500 \, \text{psia}}\right)$$

$$\boxed{= 0.15352 \, \text{BTU/lbm-}^\circ\text{R}}$$

Since the process is steady flow polytropic,

$$w = h_1 - h_2$$

From the air tables,

$$w = (124.27 - 85.97) \, \frac{\text{BTU}}{\text{lbm}}$$

$$\boxed{= 38.3 \, \text{BTU/lbm}}$$

If the process had occurred isentropically, from the air tables,

$$T_1 = 520^\circ\text{R}$$

$$p_{r1} = 1.2147$$

$$h_1 = 124.27 \, \text{BTU/lbm}$$

$$\frac{p_{r2,\Delta s=0}}{p_{r1}} = \frac{p_2}{p_1}$$

$$p_{r2,\Delta s=0} = p_{r1}\left(\frac{p_2}{p_1}\right)$$

$$= (1.2147)\left(\frac{14.7 \, \text{psia}}{500 \, \text{psia}}\right)$$

$$= 0.03571$$

This corresponds to

$$h_{2,\Delta s=0} = 45.04 \, \text{BTU/lbm}$$

$$w|_s = h_1 - h_{2,\Delta s=0}$$

$$= 124.27 \, \frac{\text{BTU}}{\text{lbm}} - 45.04 \, \frac{\text{BTU}}{\text{lbm}}$$

$$= 79.23 \, \text{BTU/lbm}$$

For expansion,

$$\eta = \frac{w}{w|_s} = \frac{38.3 \, \frac{\text{BTU}}{\text{lbm}}}{79.23 \, \frac{\text{BTU}}{\text{lbm}}} = 0.483 = \boxed{48.3\%}$$

16. Throttling steady-flow system:

$$s_2 - s_1 = R \ln\left(\frac{p_1}{p_2}\right)$$

$$S_2 - S_1 = nR^* \ln\left(\frac{p_1}{p_2}\right)$$

$$= \left(3 \, \frac{\text{lbmole}}{\text{sec}}\right)\left(1.986 \, \frac{\text{BTU}}{\text{lbmole-}^\circ\text{R}}\right)$$

$$\times \ln\left(\frac{9.2 \, \text{psia}}{6.4 \, \text{psia}}\right)$$

$$\boxed{2.162 \, \text{BTU/sec-}^\circ\text{R}}$$

17. $\boxed{\Delta T = 0 \text{ (throttled ideal gas)}}$

18. The Mollier diagram should be used. The constant $p_1 = 100$ psia line and $T_1 = 700^\circ\text{F}$ line intersect at the $s = 1.8$ BTU/lbm-$^\circ$R line, so follow this line down to its intersection with the $h = 1100$ BTU/lbm line and find $\boxed{p_2 \approx 5 \text{ psia.}}$

19. Enthalpy is constant during an ideal throttling process.

Find the intersection of the 900°F constant temperature line and the $p_2 = 500$ psia constant pressure line on the Mollier diagram. Follow a horizontal (i.e., constant h) line to the right to the $p = 30$ psia line.

$$\boxed{s_2 \approx 2.0 \, \text{BTU/lbm-}^\circ\text{R}}$$

20. (a) From superheated steam tables at $T_1 = 800^\circ$, $p_1 = 100$ psia,

$$h_1 = 1429.6 \, \text{BTU/lbm}$$

$$s_1 = \boxed{1.8449 \, \text{BTU/lbm-}^\circ\text{R}}$$

THERMO
Prop Changes

Since this is an isentropic process, $s_2 = s_1$.

(b) For an isentropic expansion to $p_2 = 5$ psia,

$$x_2 = \frac{s_2 - s_{F(p_{sat}=p_2)}}{s_{FG(p_{sat}=p_2)}}$$

$$= \frac{1.8449 \, \frac{\text{BTU}}{\text{lbm-}^\circ\text{R}} - 0.2349 \, \frac{\text{BTU}}{\text{lbm-}^\circ\text{R}}}{1.6093 \, \frac{\text{BTU}}{\text{lbm-}^\circ\text{R}}} = 1$$

The steam is entirely gas, so,

$$h_2 = h_{G(p_{sat}=p_2)} = 1131.0 \, \text{BTU/lbm}$$

The actual final enthalpy is given by

$$h_2' = h_1 - \eta(h_1 - h_2)$$

$$= 1429.6 \, \frac{\text{BTU}}{\text{lbm}}$$

$$- (0.80) \left(1429.6 \, \frac{\text{BTU}}{\text{lbm}} - 1131.0 \, \frac{\text{BTU}}{\text{lbm}} \right)$$

$$= 1190.7 \, \text{BTU/lbm}$$

The actual final entropy, s_2', is found by interpolation at h_2' from saturated steam tables for $p_{sat} = p_2$.

$$\boxed{s_2' = 1.93 \, \text{BTU/lbm-}^\circ\text{R}}$$

21. (a) Assume the adiabatic compression to be reversible; then, this is an isentropic process.

$$\boxed{Q = 0 \, \text{(adiabatic)}}$$

$$T_2 = T_1 \left(\frac{v_1}{v_2} \right)^{k-1}$$

$$= (20 + 460)^\circ\text{R} \left(\frac{2 \, \text{ft}^3}{1 \, \text{ft}^3} \right)^{1.4-1}$$

$$= 633.36^\circ\text{R}$$

$$\Delta U = m\Delta u = mc_v(T_2 - T_1)$$

$$= (1 \, \text{lbm}) \left(0.1714 \, \frac{\text{BTU}}{\text{lbm-}^\circ\text{R}} \right)$$

$$\left(778.2 \, \frac{\text{ft-lbf}}{\text{BTU}} \right) (633.36^\circ\text{R} - 480^\circ\text{R})$$

$$= \boxed{20,456 \, \text{ft-lbf}}$$

$$Q = \Delta U + W$$

$$W = Q - \Delta U = -\Delta U$$

$$= \boxed{-20,456 \, \text{ft-lbf}}$$

(b) Constant pressure process:

$$T_3 = T_2 \left(\frac{v_3}{v_2} \right)$$

$$= (633.36^\circ\text{R}) \left(\frac{2 \, \text{ft}^3}{1 \, \text{ft}^3} \right)$$

$$= 1266.72^\circ\text{R}$$

$$Q = mc_p(T_3 - T_2)$$

$$= (1 \, \text{lbm}) \left(0.24 \, \frac{\text{BTU}}{\text{lbm-}^\circ\text{R}} \right) \left(778.2 \, \frac{\text{ft-lbf}}{\text{BTU}} \right)$$

$$\times (1266.72^\circ\text{R} - 633.36^\circ\text{R})$$

$$= \boxed{118,291 \, \text{ft-lbf}}$$

$$W = mR(T_3 - T_2)$$

$$= (1 \, \text{lbm}) \left(53.3 \, \frac{\text{ft-lbf}}{\text{lbm-}^\circ\text{R}} \right)$$

$$\times (1266.72^\circ\text{R} - 633.36^\circ\text{R})$$

$$= \boxed{33,758.1 \, \text{ft-lbf}}$$

$$\Delta U = Q - W$$

$$= 118,291 \, \text{ft-lbf} - 33,758.1 \, \text{ft-lbf}$$

$$= \boxed{84,533 \, \text{ft-lbf}}$$

(c) Since $v_3 = v_1$, and since this is a constant volume process when carried out reversibly,

$$\boxed{W = 0}$$

$$Q = \Delta U = mc_v(T_1 - T_3)$$

$$= (1 \, \text{lbm}) \left(0.1714 \, \frac{\text{BTU}}{\text{lbm-}^\circ\text{R}} \right) \left(778 \, \frac{\text{ft-lbf}}{\text{BTU}} \right)$$

$$\times (480^\circ\text{R} - 1266.72^\circ\text{R})$$

$$= \boxed{-104,908 \, \text{ft-lbf}}$$

THERMO
Prop Changes

22. $$w_{max} = h_1 - h_2 - T_0(s_1 - s_2)$$

$T_1 = 1000°F \qquad p_1 = 100 \text{ psia}$

$T_2 = T_0 = 60°F \qquad p_2 = 14.7 \text{ psia}$

h_1 and s_1 can be obtained from superheat steam tables. From the saturated steam tables indexed by temperature: at T_2, $p_2 > p_{2sat}$. From the saturated steam tables indexed by pressure, $T_2 < T_{2sat}$. Therefore, the steam is entirely liquid. h_2, s_2 may be approximated (in the absence of sub-cooled liquid tables) by h_f, s_f listed for $T_{sat} = T_2$. (Using the temperature tables is more accurate than using the pressure tables.)

$$w_{max} = 1532.1 \frac{BTU}{lbm} - 28.06 \frac{BTU}{lbm} - (60 + 460)°R$$

$$\times \left(1.92 \frac{BTU}{lbm\text{-}°R} - 0.0555 \frac{BTU}{lbm\text{-}°R} \right)$$

$$= \boxed{534.5 \text{ BTU/lbm}}$$

23. Assume the feedwater is at the ambient temperature and pressure.

$$T_1 = T_{sat} = 400°F$$

From the saturated steam tables,

$h_1 = h_g = 1202.0 \text{ BTU/lbm}$

$s_1 = s_g = 1.5284 \text{ BTU/lbm-°R}$

$T_2 = T_0 = 80°F$

$p_2 = 1 \text{ atm}$

At this temperature and pressure, water would be entirely liquid. Approximate values for h_2, s_2 may be obtained by using the saturated steam tables at $T_{sat} = T_2 = 80°F$.

$h_2 = h_f = 48.09 \text{ BTU/lbm}$

$s_2 = s_f = 0.09332 \text{ BTU/lbm-°R}$

$$w_{max} = h_1 - h_2 - T_0(s_1 - s_2)$$

$$= 1202.0 \frac{BTU}{lbm} - 48.09 \frac{BTU}{lbm} - (460 + 80)°R$$

$$\times \left(1.5284 \frac{BTU}{lbm\text{-}°R} - 0.09332 \frac{BTU}{lbm\text{-}°R} \right)$$

$$= \boxed{378.97 \text{ BTU/lbm}}$$

Notice that the 900°F flame temperature is irrelevant.

The input energy is

$$h_1 - h_0 = 1202.0 \frac{BTU}{lbm} - 48.09 \frac{BTU}{lbm}$$

$$= 1153.91 \text{ BTU/lbm}$$

The useful work per BTU of input energy is

$$W = (1 \text{ BTU}) \left(\frac{378.97 \frac{BTU}{lbm}}{1153.91 \frac{BTU}{lbm}} \right) = \boxed{0.328 \text{ BTU}}$$

24. $T_1 = (300 + 460)°R = 760°R$

$p_1 = 40 \text{ psia}$

From air tables,

$h_1 = 182.08 \text{ BTU/lbm}$

$\phi_1 = 0.68312 \text{ BTU/lbm-°R}$

$T_0 = T_2 = (80 + 460)°R$

$= 540°R$

$p_2 = 1 \text{ atm} = 14.696 \text{ psia}$

From air tables,

$h_2 = 129.06 \text{ BTU/lbm}$

$\phi_2 = 0.60078 \text{ BTU/lbm-°R}$

$$s_2 - s_1 = \phi_2 - \phi_1 - R \ln \left(\frac{p_2}{p_1} \right)$$

$$= 0.60078 \frac{BTU}{lbm\text{-}°R} - 0.68312 \frac{BTU}{lbm\text{-}°R}$$

$$- \left(\frac{53.35 \frac{ft\text{-}lbf}{lbm\text{-}°R}}{778.2 \frac{ft\text{-}lbf}{BTU}} \right) \ln \left(\frac{14.696 \text{ psia}}{40 \text{ psia}} \right)$$

$$= -0.01369 \text{ BTU/lbm-°R}$$

$$w_{max} = h_1 - h_2 - T_0(s_1 - s_2)$$

$$= h_1 - h_2 + T_0(s_2 - s_1)$$

$$= 182.08 \frac{BTU}{lbm} - 129.06 \frac{BTU}{lbm}$$

$$+ (540°R) \left(-0.01369 \frac{BTU}{lbm\text{-}°R} \right)$$

$$= 45.63 \text{ BTU/lbm}$$

$$w_{max} = \left(45.63 \frac{BTU}{lbm} \right) \left(778.2 \frac{ft\text{-}lbf}{BTU} \right)$$

$$= \boxed{35{,}509.3 \text{ ft-lbf/lbm}}$$

THERMO
Prop Changes

PSYCHROMETRICS

1. (a) From the psychrometric chart,

$$T_{dp} = \boxed{63°F}$$

(b)
$$\phi = \boxed{48\%}$$

(c) $\omega = \boxed{86 \text{ gr/lbm } (0.0123 \text{ lbm/lbm})}$

2. (a) From the steam table,

$$p_{\text{sat}|50°F} = p_w = 0.1780 \text{ psia}$$

$$p_{\text{sat}|70°F} = 0.3632 \text{ psia}$$

$$p_a = p - p_w$$

$$= 10 \text{ psia} - 0.1780 \text{ psia}$$

$$= 9.8220 \text{ psia}$$

$$\omega = 0.621 \frac{p_w}{p_a} = (0.621)\left(\frac{0.1780 \text{ psia}}{9.8220 \text{ psia}}\right)$$

$$= 0.0113 \frac{\text{lbm}}{\text{lbm}}$$

$$\omega = \left(0.0113 \frac{\text{lbm}}{\text{lbm}}\right)\left(7000 \frac{\text{gr}}{\text{lbm}}\right)$$

$$= \boxed{79.1 \text{ gr/lbm}}$$

(b) $\phi = \dfrac{p_w}{p_{\text{sat}|70°F}} = \dfrac{0.1780 \text{ psia}}{0.3632 \text{ psia}}$

$$= 0.49 \quad \boxed{49\%}$$

(c)
$$pV = mRT$$

$$\frac{V}{m} = v = \frac{RT}{p}$$

$$v = \frac{\left(53.3 \frac{\text{ft-lbf}}{\text{lbm-°R}}\right)(460 + 70)°R}{\left(9.8220 \frac{\text{lbf}}{\text{in}^2}\right)\left(144 \frac{\text{in}^2}{\text{ft}^2}\right)}$$

$$= \boxed{19.97 \text{ ft}^3/\text{lbm}}$$

3. (a) From the psychrometric chart,

$$h_1 = 41.6 \text{ BTU/lbm of dry air}$$

$$h_2 = 20.8 \text{ BTU/lbm of dry air}$$

$$\omega_1 = 0.0168 \text{ lbm/lbm}$$

$$m_a = \text{mass of dry air} = \frac{m_{\text{moist air}}}{1 + \omega}$$

$$= \frac{1000 \text{ lbm}}{1 + 0.0168} = 983.5 \text{ lbm}$$

total heat removed

$$= m_a(h_1 - h_2)$$

$$= (983.5 \text{ lbm})\left(41.6 \frac{\text{BTU}}{\text{lbm}} - 20.8 \frac{\text{BTU}}{\text{lbm}}\right)$$

$$= \boxed{20,457 \text{ BTU}}$$

(b) From the psychrometric chart,

sensible heat factor $= 0.475$

sensible heat removed $= (0.475)(20,457 \text{ BTU})$

$$= \boxed{9717 \text{ BTU}}$$

(c) Latent heat removed

$$= 20,457 \text{ BTU} - 9717 \text{ BTU}$$

$$= \boxed{10,740 \text{ BTU}}$$

(d) From the psychrometric chart,

$$T_{\text{wb}|\text{final}} = \boxed{50.5°F}$$

(e)
$$T_{\text{dp}|\text{initial}} = \boxed{71.8°F}$$

(f)
$$T_{\text{dp}|\text{final}} = \boxed{47°F}$$

(g) From the psychrometric chart,

$$\omega_1 = 0.0168 \text{ lbm/lbm dry air}$$

$$\omega_2 = 0.0068 \text{ lbm/lbm dry air}$$

mass of moisture condensed

$$= m_a(\omega_1 - \omega_2)$$

$$= (983.5 \text{ lbm air})$$

$$\times \left(0.0168 \frac{\text{lbm}}{\text{lbm}} - 0.0068 \frac{\text{lbm}}{\text{lbm}}\right)$$

$$= \boxed{9.835 \text{ lbm of water}}$$

GAS DYNAMICS

1. $p_0 = 30$ psia

$$T_0 = (200 + 460)°\text{R} = 660°\text{R}$$

$$\rho_0 = \frac{p_0}{RT_0} = \frac{\left(30\,\frac{\text{lbf}}{\text{in}^2}\right)\left(\frac{12\,\text{in}}{\text{ft}}\right)^2}{\left(53.35\,\frac{\text{ft-lbf}}{\text{lbm-}°\text{R}}\right)(660°\text{R})}$$

$$= 0.1227\,\text{lbm/ft}^3$$

$$\text{v} = \sqrt{2g_c J c_p (T_0 - T)}$$

$$T = T_0 - \frac{\text{v}^2}{2g_c J c_p}$$

$$= 660°\text{R} - \frac{\left(1400\,\frac{\text{ft}}{\text{sec}}\right)^2}{(2)\left(32.2\,\frac{\text{lbm-ft}}{\text{lbf-sec}^2}\right)\left(778.2\,\frac{\text{ft-lbf}}{\text{BTU}}\right)\left(0.240\,\frac{\text{BTU}}{\text{lbm-}°\text{R}}\right)}$$

$$= 497°\text{R}$$

$$a = \sqrt{kg_c RT}$$

$$= \sqrt{(1.4)\left(32.2\,\frac{\text{lbm-ft}}{\text{lbf-sec}^2}\right)\left(53.35\,\frac{\text{ft-lbf}}{\text{lbm-}°\text{R}}\right)(497°\text{R})}$$

$$= 1093.3\,\text{ft/sec}$$

$$M = \frac{\text{v}}{a} = \frac{1400\,\frac{\text{ft}}{\text{sec}}}{1093.3\,\frac{\text{ft}}{\text{sec}}} = 1.28$$

(a) $$\frac{\dot{m}}{A^*} = \frac{\rho_0 \sqrt{kg_c RT_0}}{\left[\frac{1}{2}(k-1)+1\right]^{\frac{k+1}{2(k-1)}}}$$

$$A^* = \frac{\dot{m}\left[\frac{1}{2}(k-1)+1\right]^{\frac{k+1}{2(k-1)}}}{\rho_0 \sqrt{kg_c RT_0}}$$

$$= \frac{\left(10\,\frac{\text{lbm}}{\text{sec}}\right)\left[\left(\frac{1}{2}\right)(1.4-1)+1\right]^{\frac{1.4+1}{(2)(1.4-1)}}}{\left(0.1227\,\frac{\text{lbm}}{\text{ft}^3}\right)\sqrt{(1.4)\left(32.2\,\frac{\text{lbm-ft}}{\text{lbf-sec}^2}\right)\left(53.35\,\frac{\text{ft-lbf}}{\text{lbm-}°\text{R}}\right)(660°\text{R})}}$$

$$= 0.1118\,\text{ft}^2$$

$$A = A^*\left[\frac{A}{A^*}\right]_{M=1.28}$$

$$= (0.118\,\text{ft}^2)(1.0591) = \boxed{0.1184\,\text{ft}^2}$$

(b) $$T = T_0\left[\frac{T}{T_0}\right]_{M=1.28}$$

$$= (660°\text{R})(0.7532) = \boxed{497.1°\text{R}\ (37.1°\text{F})}$$

(c) $$p = p_0\left[\frac{p}{p_0}\right]_{M=1.28}$$

$$= (30\,\text{psia})(0.3712) = \boxed{11.14\,\text{psia}}$$

(d) $$M = \boxed{1.28}$$

2. The velocity is effectively zero at the point of combustion.

$$T_0 = 1700°\text{F} = (1700 + 460)°\text{R} = 2160°\text{R}$$

At exit, $\text{v}_e = 8207$ ft/sec

$$T_e = 274°\text{F} = (274 + 460)°\text{R} = 734°\text{R}$$

From the steady-flow energy equation,

$$Jh_0 + \frac{\text{v}_0^2}{2g_c} = Jh_e + \frac{\text{v}_e^2}{2g_c}$$

$$(\Delta h)_{\text{actual}} = h_0 - h_e = \frac{\text{v}_e^2}{2Jg_c}$$

Also, $$(\Delta h)_{\text{ideal}} = c_p(T_0 - T_e)$$

$$\eta = \frac{(\Delta h)_{\text{actual}}}{(\Delta h)_{\text{ideal}}} = \frac{\left(\frac{\text{v}_e^2}{2Jg_c}\right)}{c_p(T_0 - T_e)}$$

$$= \frac{\left(8207\,\frac{\text{ft}}{\text{sec}}\right)^2}{(2)\left(778.2\,\frac{\text{ft-lbf}}{\text{BTU}}\right)\left(32.2\,\frac{\text{lbm-ft}}{\text{lbf-sec}^2}\right)}$$
$$\overline{\left(1.9\,\frac{\text{BTU}}{\text{lbm-}°\text{R}}\right)(2160°\text{R} - 734°\text{R})}$$

$$= \boxed{0.496\ (49.6\%)}$$

THERMO
Gas Dynamics

3. (a) Consider air to be an ideal gas.

$$c_p = \frac{Rk}{J(k-1)}$$

$$= \frac{\left(53.35 \frac{\text{ft-lbf}}{\text{lbm-}^\circ\text{R}}\right)(1.4)}{\left(778.17 \frac{\text{ft-lbf}}{\text{BTU}}\right)(0.4)} = 0.240 \text{ BTU/lbm-}^\circ\text{R}$$

$$T_0 = T_1 + \frac{v_1^2}{2g_c J c_p} = 540^\circ\text{R}$$

$$+ \frac{\left(500 \frac{\text{ft}}{\text{sec}}\right)^2}{(2)\left(32.2 \frac{\text{lbm-ft}}{\text{lbf-sec}^2}\right)\left(778.17 \frac{\text{ft-lbf}}{\text{BTU}}\right)\left(0.240 \frac{\text{BTU}}{\text{lbm-}^\circ\text{R}}\right)}$$

$$= \boxed{560.8^\circ\text{R}}$$

(b) $v_{max} = \sqrt{2g_c J c_p T_0}$

$$= \sqrt{(2)\left(32.2 \frac{\text{lbm-ft}}{\text{lbf-sec}^2}\right)\left(778.17 \frac{\text{ft-lbf}}{\text{BTU}}\right)\left(0.240 \frac{\text{BTU}}{\text{lbm-}^\circ\text{R}}\right)(560.8^\circ\text{R})}$$

$$= \boxed{2597 \text{ ft/sec}}$$

(c) $\rho_1 = \frac{p_1}{RT_1} = \frac{\left(44 \frac{\text{lbf}}{\text{in}^2}\right)\left(144 \frac{\text{in}^2}{\text{ft}^2}\right)}{\left(53.35 \frac{\text{ft-lbf}}{\text{lbm-}^\circ\text{R}}\right)(540^\circ\text{R})}$

$$= 0.220 \text{ lbm/ft}^3$$

$$\rho_2 = \frac{p_2}{RT_2} = \frac{\left(29 \frac{\text{lbf}}{\text{in}^2}\right)\left(144 \frac{\text{in}^2}{\text{ft}^2}\right)}{\left(53.35 \frac{\text{ft-lbf}}{\text{lbm-}^\circ\text{R}}\right)(500^\circ\text{R})}$$

$$= 0.157 \text{ lbm/ft}^3$$

$$\dot{m} = \rho_1 A_1 v_1 = \rho_2 A_2 v_2$$

$$= \left(0.220 \frac{\text{lbm}}{\text{ft}^3}\right)(78 \text{ in}^2)\left(\frac{\text{ft}^2}{144 \text{ in}^2}\right)\left(500 \frac{\text{ft}}{\text{sec}}\right)$$

$$= \boxed{59.58 \text{ lbm/sec}}$$

(d) $v_2 = \sqrt{2g_c J c_p (T_0 - T_2)} =$

$$\sqrt{(2)\left(32.2 \frac{\text{lbm-ft}}{\text{lbf-sec}^2}\right)\left(778.17 \frac{\text{ft-lbf}}{\text{BTU}}\right)\left(0.240 \frac{\text{BTU}}{\text{lbm-}^\circ\text{R}}\right)(560.8^\circ\text{R}-500^\circ\text{R})}$$

$$= 855.1 \text{ ft/sec}$$

$$A_2 = \left(\frac{\rho_1 v_1}{\rho_2 v_2}\right) A_1$$

$$= \left[\frac{\left(0.220 \frac{\text{lbm}}{\text{ft}^3}\right)\left(500 \frac{\text{ft}}{\text{sec}}\right)}{\left(0.157 \frac{\text{lbm}}{\text{ft}^3}\right)\left(855.1 \frac{\text{ft}}{\text{sec}}\right)}\right](78 \text{ in}^2)$$

$$= \boxed{63.9 \text{ in}^2}$$

4. (a) From the superheated steam tables,

$$h_1 = h_0 = 1329.3 \text{ BTU/lbm}$$
$$h_2 = 1283.0 \text{ BTU/lbm}$$
$$v_2 = \sqrt{2g_c J(h_0 - h_2)} = (223.8)\sqrt{h_0 - h_2}$$
$$= (223.8)\sqrt{1329.3 \frac{\text{BTU}}{\text{lbm}} - 1283.0 \frac{\text{BTU}}{\text{lbm}}}$$
$$= \boxed{1523 \text{ ft/sec}}$$

(b) $v_2 = \sqrt{\frac{2g_c Rk(T_0 - T_2)}{k - 1}}$

$$= \sqrt{\frac{(2)\left(32.2 \frac{\text{lbm-ft}}{\text{lbf-sec}^2}\right)\left(85.5 \frac{\text{ft-lbf}}{\text{lbm-}^\circ\text{R}}\right)(1.3)(600-500)^\circ\text{R}}{1.3-1}}$$

$$= \boxed{1545 \text{ ft/sec}}$$

5. (a) $\frac{T_0}{T_1} = 1 + \left(\frac{1}{2}\right)(k-1)M_1^2$

$$= 1 + \left(\frac{1}{2}\right)(1.3-1)(1.5)^2 = 1.3375$$

$$T_0 = (1.3375)(460 + 80)^\circ\text{R} = \boxed{722.3^\circ\text{R}}$$

(b) $\frac{T_0}{T_2} = 1 + \left(\frac{1}{2}\right)(k-1)M_2^2$

$$= 1 + \left(\frac{1}{2}\right)(1.3-1)(2.5)^2$$

$$= 1.9375$$

$$T_2 = \frac{722.3^\circ\text{R}}{1.9375} = \boxed{372.8^\circ\text{R}}$$

$$v_2 = M_2 a_2 = M_2 \sqrt{k g_c R T_2}$$

$$= (2.5)\sqrt{(1.3)\left(32.2 \frac{\text{lbm-ft}}{\text{lbf-sec}^2}\right)\left(87.4 \frac{\text{ft-lbf}}{\text{lbm-}^\circ\text{R}}\right)(372.8^\circ\text{R})}$$

$$= \boxed{2920 \text{ ft/sec}}$$

THERMO
Gas Dynamics

(c) $\dfrac{\dot{m}}{A_1} = \rho_1 v_1$

$\rho_1 = \dfrac{p_1}{RT_1}$

$= \dfrac{\left(36\,\frac{\text{lbf}}{\text{in}^2}\right)\left(\frac{144\,\text{in}^2}{\text{ft}^2}\right)}{\left(87.4\,\frac{\text{ft-lbf}}{\text{lbm-}^\circ\text{R}}\right)(460+80)^\circ\text{R}}$

$= 0.1098\ \text{lbm/ft}^3$

$v_1 = M_1 a_1 = M_1\sqrt{kg_cRT_1}$

$= (1.5)\sqrt{(1.3)\left(32.2\,\frac{\text{lbm-ft}}{\text{lbf-sec}^2}\right)\left(87.4\,\frac{\text{ft-lbf}}{\text{lbm-}^\circ\text{R}}\right)(540^\circ\text{R})}$

$= 2108\ \text{ft/sec}$

$\dfrac{\dot{m}}{A_1} = \left(0.1098\,\frac{\text{lbm}}{\text{ft}^3}\right)\left(2108\,\frac{\text{ft}}{\text{sec}}\right)$

$= \boxed{231.5\ \text{lbm/sec-ft}^2}$

6. $a = \sqrt{kg_cRT_1}$

$= \sqrt{(1.4)\left(32.2\,\frac{\text{lbm-ft}}{\text{lbf-sec}^2}\right)\left(53.35\,\frac{\text{ft-lbf}}{\text{lbm-}^\circ\text{R}}\right)(460+100)^\circ\text{R}}$

$= 1161\ \text{ft/sec}$

Since $M_1 = \dfrac{v_1}{a_1}$,

$\dfrac{T_0}{T_1} = 1 + \left(\dfrac{k-1}{2}\right)\left(\dfrac{v_1}{a_1}\right)^2$

$= 1 + \left(\dfrac{1.4-1}{2}\right)\left(\dfrac{980\,\frac{\text{ft}}{\text{sec}}}{1161\,\frac{\text{ft}}{\text{sec}}}\right)^2 = 1.143$

$T_0 = (1.143)(560^\circ\text{R}) = 640^\circ\text{R}$

$\dfrac{T_0}{T_2} = 1 + \left(\dfrac{k-1}{2}\right) = M_2^2$

$= 1 + \left(\dfrac{1.4-1}{2}\right)(2)^2 = 1.8$

$T_2 = \dfrac{640^\circ\text{R}}{1.8} = \boxed{355.6^\circ\text{R} \quad (-104.4^\circ\text{F})}$

7. (a) $M = \sqrt{\left(\dfrac{2}{k-1}\right)\left(\dfrac{T_0}{T}-1\right)}$

$= \sqrt{\left(\dfrac{2}{1.4-1}\right)\left(\dfrac{580^\circ\text{R}}{531.5^\circ\text{R}}-1\right)}$

$= \boxed{0.675}$

(b) $v^2 = \left(\dfrac{2k}{k-1}\right)g_cR(T_0-T)$

$= \left(\dfrac{2k}{k-1}\right)g_c\left(RT_0-\dfrac{p}{\rho}\right)$

$\rho = \dfrac{\dot{m}}{Av}$

$v^2 = \left(\dfrac{2k}{k-1}\right)g_c\left(RT_0-\dfrac{pAv}{\dot{m}}\right)$

$v^2 + \left(\dfrac{2k}{k-1}\right)g_c\left(\dfrac{pA}{\dot{m}}\right)v - \left(\dfrac{2k}{k-1}\right)Rg_cT_0 = 0$

$v^2 + \left[\dfrac{(2)(1.4)}{1.4-1}\right]\left(32.2\,\dfrac{\text{lbm-ft}}{\text{lbf-sec}^2}\right)$

$\times \left[\dfrac{\left(7.3\,\frac{\text{lbf}}{\text{in}^2}\right)\left(\frac{144\,\text{in}^2}{\text{ft}^2}\right)(\pi)(0.3\,\text{ft})^2}{\left(2\,\frac{\text{lbm}}{\text{sec}}\right)(4)}\right]v - \left[\dfrac{(2)(1.4)}{1.4-1}\right]$

$\times \left(53.35\,\dfrac{\text{ft-lbf}}{\text{lbm-}^\circ\text{R}}\right)\left(32.2\,\dfrac{\text{lbm-ft}}{\text{lbf-sec}^2}\right)(460+120)^\circ\text{R} = 0$

$v^2 + 8374.2\,v - 6.9746\times10^6 = 0$

$v = \boxed{763.3\ \text{ft/sec}}$

$T = \dfrac{p}{\rho R} = \left(\dfrac{p}{\dot{m}R}\right)Av$

$= \left[\dfrac{\left(7.3\,\frac{\text{lbf}}{\text{in}^2}\right)\left(\frac{144\,\text{in}^2}{\text{ft}^2}\right)}{\left(2\,\frac{\text{lbm}}{\text{sec}}\right)\left(53.35\,\frac{\text{ft-lbf}}{\text{lbm-}^\circ\text{R}}\right)}\right]\left[\dfrac{(\pi)(0.3\,\text{ft})^2}{4}\right]\left(763.3\,\dfrac{\text{ft}}{\text{sec}}\right)$

$= 531.5^\circ\text{R}$

(c) $\dfrac{p_0}{p} = \left(\dfrac{T_0}{T}\right)^{\frac{k}{k-1}}$

$p_0 = p\left(\dfrac{T_0}{T}\right)^{\frac{k}{k-1}}$

$= (7.3\,\text{psi})\left(\dfrac{580^\circ\text{R}}{531.5^\circ\text{R}}\right)^{\frac{1.4}{1.4-1}}$

$= \boxed{9.91\ \text{psi}}$

8. Air flow is sonic at the throat.

$\dfrac{T^*}{T_0} = \dfrac{2}{k+1}$

$T^* = \left(\dfrac{2}{k+1}\right)T_0 = \dfrac{(2)(537^\circ\text{R})}{1.4+1} = 447.5^\circ\text{R}$

$$v^* = a^* = \sqrt{kg_c RT^*}$$

$$= \sqrt{(1.4)\left(32.2 \ \frac{\text{lbm-ft}}{\text{lbf-sec}^2}\right)\left(53.35 \ \frac{\text{ft-lbf}}{\text{lbm-}^\circ\text{R}}\right)(447.5^\circ\text{R})}$$

$$= 1037.4 \ \text{ft/sec}$$

$$\frac{p^*}{p_0} = \left(\frac{2}{k+1}\right)^{\frac{k}{k-1}}$$

$$p^* = p_0 \left(\frac{2}{k+1}\right)^{\frac{k}{k-1}}$$

$$= (29 \ \text{psi}) \left(\frac{2}{1.4+1}\right)^{\frac{1.4}{1.4-1}} = 15.32 \ \text{psi}$$

$$\rho^* = \frac{p^*}{RT^*} = \frac{\left(15.32 \ \frac{\text{lbf}}{\text{in}^2}\right)\left(\frac{144 \ \text{in}^2}{\text{ft}^2}\right)}{\left(53.35 \ \frac{\text{ft-lbf}}{\text{lbm-}^\circ\text{R}}\right)(447.5^\circ\text{R})}$$

$$= 0.0924 \ \text{lbm/ft}^3$$

$$\dot{m} = \rho^* A^* v^*$$

$$= \left(0.0924 \ \frac{\text{lbm}}{\text{ft}^3}\right)(78 \ \text{in}^2)\left(\frac{\text{ft}^2}{144 \ \text{in}^2}\right)\left(1037.6 \ \frac{\text{ft}}{\text{sec}}\right)$$

$$= \boxed{51.92 \ \text{lbm/sec}}$$

The mass flow rate is the same everywhere.

9. (a) The Mach number at the exit is unknown.

$$\frac{p_{\text{exit}}}{p_0} = \frac{63 \ \text{psi}}{360 \ \text{psi}} = 0.175$$

From the isentropic flow factor tables for $M = 1.8$,

$$\frac{A}{A^*} = 1.439$$

$$A^* = \frac{A}{1.439} = \frac{15.5 \ \text{in}^2}{1.439} = \boxed{10.8 \ \text{in}^2}$$

(b) $\dfrac{T^*}{T_0} = \dfrac{2}{k+1}$

$$T^* = \left(\frac{2}{k+1}\right) T_0 = \left(\frac{2}{1.4+1}\right)(1890^\circ\text{R})$$

$$= \boxed{1575^\circ\text{R}}$$

$$\frac{p^*}{p_0} = \left(\frac{2}{k+1}\right)^{\frac{k}{k-1}}$$

$$p^* = \left(\frac{2}{k+1}\right)^{\frac{k}{k-1}} p_0$$

$$= \left(\frac{2}{1.4+1}\right)^{\frac{1.4}{1.4-1}} (360 \ \text{psi})$$

$$= \boxed{190.2 \ \text{psi}}$$

(c) From the isentropic flow factor tables for $M = 1.8$,

$$\frac{T}{T_0} = 0.6068$$

$$T = (0.6068)(1890^\circ\text{R}) = \boxed{1147^\circ\text{R}}$$

(d) The theoretical maximum velocity is achieved when all thermal energy possessed by the gas is converted to kinetic energy.

$$h_0 = \frac{1}{2} v_{\text{max}}^2$$

$$v_{\text{max}} = \sqrt{2 g_c J h_0} = \sqrt{\frac{2 g_c R k T_0}{k-1}}$$

At the exit,

$$v = Ma = M\sqrt{kg_c RT}$$

$$\frac{v}{v_{\text{max}}} = M\sqrt{\frac{(k-1)T}{2T_0}}$$

$$= (1.8)\sqrt{\frac{(1.4-1)(1147^\circ\text{R})}{(2)(1890^\circ\text{R})}}$$

$$= \boxed{0.63}$$

(e) $v^* = \sqrt{kg_c RT^*}$

$$= \sqrt{(1.4)\left(32.2 \ \frac{\text{lbm-ft}}{\text{lbf-sec}^2}\right)\left(53.35 \ \frac{\text{ft-lbf}}{\text{lbm-}^\circ\text{R}}\right)(1575^\circ\text{R})}$$

$$= 1946 \ \text{ft/sec}$$

$$\dot{m} = \rho^* A^* v^* = \left(\frac{p^*}{RT^*}\right) A^* v^*$$

$$= \left[\frac{190.2 \ \frac{\text{lbf}}{\text{in}^2}}{\left(53.35 \ \frac{\text{ft-lbf}}{\text{lbm-}^\circ\text{R}}\right)(1575^\circ\text{R})}\right](10.8 \ \text{in}^2)\left(1946 \ \frac{\text{ft}}{\text{sec}}\right)$$

$$= \boxed{47.57 \ \text{lbm/sec}}$$

THERMO
Gas Dynamics

10. (a) From the isentropic flow factor tables for $M_1 = 0.9$,

$$\frac{A_1}{A^*} = 1.0089$$

For $M_2 = 0.2$,

$$\frac{A_2}{A^*} = 2.9635$$

Therefore,

$$\frac{A_2}{A_1} = \frac{\dfrac{A_2}{A^*}}{\dfrac{A_1}{A^*}} = \frac{2.9635}{1.0089} = \boxed{2.937}$$

(b) Similarly, from the tables,

$$\frac{p_1}{p_0} = 0.5913$$

$$\frac{p_2}{p_0} = 0.9725$$

$$\frac{p_2}{p_1} = \frac{\dfrac{p_2}{p_0}}{\dfrac{p_1}{p_0}} = \frac{0.9725}{0.5913} = 1.645$$

$$= p_2 - p_1 = (1.645 - 1)p_1$$

$$= (0.645)(60.5 \text{ psi}) = \boxed{39.0 \text{ psi}}$$

11. (a) $$\frac{p}{p_0} = \frac{160 \text{ psi}}{290 \text{ psi}} = 0.55$$

Searching the tables for this pressure ratio, the exit Mach number, M, is 0.965.

The nozzle can be either divergent or convergent-divergent. If the nozzle is divergent, the flow is entirely subsonic. If the nozzle is convergent-divergent, the flow is subsonic except at the throat.

(b) $$\frac{A}{A^*} = 1.0011$$
$$A^* = 10.788 \text{ ft}^2$$

In the case of a convergent-divergent nozzle, the flow would be sonic at the throat if $A_{\text{throat}} = A^* = 10.788 \text{ ft}^2$.

12. (a) $$v_e = M\sqrt{kg_c R T_e}$$

$$= (1.8)\sqrt{(1.4)\left(32.2 \frac{\text{lbm-ft}}{\text{lbf-sec}^2}\right)\left(53.35 \frac{\text{ft-lbf}}{\text{lbm-}^\circ\text{R}}\right)(1147^\circ\text{R})}$$

$$= 2990 \text{ ft/sec}$$

$$F = \frac{\dot{m}v_e}{g_c} + A_e(p_e - p_a)$$

$$= \frac{\left(47.57 \dfrac{\text{lbm}}{\text{sec}}\right)\left(2990 \dfrac{\text{ft}}{\text{sec}}\right)}{32.2 \dfrac{\text{lbm-ft}}{\text{lbf-sec}^2}}$$

$$+ (15.5 \text{ in}^2)\left(63 \frac{\text{lbf}}{\text{in}^2} - 7.3 \frac{\text{lbf}}{\text{in}^2}\right)$$

$$= \boxed{5281 \text{ lbf}}$$

(b) $$C_f = \frac{F}{p_0 A_t} = \frac{5281 \text{ lbf}}{\left(360 \dfrac{\text{lbf}}{\text{in}^2}\right)(10.788 \text{ in}^2)}$$

$$= \boxed{1.42}$$

(c) $$v_{\text{eff}} = \frac{F g_c}{\dot{m}}$$

$$= \frac{(5281 \text{ lbf})\left(32.2 \dfrac{\text{lbm-ft}}{\text{lbf-sec}^2}\right)}{47.57 \dfrac{\text{lbm}}{\text{sec}}}$$

$$= \boxed{3575 \text{ ft/sec}}$$

(d) $$I_{\text{sp}} = \frac{v_{\text{eff}}}{g} = \frac{3575 \dfrac{\text{ft}}{\text{sec}}}{32.2 \dfrac{\text{ft}}{\text{sec}^2}} = \boxed{111.0 \text{ sec}}$$

13. (a) $$\eta = \left(\frac{v_{\text{actual}}}{v_{\text{ideal}}}\right)^2 = \left(\frac{2630 \dfrac{\text{ft}}{\text{sec}}}{2920 \dfrac{\text{ft}}{\text{sec}}}\right)^2 = \boxed{0.811}$$

(b) $$C_v = \sqrt{\eta} = \boxed{0.901}$$

(c) $$\dot{m}_{\text{ideal}} = \left(\frac{\dot{m}}{A_1}\right)A$$

$$= \left(231.5 \frac{\text{lbm}}{\text{sec-ft}^2}\right)(1.1 \text{ ft}^2)$$

$$= 254.65 \text{ lbm/sec}$$

$$\dot{m}_{\text{actual}} = C_d \dot{m}_{\text{ideal}}$$

$$= (0.85)\left(254.65 \,\frac{\text{lbm}}{\text{sec}}\right)$$

$$= \boxed{216.5 \text{ lbm/sec}}$$

14. From the normal shock parameter tables for $M_y = 0.5130$,

$$\frac{p_y}{p_x} = 7.125$$

$$\frac{T_y}{T_x} = 2.137$$

$$\frac{T_y}{T_{0y}} = 0.950$$

(a) $\qquad T_y = (2.137)(900°\text{R}) = \boxed{1923°\text{R}}$

(b) $\qquad T_{0y} = \dfrac{1923°\text{R}}{0.950} = \boxed{2024°\text{R}}$

(c) $\qquad p_y = (7.125)(14.5 \text{ psi}) = \boxed{103.3 \text{ psi}}$

15. (a) $\rho_x = \dfrac{p_x}{RT_x} = \dfrac{\left(11.6 \,\frac{\text{lbf}}{\text{in}^2}\right)\left(144 \,\frac{\text{in}^2}{\text{ft}^2}\right)}{\left(53.35 \,\frac{\text{ft-lbf}}{\text{lbm-}°\text{R}}\right)(670°\text{R})}$

$$= 0.0467 \text{ lbm/ft}^3$$

From the normal shock tables at $M_x = 1.8$,

$$\frac{p_y}{p_x} = 3.613$$

$$\frac{T_y}{T_x} = 1.532$$

$$\frac{\rho_y}{\rho_x} = \left(\frac{p_y}{p_x}\right)\left(\frac{T_x}{T_y}\right) = \frac{3.613}{1.532} = 2.358$$

$$\rho_y = (2.358)\left(0.0467 \,\frac{\text{lbm}}{\text{ft}^3}\right)$$

$$= \boxed{0.1101 \text{ lbm/ft}^3}$$

(b) For isentropic flow,

$$\frac{\rho_y}{\rho_x} = \left(\frac{p_y}{p_x}\right)^{\frac{1}{k}} = (3.613)^{1/1.4} = 2.503$$

$$\rho_{y,\text{isentropic}} = (2.503)\left(0.0467 \,\frac{\text{lbm}}{\text{ft}^3}\right)$$

$$= \boxed{0.1169 \text{ lbm/ft}^3}$$

Isentropic compression is greater than shock compression.

16. From the normal shock tables at $M_y = 0.5774$,

$$\frac{p_y}{p_x} = 4.5$$

$$\frac{T_y}{T_x} = 1.687$$

(a) $\boxed{M_y = 0.5774}$

(b) $p_y = (4.5)(10 \text{ psi}) = \boxed{45 \text{ psi}}$

(c) $T_y = (1.687)(495°\text{R}) = \boxed{835.1°\text{R}}$

(d) $\rho_y = \dfrac{p_y}{RT_y} = \dfrac{\left(45 \,\frac{\text{lbf}}{\text{in}^2}\right)\left(144 \,\frac{\text{in}^2}{\text{ft}^2}\right)}{\left(53.35 \,\frac{\text{ft-lbf}}{\text{lbm-}°\text{R}}\right)(835.1°\text{R})}$

$$= \boxed{0.1454 \text{ lbm/ft}^3}$$

(e) $\qquad a = \sqrt{kg_c RT}$

$$= \sqrt{(1.4)\left(32.2 \,\frac{\text{lbm-ft}}{\text{lbf-sec}^2}\right)\left(53.35 \,\frac{\text{ft-lbf}}{\text{lbm-}°\text{R}}\right)(835.1°\text{R})}$$

$$= \boxed{1417 \text{ ft/sec}}$$

(f) $\qquad v = Ma = (0.5774)\left(1417 \,\frac{\text{ft}}{\text{sec}}\right)$

$$= \boxed{818.2 \text{ ft/sec}}$$

THERMO
Gas Dynamics

VAPOR POWER CYCLES

Subscripts F and G denote saturated fluid and saturated steam properties, respectively. Subscripts f and g denote properties at points f and g in the cycle, respectively.

1. From the superheated steam table,

$$h_1 = 1279.1 \text{ BTU/lbm}$$

Find the T_1, p_1 intersection on the Mollier diagram by following the constant s line vertically downward to p_2.

$$h_2 = 1242.1 \text{ BTU/lbm}$$

Isentropic flow means $Q = 0$.

$$h_1 + \frac{\mathrm{v}_1^2}{2g_c J} = h_2 + \frac{\mathrm{v}_2^2}{2g_c J}$$

$$\mathrm{v}_2 = \sqrt{2g_c J(h_1 - h_2)}$$

$$\mathrm{v}_2 =$$

$$\sqrt{(2)\left(32.2\,\frac{\text{lbm-ft}}{\text{lbf-sec}^2}\right)\left(778.2\,\frac{\text{ft-lbf}}{\text{BTU}}\right)\left(1279.1\,\frac{\text{BTU}}{\text{lbm}} - 1242.1\,\frac{\text{BTU}}{\text{lbm}}\right)}$$

$$= \boxed{1361.7 \text{ ft/sec}}$$

2. From the superheated steam table,

$$h_1 = 1279.1 \text{ BTU/lbm}$$

$$s_2 = s_1 = 1.7085 \text{ BTU/lbm-}^\circ\text{F}$$

From the Mollier diagram it is seen that the steam is part vapor and part liquid at p_2.

$$x_2 = \frac{s_2 - s_{F(p_{\text{sat}}=p_2)}}{s_{FG(p_{\text{sat}}=p_2)}}$$

$$\cong \frac{1.7085\,\frac{\text{BTU}}{\text{lbm-}^\circ\text{F}} - 0.2009\,\frac{\text{BTU}}{\text{lbm-}^\circ\text{F}}}{1.6852\,\frac{\text{BTU}}{\text{lbm-}^\circ\text{F}}}$$

$$= 0.8946$$

$$h_2 = h_{F(p_{\text{sat}}=p_2)} + x_2 h_{FG(p_{\text{sat}}=p_2)}$$

$$= 109.39\,\frac{\text{BTU}}{\text{lbm}} + (0.8946)\left(1013.1\,\frac{\text{BTU}}{\text{lbm}}\right)$$

$$= \boxed{1015.7 \text{ BTU/lbm}}$$

3. (a) From the superheated steam table,

$$h_1 = 1473.6 \text{ BTU/lbm}$$

$$s_1 = 1.7589 \text{ BTU/lbm-}^\circ\text{F}$$

$$h_2 = h_{G,p_{\text{sat}}=p_2} = 1131.0 \text{ BTU/lbm}$$

$$h_1 - h_2 = 1473.6\,\frac{\text{BTU}}{\text{lbm}} - 1131.0\,\frac{\text{BTU}}{\text{lbm}}$$

$$= 342.6 \text{ BTU/lbm}$$

$$\dot{m} = \frac{P}{W} = \frac{(4000 \text{ hp})\left(\dfrac{1\,\frac{\text{BTU}}{\text{hr}}}{3.929 \times 10^{-4}\,\text{hp}}\right)}{342.6\,\frac{\text{BTU}}{\text{lbm}}}$$

$$= \boxed{29{,}716 \text{ lbm/hr}}$$

(b) If the expansion is isentropic, $s_2 = s_1$.

$$x_2 = \frac{s_2 - s_{F(p_{\text{sat}}=p_2)}}{s_{FG(p_{\text{sat}}=p_2)}}$$

$$= \frac{1.7589\,\frac{\text{BTU}}{\text{lbm-}^\circ\text{F}} - 0.2349\,\frac{\text{BTU}}{\text{lbm-}^\circ\text{F}}}{1.6093\,\frac{\text{BTU}}{\text{lbm-}^\circ\text{F}}}$$

$$= 0.9470$$

$$h_2 = h_{F(p_2=p_{\text{sat}})} + x_2 h_{FG(p_2=p_{\text{sat}})}$$

$$= 130.17\,\frac{\text{BTU}}{\text{lbm}} + (0.9470)\left(1000.9\,\frac{\text{BTU}}{\text{lbm}}\right)$$

$$= 1078.02 \text{ BTU/lbm}$$

$$W = h_1 - h_2$$

$$= 1473.6\,\frac{\text{BTU}}{\text{lbm}} - 1078.02\,\frac{\text{BTU}}{\text{lbm}} = 395.6\,\frac{\text{BTU}}{\text{lbm}}$$

$$\eta_{\text{isentropic}} = \frac{W}{W_{(\Delta s = 0)}} = \frac{342.6\,\frac{\text{BTU}}{\text{lbm}}}{395.6\,\frac{\text{BTU}}{\text{lbm}}} = \boxed{0.866}$$

THERMO
Vapor Cycles

4. $\eta_{max} = 1 - \dfrac{T_{low}}{T_{high}}$

$= 1 - \dfrac{(100 + 460)°R}{(650 + 460)°R} = \boxed{0.495 \ (49.5\%)}$

5. $h_a = h_{F(T_{sat}=T_{high})} = 696.5 \ BTU/lbm$

$s_a = s_{F(T_{sat}=T_{high})} = 0.8836 \ BTU/lbm$

$h_b = h_{G(T_{sat}=T_{high})} = 1118.7 \ BTU/lbm$

$s_b = s_{G(T_{sat}=T_{high})} = 1.2643 \ BTU/lbm$

Assuming an isentropic turbine and compressor,

$$s_d = s_a \qquad s_c = s_b$$

$x_c = \dfrac{s_c - s_{F(T_{sat}=T_{low})}}{s_{G(T_{sat}=T_{low})} - s_{F(T_{sat}=T_{low})}}$

$= \dfrac{1.2643 \dfrac{BTU}{lbm} - 0.1296 \dfrac{BTU}{lbm}}{1.9822 \dfrac{BTU}{lbm} - 0.1296 \dfrac{BTU}{lbm}} = 0.6125$

$x_d = \dfrac{s_d - s_{F(T_{sat}=T_{low})}}{s_{G(T_{sat}=T_{low})} - s_{F(T_{sat}=T_{low})}}$

$= \dfrac{0.8836 \dfrac{BTU}{lbm} - 0.1296 \dfrac{BTU}{lbm}}{1.9822 \dfrac{BTU}{lbm} - 0.1296 \dfrac{BTU}{lbm}} = 0.4070$

$h_c = h_{F(T_{sat}=T_{low})} + x_c h_{FG(T_{sat}=T_{low})}$

$= 68.05 \dfrac{BTU}{lbm} + (0.6125)\left(1037 \dfrac{BTU}{lbm}\right)$

$= 703.21 \ BTU/lbm$

$h_d = h_{F(T_{sat}=T_{low})} + x_d h_{FG(T_{sat}=T_{low})}$

$= 68.05 \dfrac{BTU}{lbm} + (0.4070)\left(1037 \dfrac{BTU}{lbm}\right)$

$= 490.11 \ BTU/lbm$

$h'_c = h_b - \eta(h_b - h_c)$

$= 1118.7 \dfrac{BTU}{lbm}$

$\qquad - (0.90)\left(1118.7 \dfrac{BTU}{lbm} - 703.21 \dfrac{BTU}{lbm}\right)$

$= 744.76 \ BTU/lbm$

$h'_a = h_d + \dfrac{h_a - h_d}{\eta_{pump}}$

$= 490.11 \dfrac{BTU}{lbm} + \dfrac{696.5 \dfrac{BTU}{lbm} - 490.11 \dfrac{BTU}{lbm}}{0.80}$

$= 748.1 \ BTU/lbm$

$\eta_{th} = \dfrac{(h_b - h'_c) - (h'_a - h_d)}{h_b - h_a}$

$= \dfrac{\left(1118.7 \dfrac{BTU}{lbm} - 744.76 \dfrac{BTU}{lbm}\right) - \left(748.1 \dfrac{BTU}{lbm} - 490.11 \dfrac{BTU}{lbm}\right)}{1118.7 \dfrac{BTU}{lbm} - 748.1 \dfrac{BTU}{lbm}}$

$= \boxed{0.313 \ (31.3\%)}$

6. $T_{high} = 100°C \ (or \ 672°R)$

$T_{low} = 0°C \ (or \ 492°R)$

$Q_{in} = 100 \ BTU$

$\eta = 1 - \dfrac{T_{low}}{T_{high}}$

$= 1 - \dfrac{(0 + 273.15)K}{(100 + 273.15)K} = 0.268$

$W = \eta Q_{in} = (0.268)(100 \ BTU)$

$= \boxed{26.8 \ BTU}$

$\eta = \dfrac{Q_{in} - Q_{out}}{Q_{in}}$

$Q_{out} = Q_{in}(1 - \eta) = (100 \ BTU)(1 - 0.268)$

$= \boxed{73.2 \ BTU}$

7. $h_b = h_{G(T_{sat}=T_{high})} = 1195.2 \ BTU/lbm$

$s_b = s_{G(T_{sat}=T_{high})} = 1.5688 \ BTU/lbm\text{-}°F$

$h_d = h_{F(T_{sat}=T_{low})} = 68.05 \ BTU/lbm$

$s_d = s_{F(T_{sat}=T_{low})} = 0.1296 \ BTU/lbm\text{-}°F$

For an isentropic turbine and pump, $s_c = s_b$, and $s_e = s_d$.

$x_c = \dfrac{s_c - s_{F(T_{sat}=T_{low})}}{s_{G(T_{sat}=T_{low})} - s_{F(T_{sat}=T_{low})}}$

$= \dfrac{1.5688 \dfrac{BTU}{lbm\text{-}°F} - 0.1296 \dfrac{BTU}{lbm\text{-}°F}}{1.9822 \dfrac{BTU}{lbm\text{-}°F} - 0.1296 \dfrac{BTU}{lbm\text{-}°F}} = 0.7769$

$h_c = h_{F(T_{sat}=T_{low})} + x_c h_{FG(T_{sat}=T_{low})}$

$= 68.05 \dfrac{BTU}{lbm} + (0.7769)\left(1037.0 \dfrac{BTU}{lbm}\right)$

$= 873.70 \ BTU/lbm$

$$p_e = p_a = p_{\text{sat}(T_{\text{sat}}=T_{\text{high}})} = 152.92 \, \text{lbf/in}^2$$

$$p_d = p_{\text{sat}(T_{\text{sat}}=T_{\text{low}})} = 0.9503 \, \text{lbf/in}^2$$

$$v_d = v_{F(T_{\text{sat}}=T_{\text{low}})} = 0.01613 \, \text{ft}^3/\text{lbm}$$

$$
\begin{aligned}
h_e &= h_d + \frac{v_d(p_e - p_d)}{J} \\
&= 68.05 \, \frac{\text{BTU}}{\text{lbm}} \\
&+ \frac{\left(0.01613 \, \frac{\text{ft}^3}{\text{lbm}}\right)\left(152.92 \, \frac{\text{lbf}}{\text{in}^2} - 0.9503 \, \frac{\text{lbf}}{\text{in}^2}\right)\left(\frac{12 \, \text{in}}{\text{ft}}\right)^2}{778.2 \, \frac{\text{ft-lbf}}{\text{BTU}}} \\
&= 68.50 \, \text{BTU/lbm}
\end{aligned}
$$

$$\eta_{\text{th}} = \frac{(h_b - h_c) - (h_e - h_d)}{h_b - h_e} =$$

$$\frac{\left(1195.2 \, \frac{\text{BTU}}{\text{lbm}} - 873.70 \, \frac{\text{BTU}}{\text{lbm}}\right) - \left(68.50 \, \frac{\text{BTU}}{\text{lbm}} - 68.05 \, \frac{\text{BTU}}{\text{lbm}}\right)}{1195.2 \, \frac{\text{BTU}}{\text{lbm}} - 68.50 \, \frac{\text{BTU}}{\text{lbm}}}$$

$$\boxed{= 0.285 \, (28.5\%)}$$

8. At the entrance to the pump, $T_d = 80°\text{F}$, $p_d = 1 \, \text{atm}$. For $p_{\text{sat}} = 1 \, \text{atm}$, $T_{\text{sat}} = 212$, so this is subcooled liquid. The modified Rankine diagram is

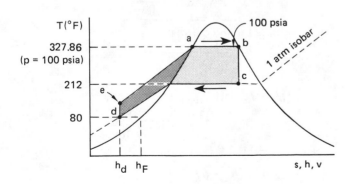

For saturated steam at $p_b = 100 \, \text{lbf/in}^2$,

$$h_b = h_{G(p_{\text{sat}}=p_b)} = 1187.8 \, \text{BTU/lbm}$$

$$s_b = s_{G(p_{\text{sat}}=p_b)} = 1.6034 \, \frac{\text{BTU}}{\text{lbm-}°\text{F}}$$

For the vapor/water mixture at $p_c = 14.696 \, \text{lbf/in}^2 = 1 \, \text{atm}$,

$$s_c = s_b$$

$$
\begin{aligned}
x_c &= \frac{s_c - s_{F(p_{\text{sat}}=p_c)}}{s_{FG(p_{\text{sat}}=p_c)}} \\
&= \frac{1.6034 \, \frac{\text{BTU}}{\text{lbm-}°\text{F}} - 0.3121 \, \frac{\text{BTU}}{\text{lbm-}°\text{F}}}{1.4446 \, \frac{\text{BTU}}{\text{lbm-}°\text{F}}} \\
&= 0.894
\end{aligned}
$$

$$
\begin{aligned}
h_c &= h_{F(p_{\text{sat}}=p_c)} + x_c h_{FG(p_{\text{sat}}=p_c)} \\
&= 180.15 \, \frac{\text{BTU}}{\text{lbm}} + (0.894)\left(970.4 \, \frac{\text{BTU}}{\text{lbm}}\right) \\
&= 1047.7 \, \text{BTU/lbm}
\end{aligned}
$$

$p_d = p_c$ and $T_d = 80°\text{F}$.

$$
\begin{aligned}
h_d &= h_{F(p_{\text{sat}}=1 \, \text{atm})} - c_p(T_c - T_d) \\
&= 180.15 \, \frac{\text{BTU}}{\text{lbm}} - \left(1 \, \frac{\text{BTU}}{\text{lbm-}°\text{R}}\right)(212°\text{F} - 80°\text{F}) \\
&= 48.15 \, \frac{\text{BTU}}{\text{lbm}}
\end{aligned}
$$

$$p_e = p_a = p_b = 100 \, \text{psia}$$

Approximate $v_e = v_d$ by $v_{F(T_{\text{sat}}=Td)} = 0.01607 \, \text{ft}^3/\text{lbm}$ since the liquid is incompressible.

$$
\begin{aligned}
h_e &= h_d + \frac{v_d(p_e - p_d)}{J} \\
&= 48.15 \, \frac{\text{BTU}}{\text{lbm}} \\
&+ \frac{\left(0.01607 \, \frac{\text{ft}^3}{\text{lbm}}\right)\left(100 \, \frac{\text{lbf}}{\text{in}^2} - 14.7 \, \frac{\text{lbf}}{\text{in}^2}\right)\left(\frac{12 \, \text{in}}{\text{ft}}\right)^2}{778.2 \, \frac{\text{ft-lbf}}{\text{BTU}}} \\
&= 48.40 \, \frac{\text{BTU}}{\text{lbm}}
\end{aligned}
$$

Since pump, turbine are actually not isentropic,

$$
\begin{aligned}
h_c' &= h_b - \eta_{\text{turb}}(h_b - h_c) \\
&= 1187.8 \, \frac{\text{BTU}}{\text{lbm}} \\
&- (0.80)\left(1187.8 \, \frac{\text{BTU}}{\text{lbm}} - 1047.7 \, \frac{\text{BTU}}{\text{lbm}}\right) \\
&= 1075.7 \, \text{BTU/lbm}
\end{aligned}
$$

$$h'_e = h_d + \frac{h_e - h_d}{\eta_{\text{pump}}}$$

$$= 48.15 \frac{\text{BTU}}{\text{lbm}} + \frac{48.40 \frac{\text{BTU}}{\text{lbm}} - 48.15 \frac{\text{BTU}}{\text{lbm}}}{0.60}$$

$$= 48.57 \text{ BTU/lbm}$$

$$\eta_{\text{th}} = \frac{(h_b - h'_e) - (h'_c - h_d)}{h_b - h'_e} = 1 - \frac{h'_c - h_d}{h_b - h'_e}$$

$$= 1 - \frac{1075.7 \frac{\text{BTU}}{\text{lbm}} - 48.15 \frac{\text{BTU}}{\text{lbm}}}{1187.8 \frac{\text{BTU}}{\text{lbm}} - 48.57 \frac{\text{BTU}}{\text{lbm}}}$$

$$= \boxed{0.098 \ (9.8\%)}$$

9. From the superheated steam tables for T_d and p_d,

$$p_f = p_e = 1 \text{ lbf/in}^2$$
$$h_f = h_{F(p_{\text{sat}}=p_f)} = 69.74 \text{ BTU/lbm}$$
$$h_d = 1368.3 \text{ BTU/lbm}$$
$$s_e = s_d = 1.6751 \text{ BTU/lbm-}°\text{F}$$
$$x_e = \frac{s_e - s_{F(p_{\text{sat}}=p_e)}}{s_{FG(p_{\text{sat}}=p_e)}}$$

$$= \frac{1.6751 \frac{\text{BTU}}{\text{lbm-}°\text{F}} - 0.1327 \frac{\text{BTU}}{\text{lbm-}°\text{F}}}{1.8453 \frac{\text{BTU}}{\text{lbm-}°\text{F}}} = 0.8359$$

$$h_e = h_{F(p_{\text{sat}}=p_f)} + x_e h_{FG(p_{\text{sat}}=p_f)}$$

$$= 69.74 \frac{\text{BTU}}{\text{lbm}} + (0.8359)\left(1036.0 \frac{\text{BTU}}{\text{lbm}}\right)$$

$$= 935.73 \text{ BTU/lbm}$$
$$v_f = v_{F(p_{\text{sat}}=p_f)} = 0.01614 \text{ ft}^3/\text{lbm}$$
$$p_a = p_d = 300 \text{ lbf/in}^2$$
$$h_a = h_f + \frac{v_f(p_a - p_f)}{J}$$

$$= 69.74 \text{ BTU/lbm}$$

$$+ \frac{\left(0.01614 \frac{\text{ft}^3}{\text{lbm}}\right)\left(300 \frac{\text{lbf}}{\text{in}^2} - 1 \frac{\text{lbf}}{\text{in}^2}\right)\left(\frac{12 \text{ in}}{\text{ft}}\right)^2}{778.2 \frac{\text{ft-lbf}}{\text{BTU}}}$$

$$= 70.63 \text{ BTU/lbm}$$

$$\eta_{\text{th}} = \frac{(h_d - h_a) - (h_e - h_f)}{h_d - h_a}$$

$$= \frac{(1368.3 - 70.63) - (935.73 - 69.74)}{1368.3 - 70.63}$$

$$= \boxed{0.333 \ (33.3\%)}$$

10. $p_d = 500 \text{ lbf/in}^2$
    $T_d = 1000°\text{F}$

From the superheated steam tables,

$$h_d = 1520.7 \text{ BTU/lbm}$$
$$s_e = s_d = 1.7471 \text{ BTU/lbm-}°\text{F}$$
$$p_e = p_f = 5 \text{ lbf/in}^2$$
$$x_e = \frac{s_e - s_{F(p_{\text{sat}}=p_f)}}{s_{FG(p_{\text{sat}}=p_f)}}$$

$$= \frac{1.7471 \frac{\text{BTU}}{\text{lbm-}°\text{F}} - 0.2349 \frac{\text{BTU}}{\text{lbm-}°\text{F}}}{1.6093 \frac{\text{BTU}}{\text{lbm-}°\text{F}}}$$

$$= 0.94$$
$$h_e = h_{F(p_{\text{sat}}=p_f)} + x_e h_{FG(p_{\text{sat}}=p_f)}$$

$$= 130.17 \frac{\text{BTU}}{\text{lbm}} + (0.94)\left(1000.9 \frac{\text{BTU}}{\text{lbm}}\right)$$

$$= 1071.02 \text{ BTU/lbm}$$
$$h_f = h_{F(p_{\text{sat}}=p_f)} = 130.17 \text{ BTU/lbm}$$

(a) $W_{\text{turbine}} = (h_d - h_e)\eta_{\text{turbine}}$

$$= \left(1520.7 \frac{\text{BTU}}{\text{lbm}} - 1070.68 \frac{\text{BTU}}{\text{lbm}}\right)(0.75)$$

$$= 337.5 \text{ BTU/lbm}$$

$$\dot{m} = \frac{P_{\text{turbine}}}{W_{\text{turbine}}}$$

$$= \frac{(2 \times 10^5 \text{ kW})\left(3412.9 \frac{\frac{\text{BTU}}{\text{hr}}}{\text{kW}}\right)}{337.5 \frac{\text{BTU}}{\text{lbm}}}$$

$$= \boxed{2.022 \times 10^6 \text{ lbm/hr}}$$

(b) $h'_e = h_d - W_{\text{turbine}}$

$$= 1520.7 \frac{\text{BTU}}{\text{lbm}} - 337.5 \frac{\text{BTU}}{\text{lbm}}$$

$$= 1183.2 \text{ BTU/lbm}$$

$$Q_{\text{out}} = \dot{m}(h'_e - h_f)$$

$$= \left(2.022 \times 10^6 \frac{\text{lbm}}{\text{hr}}\right)$$

$$\times \left(1183.2 \frac{\text{BTU}}{\text{lbm}} - 130.17 \frac{\text{BTU}}{\text{lbm}}\right)$$

$$= \boxed{2.13 \times 10^9 \text{ BTU/hr}}$$

11. $p_f = p_e = (1.5 \text{ in Hg}) \left(0.49116 \dfrac{\frac{\text{lbf}}{\text{in}^2}}{\text{in Hg}} \right)$

$\qquad = 0.7367 \text{ lbf/in}^2$

This is close to p_{sat} for 92°F, so use 92°F data. From the superheated steam tables at $p_a = p_d = 200 \text{ lbf/in}^2$, $T_d = 400°F$,

$h_d = 1210.8 \text{ BTU/lbm}$

$s_e = s_d = 1.5600 \text{ BTU/lbm-°F}$

$h_f = h_{F(p_{\text{sat}}=p_f)} = 60.06 \text{ BTU/lbm}$

$v_f = v_{F(p_{\text{sat}}=p_f)} = 0.01611 \text{ ft}^3/\text{lbm}$

$x_e = \dfrac{s_e - s_{F(p_{\text{sat}}=p_e)}}{s_{FG(p_{\text{sat}}=p_e)}} = \dfrac{s_e - s_F}{s_G - s_F}$

$\quad = \dfrac{1.5600 \frac{\text{BTU}}{\text{lbm-°F}} - 0.1153 \frac{\text{BTU}}{\text{lbm-°F}}}{2.0030 \frac{\text{BTU}}{\text{lbm-°F}} - 0.1153 \frac{\text{BTU}}{\text{lbm}}} = 0.7653$

$h_e = h_{F(p_{\text{sat}}=p_f)} + x_e h_{FG(p_{\text{sat}}=p_f)}$

$\quad = 60.06 \dfrac{\text{BTU}}{\text{lbm}} + (0.7653) \left(1041.5 \dfrac{\text{BTU}}{\text{lbm}} \right)$

$\quad = 857.12 \text{ BTU/lbm}$

$h_a = h_f + \dfrac{v_f(p_a - p_f)}{J}$

$\quad = 60.06 \dfrac{\text{BTU}}{\text{lbm}}$

$\qquad + \dfrac{\left(0.01611 \frac{\text{ft}^3}{\text{lbm}} \right) \left(200 \frac{\text{lbf}}{\text{in}^2} - 0.7367 \frac{\text{lbf}}{\text{in}^2} \right) \left(\frac{12 \text{ in}}{\text{ft}} \right)^2}{778.2 \frac{\text{ft-lbf}}{\text{BTU}}}$

$\quad = 60.65 \text{ BTU/lbm}$

$\eta_{\text{th}} = \dfrac{(h_d - h_a) - (h_e - h_f)}{h_d - h_a}$

$\quad = \dfrac{(1210.8 - 60.65) - (857.12 - 60.06)}{1210.8 - 60.65}$

$\quad = \boxed{0.307 \ (30.7\%)}$

12. $T_c = T_d = T_f = 700°F$

$\quad p_a = p_d = 600 \text{ lbf/in}^2$

$\quad p_e = p_f = 200 \text{ lbf/in}^2$

$\quad T_h = T_g = 70°F$

From the superheated steam tables,

$h_d = 1350.6 \text{ BTU/lbm}$

$h_f = 1373.8 \text{ BTU/lbm}$

$s_g = s_f = 1.7234 \text{ BTU/lbm-°R}$

$s_e = s_d = 1.5872 \text{ BTU/lbm-°R}$

$x_g = \dfrac{s_f - s_{F(T_{\text{sat}}=T_g)}}{s_{G(T_{\text{sat}}=T_g)} - s_{F(T_{\text{sat}}=T_g)}}$

$\quad = \dfrac{1.7234 \frac{\text{BTU}}{\text{lbm-°R}} - 0.07463 \frac{\text{BTU}}{\text{lbm-°R}}}{2.0642 \frac{\text{BTU}}{\text{lbm-°R}} - 0.07463 \frac{\text{BTU}}{\text{lbm-°R}}} = 0.829$

$h_g = h_{F(T_{\text{sat}}=T_g)} + x_g h_{FG(T_{\text{sat}}=T_g)}$

$\quad = 38.09 \dfrac{\text{BTU}}{\text{lbm}} + (0.829) \left(1054.0 \dfrac{\text{BTU}}{\text{lbm}} \right)$

$\quad = 911.86 \text{ BTU/lbm}$

$h_h = h_{F(T_{\text{sat}}=T_h)} = 38.09 \text{ BTU/lbm}$

$p_h = p_{sat(T_{\text{sat}}=T_h)} = 0.3632 \text{ lbf/in}^2$

$v_h = v_{F(T_{\text{sat}}=T_h)} = 0.01605 \text{ ft}^3/\text{lbm}$

$h_a = h_h + \dfrac{v_h(p_a - p_h)}{J}$

$\quad = 38.09 \dfrac{\text{BTU}}{\text{lbm}}$

$\qquad + \dfrac{\left(0.01605 \frac{\text{ft}^3}{\text{lbm}} \right) \left(600 \frac{\text{lbf}}{\text{in}^2} - 0.3632 \frac{\text{lbf}}{\text{in}^2} \right) \left(\frac{12 \text{ in}}{\text{ft}} \right)^2}{778.2 \frac{\text{ft-lbf}}{\text{BTU}}}$

$\quad = 39.87 \text{ BTU/lbm}$

$x_e = \dfrac{s_e - s_{F(p_{\text{sat}}=p_e)}}{s_{FG(p_{\text{sat}}=p_e)}}$

$\quad = \dfrac{1.5872 \frac{\text{BTU}}{\text{lbm-°R}} - 0.5440 \frac{\text{BTU}}{\text{lbm-°R}}}{1.0025 \frac{\text{BTU}}{\text{lbm-°R}}} = 1.041$

Since steam is superheated at point e, read h_e from the Mollier diagram. Drop straight down from intersection of $T = 700°F$ and $p = 600$ psia to 200 psia.

$h_e = 1235 \text{ BTU/lbm}$

$h_e' = h_d - \eta_{\text{turbine}}(h_d - h_e)$

$\quad = 1350.6 \dfrac{\text{BTU}}{\text{lbm}} - (0.88) \left(1350.6 \dfrac{\text{BTU}}{\text{lbm}} - 1235 \dfrac{\text{BTU}}{\text{lbm}} \right)$

$\quad = 1248.9 \text{ BTU/lbm}$

$h_g' = h_f - \eta_{\text{turbine}}(h_f - h_g)$

$\quad = 1373.8 \dfrac{\text{BTU}}{\text{lbm}} - (0.88) \left(1373.8 \dfrac{\text{BTU}}{\text{lbm}} - 911.86 \dfrac{\text{BTU}}{\text{lbm}} \right)$

$\quad = 967.3 \text{ BTU/lbm}$

$h_a' = h_h + \dfrac{h_a - h_h}{\eta_{\text{pump}}}$

$\quad = 38.09 \dfrac{\text{BTU}}{\text{lbm}} + \dfrac{39.87 \frac{\text{BTU}}{\text{lbm}} - 38.09 \frac{\text{BTU}}{\text{lbm}}}{0.96}$

$\quad = 39.94 \text{ BTU/lbm}$

THERMO
Vapor Cycles

$$\eta_{\text{th}} = \frac{(h_d - h'_a) + (h_f - h'_e) - (h'_g - h_h)}{(h_d - h'_a) + (h_f - h'_e)}$$

$$= \frac{(1350.6 - 39.94) + (1373.8 - 1248.9) - (967.3 - 38.09)}{(1350.6 - 39.94) + (1373.8 - 1248.9)}$$

$$= \boxed{0.353 \ (35.3\%)}$$

13.

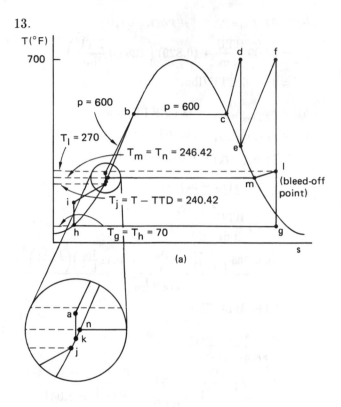

(a)

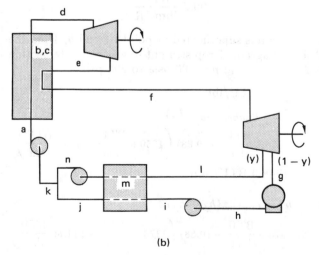

(b)

Start at f. From superheated steam tables at $T_f = 700°F$ and $p_f = 200 \ \text{lbf/in}^2$, $h_f = 1373.8 \ \text{BTU/lbm}$, $s_f = 1.7234 \ \text{BTU/lbm-°R}$.

Assume isentropic expansion from f to l (where bleed occurs).

From the Mollier diagram at $T_l = 270°F$ and $s_l = s_f$,

$$h_l \approx 1175 \ \text{BTU/lbm}$$

$$p_l = p_{\text{bleed}} \approx 28 \ \text{lbf/in}^2$$

Fraction y undergoes isobaric cooling to a saturated liquid from l to m.

$$p_m = p_l$$

$$T_m = T_{\text{sat}(p_{\text{sat}}=p_m)} = 246.42°F$$

$$h_m = h_{G(p_{\text{sat}}=p_m)} = 1163.0 \ \text{BTU/lbm}$$

Fraction y adiabatically condenses in the closed feedwater heater from m to n.

$$h_n = h_{F(p_{\text{sat}}=p_m)} = 214.95 \ \text{BTU/lbm}$$

Fraction $(1 - y)$ undergoes isentropic expansion from l to g. From problem 12,

$$h_g = 911.86 \ \text{BTU/lbm}$$

Fraction $(1 - y)$ undergoes adiabatic condensing from g to h. From problem 12,

$$h_h = 38.09 \ \text{BTU/lbm}$$

$$v_h = 0.01605 \ \text{ft}^3/\text{lbm}$$

$$p_h = 0.3632 \ \text{lbf/in}^2$$

From h to i, $(1 - y)$, as saturated fluid, is pumped to $p_i = p_j = p_{\text{sat}(T=T_j)}$ before entering the closed feedwater heater where the temperature is raised to T_j from i to j. Apply closed feedwater heater equations.

$$\text{TTD} = T_{\text{sat}(p=p_{\text{bleed}} \text{ for steam})} - T_{\text{(feedwater on exit)}}$$

$$= T_{\text{(steam on exit)}} - T_{\text{(feedwater on exit)}}$$

$$= T_m - T_j$$

$$T_j = T_m - \text{TTD} = 246.42°F - 6°F$$

$$= 240.42°F \quad (\text{say } 240°F)$$

$$h_j = h_{F(T_{\text{sat}}=T_j)} = 208.4 \ \text{BTU/lbm}$$

$$p_j = p_{\text{sat}(T_{\text{sat}}=T_j)} = 24.97 \ \text{lbf/in}^2$$

$$p_i = p_j$$

$$h_i = h_h + \frac{v_h(p_i - p_h)}{J}$$

$$= 38.09 \ \frac{\text{BTU}}{\text{lbm}}$$

$$+ \frac{\left(0.01605 \ \frac{\text{ft}^3}{\text{lbm}}\right)\left(24.97 \ \frac{\text{lbf}}{\text{in}^2} - 0.3632 \ \frac{\text{lbf}}{\text{in}^2}\right)\left(\frac{12 \ \text{in}}{\text{in}}\right)^2}{778.2 \ \frac{\text{ft-lbf}}{\text{BTU}}}$$

$$= 38.16 \ \text{BTU/lbm}$$

Heat balance for the closed feedwater heater is

$$yh_l + (1-y)h_i + W_{\text{drip pump}} = yh_n + (1-y)h_j$$

Neglecting the drip pump work,

$$y = \frac{h_j - h_i}{h_l - h_i - h_n + h_j}$$

$$= \frac{208.4 - 38.16}{1175 - 38.16 - 214.95 + 208.4}$$

$$= 0.1506$$

Mixing the products of the closed feedwater heater,

$$h_k = (1-y)h_j + yh_n$$

$$= (1 - 0.1506)\left(208.4\ \frac{\text{BTU}}{\text{lbm}}\right)$$

$$\quad + (0.1506)\left(214.95\ \frac{\text{BTU}}{\text{lbm}}\right)$$

$$= 209.39\ \text{BTU/lbm}$$

Since the water at k is saturated or very slightly sub-cooled, find a point in the steam table where $h_F = h_k$. This occurs at approximately $p_{\text{sat}} = 25$ psia.

$$T_k = T_{\text{sat}} = 240.08°\text{F}$$

$$p_k = p_{\text{sat}} = 25\ \text{lbf/in}^2$$

$$v_k = v_F = 0.01692\ \text{ft}^3/\text{lbm}$$

Assuming isentropic compression from k to a,

$$h_a = h_k + \frac{v_k(p_a - p_k)}{J}$$

$$= 209.39\ \frac{\text{BTU}}{\text{lbm}}$$

$$\quad + \frac{\left(0.01692\ \frac{\text{ft}^3}{\text{lbm}}\right)\left(600\ \frac{\text{lbf}}{\text{in}^2} - 25\ \frac{\text{lbf}}{\text{in}^2}\right)\left(\frac{12\ \text{in}}{\text{ft}}\right)^2}{778.2\ \frac{\text{ft-lbf}}{\text{BTU}}}$$

$$= 211.19\ \text{BTU/lbm}$$

From problem 12,

$$h_d = 1350.6\ \text{BTU/lbm}$$

$$h_f = 1373.8\ \text{BTU/lbm}$$

$$h'_e = 1248.9\ \text{BTU/lbm}$$

However, the pump and turbine are not isentropic.

$$h'_l = h_f - \eta_{\text{turbine}}(h_f - h_l)$$

$$= 1373.8\ \frac{\text{BTU}}{\text{lbm}} - (0.88)\left(1373.8\ \frac{\text{BTU}}{\text{lbm}} - 1175\ \frac{\text{BTU}}{\text{lbm}}\right)$$

$$= 1198.86\ \text{BTU/lbm}$$

$$h'_g = h_f - \eta_{\text{turbine}}(h_f - h_g)$$

$$= 1373.8\ \frac{\text{BTU}}{\text{lbm}} - (0.88)\left(1373.8\ \frac{\text{BTU}}{\text{lbm}} - 911.86\ \frac{\text{BTU}}{\text{lbm}}\right)$$

$$= 967.29\ \text{BTU/lbm}$$

$$h'_i = h_h + \frac{h_i - h_h}{\eta_{\text{pump}}}$$

$$= 38.09\ \frac{\text{BTU}}{\text{lbm}} + \frac{38.16\ \frac{\text{BTU}}{\text{lbm}} - 38.09\ \frac{\text{BTU}}{\text{lbm}}}{0.96}$$

$$= 38.16\ \text{BTU/lbm}$$

$$h'_a = h_k + \frac{h_a - h_k}{\eta_{\text{pump}}}$$

$$= 209.39\ \frac{\text{BTU}}{\text{lbm}} + \frac{211.19\ \frac{\text{BTU}}{\text{lbm}} - 209.39\ \frac{\text{BTU}}{\text{lbm}}}{0.96}$$

$$= 211.27\ \text{BTU/lbm}$$

$$\eta_{\text{th}} = \frac{W_{\text{out}} - W_{\text{in}}}{Q_{\text{in}}}$$

$$= \frac{\begin{array}{c}[(h_d - h'_e) + (h_f - h'_l) + (1-y)(h'_l - h'_g)]\\ -[(1-y)(h'_i - h_h) + (h'_a - h_k)]\end{array}}{(h_d - h'_a) + (h_f - h'_e)} =$$

$$\frac{\begin{array}{c}[(1350.6-1248.9)+(1373.8-1198.86)+(1-0.1506)(1198.86-967.29)]\\ -[(1-0.1506)(38.16-38.09)+(211.19-209.39)]\end{array}}{(1350.6-211.19)+(1373.8-1248.9)}$$

$$= \boxed{0.373\ (37.3\%)}$$

14.

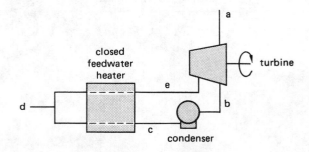

(Pumps are left out because work performed is small.)

For superheated steam at $T_a = 1100°\text{F}$,

$$p_a = 1200\ \text{lbf/in}^2$$

$$h_a = 1557.9\ \text{BTU/lbm}$$

$$s_a = 1.6682\ \text{BTU/lbm-}°\text{R}$$

For superheated steam at $T_e = 900°\text{F}$,

$$s_e = s_a$$

$$p_e = 644.2\ \text{lbf/in}^2$$

$$h_e = 1461.3\ \text{BTU/lbm}$$

For saturated liquid at $T_F = T_{\text{sat},p_e}$,

$$h_f = 480.5\ \text{BTU/lbm}$$

The vapor-liquid mixture is at

$$p_b = (1.5 \text{ in Hg}) \left(0.49116 \frac{\frac{\text{lbf}}{\text{in}^2}}{\text{in Hg}} \right) = 0.7367 \text{ lbf/in}^2$$

$$s_b = s_a$$

$$x_b = \frac{s_b - s_{F(p_{\text{sat}} = p_b)}}{s_{FG(p_{\text{sat}} = p_b)}}$$

$$= \frac{1.6682 \frac{\text{BTU}}{\text{lbm-}^\circ\text{R}} - 0.1143 \frac{\text{BTU}}{\text{lbm-}^\circ\text{R}}}{1890 \frac{\text{BTU}}{\text{lbm-}^\circ\text{R}}}$$

$$= 0.8222$$

$$h_b = h_{F(p_{\text{sat}} = p_b)} + x_b h_{FG(p_{\text{sat}} = p_b)}$$

$$= 59.56 \frac{\text{BTU}}{\text{lbm}} + (0.8222) \left(1041.82 \frac{\text{BTU}}{\text{lbm}} \right)$$

$$= 916.14 \text{ BTU/lbm}$$

The Mollier diagram can also be used to determine h_b.

$$h_c = h_{F(p_{\text{sat}} = p_b)} = 59.56 \text{ BTU/lbm}$$
$$h_d = h_{F(T_{\text{sat}} = T_d = 180^\circ\text{F})}$$
$$= 147.99 \text{ BTU/lbm}$$

The heat balance for the heater (neglecting the small amount of pump work) is

$$yh_e + (1 - y)h_c = h_d$$

$$y = \frac{h_d - h_c}{h_e - h_c}$$

$$= \frac{147.99 \frac{\text{BTU}}{\text{lbm}} - 59.56 \frac{\text{BTU}}{\text{lbm}}}{1461.3 \frac{\text{BTU}}{\text{lbm}} - 59.56 \frac{\text{BTU}}{\text{lbm}}} = 0.0631$$

$$\eta_{\text{th}} = \frac{W_{\text{out}}}{Q_{\text{in}}} = \frac{y(h_a - h_e) + (1 - y)(h_a - h_b)}{h_a - h_d}$$

$$= \frac{(0.0631)(1557.9 - 1461.3) + (1 - 0.0631)(1557.9 - 916.11)}{1557.9 - 147.99}$$

$$= \boxed{0.431 \ (43.1\%)}$$

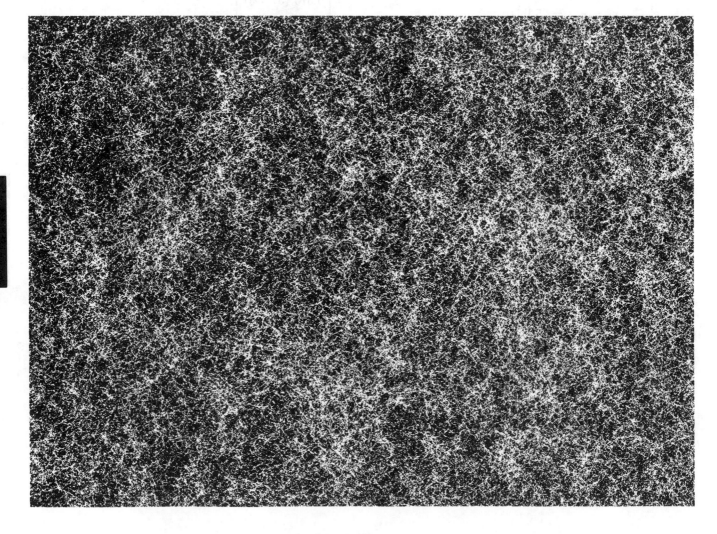

COMBUSTION POWER CYCLES

1. (a) $T_{\text{high}} = (360 + 460)^\circ\text{R} = 820^\circ\text{R}$

   $T_{\text{low}} = (100 + 460)^\circ\text{R} = 560^\circ\text{R}$

   $$\eta_{\text{th}} = \frac{T_{\text{high}} - T_{\text{low}}}{T_{\text{high}}} = \frac{820^\circ\text{R} - 560^\circ\text{R}}{820^\circ\text{R}} = 0.317$$

   $$\eta_{\text{th}} = \frac{Q_{\text{in}} - Q_{\text{out}}}{Q_{\text{in}}}$$

   $$Q_{\text{out}} = Q_{\text{in}}(1 - \eta_{\text{th}}) = \left(100\,\frac{\text{BTU}}{\text{lbm}}\right)(1 - 0.317)$$

   $$= 68.30\ \text{BTU/lbm}$$

   $$W_{a \to b} = c_{\text{v}}(T_a - T_b)$$

   $$= \left(0.171\,\frac{\text{BTU}}{\text{lbm-}^\circ\text{R}}\right)(560^\circ\text{R} - 820^\circ\text{R})$$

   $$= -44.46\ \text{BTU/lbm}$$

   $W_{b \to c} = Q_{b \to c} = 100\ \text{BTU/lbm (given)}$

   $W_{c \to d} = c_{\text{v}}(T_c - T_d)$

   $$= \left(0.171\,\frac{\text{BTU}}{\text{lbm-}^\circ\text{R}}\right)(820^\circ\text{R} - 560^\circ\text{R})$$

   $$= 44.46\ \text{BTU/lbm}$$

   $W_{d \to a} = Q_{d \to a} = -Q_{\text{out}}$

   $$= -68.30\ \text{BTU/lbm}$$

   $$W_{\text{expansion}} = W_{\text{out}} = W_{b \to c} + W_{c \to d}$$

   $$= 100\,\frac{\text{BTU}}{\text{lbm}} + 44.46\,\frac{\text{BTU}}{\text{lbm}}$$

   $$= \boxed{144.46\ \text{BTU/lbm}}$$

   $$W_{\text{compression}} = W_{\text{in}} = |W_{d \to a} + W_{a \to b}|$$

   $$= \left|-68.30\,\frac{\text{BTU}}{\text{lbm}} - 44.46\,\frac{\text{BTU}}{\text{lbm}}\right|$$

   $$= \boxed{112.76\ \text{BTU/lbm}}$$

   (b) $W_{\text{net}} = W_{\text{expansion}} - W_{\text{compression}}$

   $$= 144.46\,\frac{\text{BTU}}{\text{lbm}} - 112.76\,\frac{\text{BTU}}{\text{lbm}}$$

   $$= \boxed{31.70\ \text{BTU/lbm}}$$

2. (a) $T_a = (80 + 460)^\circ\text{R} = 540^\circ\text{R}$

   $$r_{\text{v}} = \frac{V_a}{V_b} = 10$$

   $$T_b = T_a\left(\frac{V_a}{V_b}\right)^{k-1} = (540^\circ\text{R})(10)^{1.4-1}$$

   $$= 1356.4^\circ\text{R}$$

   Assume ideal gas behavior.

   $$pV = mRT$$

   $$m = \frac{pV}{RT} = \frac{p_a V_a}{RT_a}$$

   $$= \frac{\left(14.3\,\frac{\text{lbf}}{\text{in}^2}\right)\left(\dfrac{12\ \text{in}}{\text{ft}}\right)(11\ \text{ft}^3)}{\left(53.35\,\dfrac{\text{ft-lbf}}{\text{lbm-}^\circ\text{R}}\right)(540^\circ\text{R})}$$

   $$= 0.7863\ \text{lbm}$$

   $$Q_{\text{in}} = mc_{\text{v}}(T_c - T_b)$$

   $$T_c = \frac{Q_{\text{in}}}{mc_{\text{v}}} + T_b$$

   $$= \frac{(160\ \text{BTU})\left(\dfrac{^\circ\text{R}}{^\circ\text{F}}\right)}{(0.7863\ \text{lbm})\left(0.1714\,\dfrac{\text{BTU}}{\text{lbm-}^\circ\text{F}}\right)} + 1356.4^\circ\text{R}$$

   $$= 2543.6^\circ\text{R} = (2543.6 - 460)^\circ\text{F}$$

   $$= \boxed{2083.6^\circ\text{F}}$$

   (b) $\eta_{\text{th}} = 1 - \dfrac{1}{(R)^{k-1}} = 1 - \dfrac{1}{(10)^{1.4-1}} = \boxed{0.602}$

3. $\eta_{\text{th}} = 1 - \dfrac{1}{(R)^{k-1}} = 1 - \dfrac{1}{(6)^{1.4-1}} = \boxed{0.5116}$

4. $A = \dfrac{\pi(\text{bore})^2}{4} = \dfrac{\pi(4\ \text{in})^2}{4} = 12.57\ \text{in}^2$

   $$N = \frac{(n)(\text{no. of cylinders})}{\left(\dfrac{\text{no. of strokes per cycle}}{2}\right)}$$

   $$= \frac{\left(3400\,\dfrac{\text{rev}}{\text{min}}\right)(6)}{\dfrac{4}{2}} = 10,200\ \text{min}^{-1}$$

THERMO
Combust Cyc

$$\text{BMEP} = \frac{\left(33,000 \ \frac{\text{ft-lbf}}{\text{min-hp}}\right)(\text{BHP})}{LAN}$$

$$= \frac{\left(33,000 \ \frac{\text{ft-lbf}}{\text{min-hp}}\right)(79.5 \ \text{hp})}{(3.125 \ \text{in})\left(\frac{\text{ft}}{12 \ \text{in}}\right)(12.57 \ \text{in}^2)\left(10,200 \ \frac{1}{\text{min}}\right)}$$

$$= \boxed{78.6 \ \text{lbf/in}^2 \ (\text{psig})}$$

$$T = \frac{\left(33,000 \ \frac{\text{ft-lbf}}{\text{min-hp}}\right)(\text{BHP})}{2\pi n}$$

$$= \frac{\left(33,000 \ \frac{\text{ft-lbf}}{\text{min-hp}}\right)(79.5 \ \text{hp})}{(2\pi)\left(3400 \ \frac{\text{rev}}{\text{min}}\right)}$$

$$= \boxed{122.81 \ \text{ft-lbf}}$$

5. $A = \dfrac{\pi(\text{bore})^2}{4} = \dfrac{\pi(10 \ \text{in})^2}{4} = 78.54 \ \text{in}^2$

$$\text{BHP} = \frac{Tn}{5252 \ \frac{\text{ft-lbf}}{\text{min-hp}}} = \frac{(600 \ \text{ft-lbf})\left(200 \ \frac{\text{rev}}{\text{min}}\right)}{5252 \ \frac{\text{ft-lbf}}{\text{min-hp}}}$$

$$= 22.85 \ \text{hp}$$

$$N = \frac{(n)(\text{no. of cylinders})}{\left(\dfrac{\text{no. of strokes per cylinder}}{2}\right)}$$

$$= \frac{\left(200 \ \frac{\text{rev}}{\text{min}}\right)(2)}{\dfrac{4}{2}} = 200 \ \text{min}^{-1}$$

$$\text{IHP} = \frac{(\text{IMEP})(LAN)}{33,000 \ \frac{\text{ft-lbf}}{\text{hp-min}}}$$

$$= \frac{(95 \ \text{psig})(18 \ \text{in})\left(\frac{\text{ft}}{12 \ \text{in}}\right)(78.54 \ \text{in}^2)\left(200 \ \frac{1}{\text{min}}\right)}{33,000 \ \frac{\text{ft-lbf}}{\text{hp-min}}}$$

$$= 67.83 \ \text{hp}$$

$$\text{FHP} = \text{IHP} - \text{BHP} = 67.83 \ \text{hp} - 22.85 \ \text{hp}$$

$$= \boxed{44.98 \ \text{hp}}$$

6. $T_a = (75 + 460)°\text{R} = 535°\text{R}$

$\quad T_b = (750 + 460)°\text{R} = 1210°\text{R}$

$\quad T_c = (2900 + 460)°\text{R} = 3360°\text{R}$

$$p_b = p_a\left(\frac{T_b}{T_a}\right)^{\frac{k}{k-1}} = \left(14.2 \ \frac{\text{lbf}}{\text{in}^2}\right)\left(\frac{1210°\text{R}}{535°\text{R}}\right)^{\frac{1.4}{1.4-1}}$$

$$= 247.1 \ \text{lbf/in}^2$$

$$p_c = p_b$$

$$v_c = \frac{RT_c}{p_c} = \frac{\left(53.35 \ \frac{\text{ft-lbf}}{\text{lbm-}°\text{R}}\right)(3360°\text{R})}{\left(247.1 \ \frac{\text{lbf}}{\text{in}^2}\right)\left(\frac{12 \ \text{in}}{\text{ft}}\right)^2}$$

$$= 5.04 \ \text{ft}^3/\text{lbm}$$

$$v_d = v_a = \frac{RT_a}{p_a}$$

$$= \frac{\left(53.35 \ \frac{\text{ft-lbf}}{\text{lbm-}°\text{R}}\right)(535°\text{R})}{\left(14.2 \ \frac{\text{lbf}}{\text{in}^2}\right)\left(\frac{12 \ \text{in}}{\text{ft}}\right)^2} = 13.96 \ \text{ft}^3/\text{lbm}$$

$$T_d = T_c\left(\frac{v_c}{v_d}\right)^{k-1} = (3360°\text{R})\left(\frac{5.04 \ \frac{\text{ft}^3}{\text{lbm}}}{13.96 \ \frac{\text{ft}^3}{\text{lbm}}}\right)^{1.4-1}$$

$$= 2235.4°\text{R}$$

$$\eta_{\text{th}} = 1 - \frac{T_d - T_a}{k(T_c - T_b)} = 1 - \frac{2235.4°\text{R} - 535°\text{R}}{(1.4)(3360°\text{R} - 1210°\text{R})}$$

$$= \boxed{0.435 \ (43.5\%)}$$

7. $T_a = (65 + 460)°\text{R} = 525°\text{R}$

$$v_a = \frac{RT_a}{p_a} = \frac{\left(53.35 \ \frac{\text{ft-lbf}}{\text{lbm-}°\text{R}}\right)(525°\text{R})}{\left(14.7 \ \frac{\text{lbf}}{\text{in}^2}\right)\left(\frac{12 \ \text{in}}{\text{ft}}\right)^2}$$

$$= 13.23 \ \text{ft}^3/\text{lbm}$$

$$r_v = \frac{v_a}{v_b} = 16$$

$$T_b = T_a\left(\frac{v_a}{v_b}\right)^{k-1} = (525°\text{R})(16)^{1.4-1} = 1591.5°\text{R}$$

$$T_c = (2600 + 460)°\text{R} = 3060°\text{R}$$

$$v_c = v_b\left(\frac{T_c}{T_b}\right) = \left(\frac{v_a}{r_v}\right)\left(\frac{T_c}{T_b}\right)$$

$$= \left(\frac{13.23 \ \frac{\text{ft}^3}{\text{lbm}}}{16}\right)\left(\frac{3060°\text{R}}{1591.5°\text{R}}\right) = 1.59 \ \text{ft}^3/\text{lbm}$$

$$v_d = v_a$$

THERMO
Combust Cyc

$$T_d = T_c \left(\frac{v_c}{v_d}\right)^{k-1}$$

$$= (3060°\text{R}) \left(\frac{1.59 \frac{\text{ft}^3}{\text{lbm}}}{13.23 \frac{\text{ft}^3}{\text{lbm}}}\right)^{1.4-1} = 1311.16°\text{R}$$

$$\eta_{\text{th}} = 1 - \frac{T_d - T_a}{k(T_c - T_b)}$$

$$= 1 - \frac{1311.16°\text{R} - 525°\text{R}}{(1.4)(3060°\text{R} - 1591.5°\text{R})}$$

$$= \boxed{0.618 \ (61.8\%)}$$

8. Because $\eta_{m,1} = \eta_{m,2}$, frictional losses cannot be constant and do not need to be calculated. Some steps are therefore unnecessary.

step 1:

    altitude 1: $p_1 = 14.7 \text{ psia}, \ T = 60°\text{F}$

    altitude 2: $z = 5000 \text{ ft}$

step 2:

$$\text{IHP}_1 = \frac{\text{BHP}_1}{\eta_{m,1}} = \frac{1000 \text{ hp}}{0.80} = 1250 \text{ hp}$$

step 4:

$$\rho_{a1} = \frac{p_1}{RT_1} = \frac{\left(14.7 \frac{\text{lbf}}{\text{in}^2}\right)\left(\frac{12 \text{ in}}{\text{ft}}\right)^2}{\left(53.35 \frac{\text{ft-lbf}}{\text{lbm-}°\text{R}}\right)(60 + 460)°\text{R}}$$

$$= 0.0763 \text{ lbm/ft}^3$$

Using altitude tables,

$$\rho_{a2} = \frac{p_2}{RT_2} = \frac{\left(12.225 \frac{\text{lbf}}{\text{in}^2}\right)\left(\frac{12 \text{ in}}{\text{ft}}\right)^2}{\left(53.35 \frac{\text{ft-lbf}}{\text{lbm-}°\text{R}}\right)(500.9°\text{R})}$$

$$= 0.0659 \text{ lbm/ft}^3$$

step 5:

$$\text{IHP}_2 = (\text{IHP}_1)\left(\frac{\rho_{a2}}{\rho_{a1}}\right)$$

$$= (1250 \text{ hp})\left(\frac{0.0659 \frac{\text{lbm}}{\text{ft}^3}}{0.0763 \frac{\text{lbm}}{\text{ft}^3}}\right) = 1079.6 \text{ hp}$$

step 7:

$$\text{BHP}_2 = (\text{IHP}_2)\eta_{m,2}$$

$$= (1079.6 \text{ hp})(0.80) = \boxed{863.68 \text{ hp}}$$

$$\dot{m}_{f1} = (\text{BSFC}_1)(\text{BHP}_1) = \left(0.45 \frac{\text{lbm}}{\text{hp-hr}}\right)(1000 \text{ hp})$$

$$= 450 \text{ lbm/hr}$$

steps 8,9:

$$\dot{V}_{a2} = \dot{V}_{a1} = \frac{\dot{m}_{a1}}{\rho_{a1}} = \frac{(R_{a/f})(\dot{m}_{f1})}{\rho_{a1}} = \frac{(23)\left(450 \frac{\text{lbm}}{\text{hr}}\right)}{0.0763 \frac{\text{lbm}}{\text{ft}^3}}$$

$$= 1.356 \times 10^5 \text{ ft}^3/\text{hr}$$

step 10:

$$\dot{m}_{a2} = \dot{V}_{a2}\rho_{a2} = \left(1.356 \times 10^5 \frac{\text{ft}^3}{\text{hr}}\right)\left(0.0659 \frac{\text{lbm}}{\text{ft}^3}\right)$$

$$= 8936.0 \text{ lbm/hr}$$

step 11: For a Diesel engine, $\dot{m}_{f2} = \dot{m}_{f1}$.

step 12:

$$\text{BSFC}_2 = \frac{\dot{m}_{f2}}{\text{BHP}_2} = \frac{\dot{m}_{f1}}{\text{BHP}_2} = \frac{450 \frac{\text{lbm}}{\text{hr}}}{863.68 \text{ hp}}$$

$$= \boxed{0.5210 \text{ lbm/hp-hr}}$$

9. Of the two common methods for calculating thermal efficiency, the one based on enthalpies (rather than temperatures) is more accurate for air since $c_{p,\text{air}}$ is strictly non-constant; air is not an ideal gas. Therefore, enthalpies taken from air tables should be used for greater accuracy.

Intake at a:

$$T_a = (60 + 460)°\text{R} = 520°\text{R}$$
$$p_a = 14.7 \text{ psia}$$
$$r_v = 5$$

This is low pressure (< 300 psia), so from air tables,

$$h_a = 124.27 \text{ BTU/lbm}$$

If compression from a to b had occurred isentropically,

$$T_b = T_a \left(\frac{v_a}{v_b}\right)^{k-1} = T_a r_v^{k-1}$$

$$= (520°\text{R})(5)^{1.4-1} = 989.9°\text{R}$$

$$p_b = p_a \left(\frac{v_a}{v_b}\right)^k = p_a r_v^k = \left(14.7 \frac{\text{lbf}}{\text{in}^2}\right)(5)^{1.4}$$
$$= 139.9 \text{ lbf/in}^2$$

This is still low pressure, so from air tables,

$$h_b = 238.5 \text{ BTU/lbm}$$

The actual enthalpy, h_b', is calculated from the isentropic efficiency.

$$h_b' = h_a + \frac{h_b - h_a}{\eta_{\text{compressor}}}$$
$$= 124.27 \frac{\text{BTU}}{\text{lbm}} + \frac{238.5 \frac{\text{BTU}}{\text{lbm}} - 124.27 \frac{\text{BTU}}{\text{lbm}}}{0.83}$$
$$= 261.90 \text{ BTU/lbm}$$

Air enters the turbine at c:

$$T_c = (1500 + 460)°\text{R} = 1960°\text{R}$$

From air tables,

$$h_c = 493.65 \text{ BTU/lbm}$$

Constant pressure heating from b to c:

$$p_c = p_b \text{ and } p_d = p_a$$

($p_d = p_a$ because in an actual Brayton cycle, the air does not recirculate.)

If turbine work from c to d had been performed isentropically,

$$T_d = T_c \left(\frac{p_d}{p_c}\right)^{\frac{k-1}{k}} = T_c \left(\frac{p_a}{p_b}\right)^{\frac{k-1}{k}}$$
$$= (1960°\text{R}) \left(\frac{14.7 \frac{\text{lbf}}{\text{in}^2}}{139.9 \frac{\text{lbf}}{\text{in}^2}}\right)^{\frac{1.4-1}{1.4}} = 1029.6°\text{R}$$

From air tables,

$$h_d = 248.36 \text{ BTU/lbm}$$

The actual enthalpy is

$$h_d' = h_c - \eta_{\text{turbine}}(h_c - h_d)$$
$$= 493.65 \frac{\text{BTU}}{\text{lbm}} - (0.92)(493.65 \frac{\text{BTU}}{\text{lbm}} - 248.36 \frac{\text{BTU}}{\text{lbm}})$$
$$= 267.98 \text{ BTU/lbm}$$

$$\eta_{\text{th}} = \frac{(h_c - h_b') - (h_d' - h_a)}{h_c - h_b'}$$
$$= \frac{(493.65 - 261.9) - (267.98 - 124.27)}{493.65 - 261.9}$$
$$= \boxed{0.380 \ (38.0\%)}$$

10. A regenerator reduces Q_{in} by recirculating some Q_{out}. All properties at the turbine and compressor entrance and exit points are the same.

Note the changes in nomenclature:

new	old	
h_a	h_a at compressor entrance	= 124.27 BTU/lbm
h_b'	h_b' at compressor exit	= 261.90 BTU/lbm
h_d	h_c at turbine entrance	= 493.65 BTU/lbm
h_e'	h_d' at turbine exit	= 267.98 BTU/lbm

$$\eta_{\text{regenerator}} = \frac{h_c - h_b'}{h_e' - h_b'}$$

$$h_c = (h_e' - h_b')\eta_{\text{regenerator}} + h_b'$$
$$= \left(267.98 \frac{\text{BTU}}{\text{lbm}} - 261.9 \frac{\text{BTU}}{\text{lbm}}\right)(0.65) + 261.9 \frac{\text{BTU}}{\text{lbm}}$$
$$= 265.85 \text{ BTU/lbm}$$

$$\eta_{\text{th}} = \frac{(h_d - h_e') - (h_b' - h_a)}{h_d - h_c}$$
$$= \frac{(493.65 - 267.98) - (261.90 - 124.27)}{493.65 - 265.85}$$
$$= \boxed{(0.386) \ 38.6\%}$$

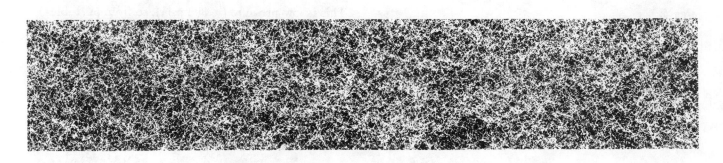

THERMO
Combust Cyc

REFRIGERATION CYCLES

1. $$\text{COP} = \frac{T_{\text{low}}}{T_{\text{high}} - T_{\text{low}}} = \frac{(460 + 10)°\text{R}}{(110°\text{F} - 10°\text{F})\left(\frac{°\text{R}}{°\text{F}}\right)}$$

    $$= \boxed{4.7}$$

    $$W_{\text{in}} = \frac{Q_{\text{in}}}{\text{COP}} = \frac{1000\,\dfrac{\text{BTU}}{\text{hr}}}{4.7}$$

    $$= \boxed{212.77\ \text{BTU/hr}}$$

    $$Q_{\text{out}} = Q_{\text{in}} + W_{\text{in}} = 1000\,\frac{\text{BTU}}{\text{hr}} + 212.77\,\frac{\text{BTU}}{\text{hr}}$$

    $$= \boxed{1212.77\ \text{BTU/hr}}$$

2. $$\text{COP} = \frac{T_{\text{high}}}{T_{\text{high}} - T_{\text{low}}} = \frac{(460 + 700)°\text{R}}{(700°\text{F} - 40°\text{F})\left(\frac{°\text{R}}{°\text{F}}\right)}$$

    $$= \boxed{1.76}$$

3. $$\dot{m} = (\dot{V})(\rho) = \left[\left(100\,\frac{\text{gal}}{\text{min}}\right)\left(0.1337\,\frac{\text{ft}^3}{\text{gal}}\right)\right]$$

    $$\times \left(62.4\,\frac{\text{lbm}}{\text{ft}^3}\right)$$

    $$= 834.5\ \text{lbm/min}$$

    $$\dot{Q}_{\text{in}} = \dot{m}c_p\Delta T$$

    $$= \left(834.5\,\frac{\text{lbm}}{\text{min}}\right)\left(1\,\frac{\text{BTU}}{\text{lbm -}°\text{F}}\right)(80°\text{F} - 20°\text{F})$$

    $$= 50{,}070\ \text{BTU/min}$$

    $$\text{COP} = \frac{T_{\text{low}}}{T_{\text{high}} - T_{\text{low}}} = \frac{(460 + 20)°\text{R}}{(80°\text{F} - 20°\text{F})\left(\frac{°\text{R}}{°\text{F}}\right)}$$

    $$= 8$$

    $$W_{\text{in,hp}} = \frac{4.715\,Q_{\text{in,tons}}}{\text{COP}}$$

    $$= \frac{(4.715)\left(\dfrac{50{,}070\,\dfrac{\text{BTU}}{\text{min}}}{200\,\dfrac{\text{BTU}}{\text{min-ton}}}\right)}{8}$$

    $$= \boxed{147.5\ \text{hp}}$$

4. At a: $p_a = 160$ psia

    Interpolating between 140 psia and 170 psia,

    $$\frac{h_a - 126}{139.3 - 126} = \frac{160 - 140}{170 - 140}$$

    $$h_a = 134.9\ \text{BTU/lbm}$$

    At b: $h_b = h_a = 134.9$ BTU/lbm

    At c: $h_c = h_b + q_{\text{in}} = 134.9\,\dfrac{\text{BTU}}{\text{lbm}} + 500\,\dfrac{\text{BTU}}{\text{lbm}}$

    $$= 634.9\ \text{BTU/lbm}$$

    From superheated ammonia table,

    $$s_c = 1.3845\ \text{BTU/lbm-}°\text{F}$$

    At d: $s_d = s_c = 1.3845$ BTU/lbm-°F

    $$p_d = p_a = 160\ \text{psia}$$

    Interpolating between 250°F and 300°F,

    $$\frac{h_d - 737.6}{1.3845 - 1.3675} = \frac{767.1 - 737.6}{1.4076 - 1.3675}$$

    $$h_d = 750.1\ \text{BTU/lbm}$$

    $$\text{COP} = \frac{q_{\text{out}}}{W_{\text{in}}} = \frac{h_c - h_b}{h_d - h_c} + 1$$

    $$= \frac{634.9\,\dfrac{\text{BTU}}{\text{lbm}} - 134.9\,\dfrac{\text{BTU}}{\text{lbm}}}{750.1\,\dfrac{\text{BTU}}{\text{lbm}} - 634.9\,\dfrac{\text{BTU}}{\text{lbm}}} + 1$$

    $$= \boxed{5.34}$$

5. At a: $T_a = 70°$F from saturated Freon-12 table

    $$h_a = 23.9\ \text{BTU/lbm}$$

    At b: $h_b = h_a = 23.9$ BTU/lbm

    At c: $T_c = T_b = -30°$F

    $$h_c = 74.7\ \text{BTU/lbm}$$

    $$v_c = 3.088\ \text{ft}^3/\text{lbm}$$

    $$q_{\text{in}} = h_c - h_b = 74.7\,\frac{\text{BTU}}{\text{lbm}} - 23.9\,\frac{\text{BTU}}{\text{lbm}}$$

    $$= 50.8\ \text{BTU/lbm}$$

    $$\dot{m} = \frac{200\,\dfrac{\text{BTU}}{\text{min-ton}}}{50.8\,\dfrac{\text{BTU}}{\text{lbm}}}$$

    $$= 3.94\ \text{lbm/min-ton}$$

$$\dot{V} = \dot{m}v_c = \left(3.94 \, \frac{\text{lbm}}{\text{min-ton}}\right)\left(3.088 \, \frac{\text{ft}^3}{\text{lbm}}\right)$$

$$= \boxed{12.16 \, \text{ft}^3/\text{min-ton}}$$

6. (a) Assuming ideal gas,

$$\frac{T_d}{T_c} = \left(\frac{p_{\text{high}}}{p_{\text{low}}}\right)^{\frac{k-1}{k}}$$

For air, $k = 1.4$.

$$T_d = T_c\left(\frac{60 \, \text{psia}}{14.7 \, \text{psia}}\right)^{\frac{1.4-1}{1.4}}$$

$$= (460 + 70)^{\circ}\text{R} \, (1.495)$$

$$= 792.1^{\circ}\text{R}$$

Temperature leaving compressor if process is not isentropic:

$$T_d' = T_c + \frac{T_d - T_c}{\eta_{\text{compressor}}}$$

$$= 530\,^{\circ}\text{R} + \frac{(792.1^{\circ}\text{R} - 530^{\circ}\text{R})}{0.7}$$

$$= \boxed{904.5^{\circ}\text{R} \, (444.5^{\circ}\text{F})}$$

$$\frac{T_a}{T_b} = \left(\frac{p_{\text{high}}}{p_{\text{low}}}\right)^{\frac{k-1}{k}}$$

$$T_b = \frac{T_a}{\left(\dfrac{60 \, \text{psia}}{14.7 \, \text{psia}}\right)^{\frac{1.4-1}{1.4}}} = \frac{(460 + 25)^{\circ}\text{R}}{1.495}$$

$$= 324.4^{\circ}\text{R}$$

Temperature leaving the turbine if process is not isentropic:

$$T_b' = T_a - \eta_{\text{turbine}}(T_a - T_b)$$

$$= 485^{\circ}\text{R} - (0.80)(485^{\circ}\text{R} - 324.4^{\circ}\text{R})$$

$$= \boxed{356.5^{\circ}\text{R} \, (-103.5^{\circ}\text{F})}$$

(b) $\text{COP} = \dfrac{T_c - T_b'}{(T_d' - T_a) - (T_c - T_b')}$

$$= \frac{530^{\circ}\text{R} - 356.5^{\circ}\text{R}}{(904.5^{\circ}\text{R} - 485^{\circ}\text{R}) - (530^{\circ}\text{R} - 356.5^{\circ}\text{R})}$$

$$= \boxed{0.705}$$

THERMO
Refrig Cyc

GAS COMPRESSION PROCESSES

1. Assume intake pressure $p_b = 14.7$ psia.

$$\eta_v = 1 - \left[\left(\frac{p_c}{p_b} \right)^{\frac{1}{n}} - 1 \right] \left(\frac{c}{100} \right)$$

$$= 1 - \left[\left(\frac{65\,\text{psia}}{14.7\,\text{psia}} \right)^{\frac{1}{1.33}} - 1 \right] \left(\frac{7}{100} \right)$$

$$= 0.856$$

$$\frac{\text{mass of air in swept volume}}{\text{min}} = \frac{48\,\dfrac{\text{lbm}}{\text{min}}}{0.856}$$

$$= 56.07\,\text{lbm}/\text{min}$$

$$\frac{\text{mass of air compressed}}{\text{min}} = (1 + 0.07) \left(56.07\,\frac{\text{lbm}}{\text{min}} \right)$$

$$= \boxed{60.0\,\text{lbm/min}}$$

2. (a) $\rho = \dfrac{p}{RT} = \dfrac{(14.5\,\text{psia}) \left(144\,\dfrac{\text{in}^2}{\text{ft}^2} \right)}{\left(53.3\,\dfrac{\text{ft-lbf}}{\text{lbm-}^\circ\text{R}} \right) (460 + 70)\,^\circ\text{R}}$

$$= 0.0739\,\text{lbm/ft}^3$$

$$\dot{m} = \rho \dot{V} = \left(0.0739\,\frac{\text{lbm}}{\text{ft}^3} \right) \left(5000\,\frac{\text{ft}^3}{\text{hr}} \right) \left(\frac{\text{hr}}{3600\,\text{sec}} \right)$$

$$= 0.1026\,\text{lbm/sec}$$

$$W = mRT_1 \ln \left(\frac{p_1}{p_2} \right)$$

$$P = \frac{W}{t} = \dot{m} R T_1 \ln \left(\frac{p_1}{p_2} \right)$$

$$= \left(0.1026\,\frac{\text{lbm}}{\text{sec}} \right) \left(53.3\,\frac{\text{ft-lbf}}{\text{lbm-}^\circ\text{R}} \right) (530\,^\circ\text{R})$$

$$\times \left[\ln \left(\frac{14.5\,\text{psia}}{100\,\text{psia}} \right) \right]$$

$$= \left(5596.77\,\frac{\text{ft-lbf}}{\text{sec}} \right) \left(\frac{\text{hp}}{550\,\dfrac{\text{ft-lbf}}{\text{sec}}} \right)$$

$$= \boxed{10.18\,\text{hp}}$$

(b)
$$W = \left(\frac{p_1 v_1}{k - 1} \right) \left[1 - \left(\frac{p_2}{p_1} \right)^{\frac{k-1}{k}} \right]$$

$$= \left(\frac{mRT_1}{k - 1} \right) \left[1 - \left(\frac{p_2}{p_1} \right)^{\frac{k-1}{k}} \right]$$

$$P = \frac{W}{t} = \left(\frac{\dot{m} R T_1}{k - 1} \right) \left[1 - \left(\frac{p_2}{p_1} \right)^{\frac{k-1}{k}} \right]$$

$$= \left[\frac{\left(0.1026\,\dfrac{\text{lbm}}{\text{sec}} \right) \left(53.3\,\dfrac{\text{ft-lbf}}{\text{lbm-}^\circ\text{R}} \right) (530\,^\circ\text{R})}{1.4 - 1} \right]$$

$$\times \left[1 - \left(\frac{100\,\text{psia}}{14.5\,\text{psia}} \right)^{\frac{1.4-1}{1.4}} \right]$$

$$= \left(5334.68\,\frac{\text{ft-lbf}}{\text{sec}} \right) \left(\frac{\text{hp}}{550\,\dfrac{\text{ft-lbf}}{\text{sec}}} \right)$$

$$= \boxed{9.7\,\text{hp}}$$

3. (a) From the air table at $(500 + 460)\,^\circ\text{R} = 960\,^\circ\text{R}$,

$$p_{r_1} = 10.61$$

$$h_1 = 231.06\,\frac{\text{BTU}}{\text{lbm}}$$

$$p_{r_2} = \left(\frac{6}{1} \right) (p_{r_1})$$

$$= (6)(10.61) = 63.66$$

For an isentropic compression to p_{r_2}, final properties are (by interpolation),

$$T_2 = 1551\,^\circ\text{R} \qquad h_2 = 383\,\frac{\text{BTU}}{\text{lbm}}$$

$$W = \frac{h_2 - h_1}{\eta}$$

$$= \frac{383\,\dfrac{\text{BTU}}{\text{lbm}} - 231.06\,\dfrac{\text{BTU}}{\text{lbm}}}{0.65}$$

$$= \boxed{233.75\,\text{BTU/lbm}}$$

(b) Enthalpy at the outlet is

$$h_2' = h_1 + W = 231.06 \, \frac{\text{BTU}}{\text{lbm}} + 233.75 \, \frac{\text{BTU}}{\text{lbm}}$$
$$= 464.81 \, \text{BTU/lbm}$$

Obtain the actual final temperature by interpolation at $h_2' = 464.81 \, \text{BTU/lbm}$,

$$T_2' = \boxed{1855°\text{R}}$$

(c) Also by interpolation, at $T_1 = 960°\text{R}$ and $p_1 = 1 \, \text{atm}$,

$$\phi_1 = 0.74030 \, \frac{\text{BTU}}{\text{lbm-°R}}$$

At T_2' and $p_2 = 6 \, \text{atm}$,

$$\phi_2' = 0.91132 \, \frac{\text{BTU}}{\text{lbm-°R}}$$

$$s_2 - s_1 = \phi_2' - \phi_1 - R \ln\left(\frac{p_2}{p_1}\right)$$
$$= 0.91132 \, \frac{\text{BTU}}{\text{lbm-°R}} - 0.74030 \, \frac{\text{BTU}}{\text{lbm-°R}}$$
$$- \left(53.3 \, \frac{\text{ft-lbf}}{\text{lbm-°R}}\right)\left(\frac{\text{BTU}}{778 \, \text{ft-lbf}}\right) \ln\left(\frac{6}{1}\right)$$
$$= \boxed{0.04827 \, \text{BTU/lbm-°R}}$$

4. From saturated steam tables at $p_{\text{sat}} = p_1 = 25 \, \text{psia}$,

$$h_1 = h_G = 1160.7 \, \frac{\text{BTU}}{\text{lbm}}$$
$$s_1 = s_G = 1.7142 \, \frac{\text{BTU}}{\text{lbm-°R}}$$

If the process had been isentropic ($s_2 = s_1$), then from the Mollier diagram at the s_2, $p_2 = 95 \, \text{psia}$ intersection,

$$h_2 = 1280 \, \frac{\text{BTU}}{\text{lbm}}$$

The actual final enthalpy is

$$h_2' = h_1 + \frac{h_2 - h_1}{\eta}$$
$$= 1160.7 \, \frac{\text{BTU}}{\text{lbm}} + \frac{1280 \, \frac{\text{BTU}}{\text{lbm}} - 1160.7 \, \frac{\text{BTU}}{\text{lbm}}}{0.70}$$
$$= \boxed{1331.1 \, \text{BTU/lbm}}$$

From superheated steam table at p_2 and h_2', the actual final temperature, entropy, and specific volume are obtained by interpolation to be

$$\boxed{T_2' = 600°\text{F}}$$

$$\boxed{s_2' = 1.77 \, \text{BTU/lbm-°R}}$$

$$\boxed{v_2' = 7.268 \, \text{ft}^3/\text{lbm}}$$

The required horsepower is

$$P = \frac{W}{t} = \dot{m}\left(h_2' - h_1\right)$$
$$= \left(200 \, \frac{\text{lbm}}{\text{min}}\right)\left(1331.1 \, \frac{\text{BTU}}{\text{lbm}} - 1160.7 \, \frac{\text{BTU}}{\text{lbm}}\right)$$
$$\times \left(\frac{60 \, \text{min}}{\text{hr}}\right)\left(\frac{\text{hp}}{2545 \, \frac{\text{BTU}}{\text{hr}}}\right)$$
$$= \boxed{804.26 \, \text{hp}}$$

INORGANIC CHEMISTRY

1. (a) Examine the previous row of the periodic table for a trend. The increase in atomic weight per element is

$$\frac{A_{83} - A_{55}}{83 - 55} = \frac{208.9\,\frac{g}{mol} - 132.9\,\frac{g}{mol}}{83 - 55}$$

$$\approx 2.71\,g/mol$$

$$A_{118} \approx 226.03\,\frac{g}{mol} + (118 - 88)\left(2.71\,\frac{g}{mol}\right)$$

$$\approx \boxed{307.33\,g/mol}$$

(Note that only non-radioactive elements were used.)

(b) Since the melting point of a given noble gas is about 40K more than the melting point of the preceding (lower atomic weight) noble gas,

$$MP_{118} \approx MP_{Rn} + 40K$$

$$= 202K + 40K = \boxed{242K}$$

(c) Similarly, the increase in boiling point is about 40K per element, so

$$BP_{118} \approx BP_{Rn} + 40K$$

$$= 211K + 40K = \boxed{251K}$$

2. Chlorine is in group VII A (the halogens), and element 119 is assumed to be in group IA (the alkali metals). Elements from these two groups form ionic compounds (e.g., LiCl, NaCl, KCl), so a hypothetical compound of chlorine and element 119 is assumed to have the characteristics of an ionic compound:

> hardness, brittleness, crystallinity, a low vapor pressure, a high melting point, nonvolatility, and electrical conductivity when dissociated.

3. Metalloids are elements with properties intermediate between metals and non-metals. They include B, Si, Ge, As, Sb, Te, and Po.

$$\boxed{(d)}$$

4. Calcium nitrate, $Ca(NO_3)_2$, has half of a mole of calcium for each mole of nitrogen.

$$m_{Ca} = n_{Ca}(A)_{Ca} = \frac{1}{2}n_N(A)_{Ca} = \frac{1}{2}\left(\frac{m_N}{(A)_N}\right)(A)_{Ca}$$

$$= \frac{1}{2}\left(\frac{20\,g}{14.007\,\frac{g}{mol}}\right)\left(40.08\,\frac{g}{mol}\right) = \boxed{28.614\,g}$$

5. $CaO + CO_2 \longrightarrow CaCO_3$

percent CaO in $CaCO_3$ = (mole fraction)(100%)

$$= \frac{(MW)_{CaO}}{(MW)_{CaCO_3}} = \frac{56.1\,\frac{g}{mol}}{100.1\,\frac{g}{mol}} \times 100\%$$

$$= \boxed{0.5604\,(56.04\,\%)}$$

6. At STP conditions, $T = 0°C = 273.15K$ and $p = 1.013 \times 10^5$ Pa. Ammonia, NH_3, boils at $-33.45K$, so it is a gas at STP. First, apply the ideal gas law for 1 mol of any gas at STP.

$$V = \frac{nR^*T}{p} = \frac{(1\,mol)\left(8.314\,\frac{J}{mol\cdot K}\right)(273.15K)}{1.013 \times 10^5\,Pa}$$

$$= (0.0224\,m^3)\left(\frac{10^3\,l}{m^3}\right) = 22.4\,l$$

Therefore, 1 mol of any ideal gas occupies 22.4 l at STP.

Assume NH_3 is an ideal gas.

$$n_{NH_3} = (125\,l)\left(\frac{1\,mol}{22.4\,l}\right) = \boxed{5.58\,mol}$$

7. Chlorine, Cl_2, boils at 239K, so it is a gas at STP. From the solution to problem 6, 1 mol of any ideal gas at STP occupies 22.4 l. Assuming Cl_2 is an ideal gas,

$$V = \left(\frac{m_{Cl_2}}{(MW)_{Cl_2}}\right)\left(\frac{22.4\,l}{mol}\right)$$

$$= \left(\frac{49\,g}{70.906\,\frac{g}{mol}}\right)\left(\frac{22.4\,l}{mol}\right) = \boxed{15.48\,l}$$

8. (a) no. of molecules

$$= \left(\frac{m_{O_2}}{(MW)_{O_2}}\right)\left(\frac{6.022 \times 10^{23} \text{ molecules}}{\text{mol}}\right)$$

$$= \left(\frac{57 \text{ g}}{32 \frac{\text{g}}{\text{mol}}}\right)\left(6.022 \times 10^{23} \frac{\text{molecules}}{\text{mol}}\right)$$

$$= \boxed{1.073 \times 10^{24} \text{ molecules}}$$

(b) Carbon dioxide, CO_2, boils at 194.7K, so it is a gas at STP.

no. of molecules

$$= n_{CO_2}\left(6.022 \times 10^{23} \frac{\text{molecules}}{\text{mol}}\right)$$

$$= V_{CO_2}\left(\frac{1 \text{ mol}}{22.4 \text{ l}}\right)\left(6.022 \times 10^{23} \frac{\text{molecules}}{\text{mol}}\right)$$

$$= (15 \text{ l})\left(\frac{1 \text{ mol}}{22.4 \text{ l}}\right)\left(6.022 \times 10^{23} \frac{\text{molecules}}{\text{mol}}\right)$$

$$= \boxed{4.033 \times 10^{23} \text{ molecules}}$$

(c) $T = 25°C = (25 + 273.15)K = 298.15K$

$$p = (770 \text{ mm Hg})\left(\frac{133.3 \text{ Pa}}{1 \text{ mm Hg}}\right)$$

$$= 1.026 \times 10^5 \text{ Pa}$$

Hydrogen, H_2, boils at $T = 20.4K$, so it is a gas at given T.

$$\text{no. of molecules} = n_{H_2}\left(6.022 \times 10^{23} \frac{\text{molecules}}{\text{mol}}\right)$$

$$= \left(\frac{pV}{R^*T}\right)\left(6.022 \times 10^{23} \frac{\text{molecules}}{\text{mol}}\right)$$

$$= \left[\frac{(1.026 \times 10^5 \text{ Pa})(9 \text{ l})\left(\frac{m^3}{10^3 \text{ l}}\right)}{\left(8.314 \frac{J}{K \cdot mol}\right)(298.15K)}\right]$$

$$\times \left(6.022 \times 10^{23} \frac{\text{molecules}}{\text{mol}}\right)$$

$$= \boxed{2.24 \times 10^{23} \text{ molecules}}$$

9. The surface water pressure is the sum of the partial pressures of the water vapor present and of the unknown gas, x.

$$p_{\text{total}} = p_{H_2O} + p_x$$

$$T = 22°C = \left[(22)\left(\frac{9}{5}\right) + 32\right]°F = 71.6°F$$

From the steam tables, the vapor pressure is

$$p_{H_2O} = 0.3631 \frac{\text{lbf}}{\text{in}^2}$$

$$= \left(0.3631 \frac{\text{lbf}}{\text{in}^2}\right)\left(\frac{51.90 \text{ mm Hg}}{\frac{\text{lbf}}{\text{in}^2}}\right)$$

$$= 18.8 \text{ mm Hg}$$

$$p_x = p_{H_2O(s)} - p_{H_2O(g)}$$

$$= 743 \text{ mm Hg} - 18.8 \text{ mm Hg}$$

$$= 724.2 \text{ mm Hg}$$

Assume gas x is ideal.

$$p_x V_x = n_x R^*T = \frac{m_x}{(MW)_x}R^*T$$

$$(MW)_x = \frac{m_x R^*T}{p_x V}$$

$$= \frac{(0.1225 \text{ g})\left(0.08206 \frac{\text{atm} \cdot \text{l}}{\text{mol} \cdot K}\right)(273 + 22)K}{(724.2 \text{ mm Hg})\left(\frac{1 \text{ atm}}{760 \text{ mm Hg}}\right)0.110 \text{ l}}$$

$$= \boxed{28.3 \text{ g/mol}}$$

10. $EW = \dfrac{MW}{\Delta \text{ oxidation number}}$

(a) $HBr \longrightarrow H^+ + Br^-$

$$(EW)_{HBr} = \frac{80.912 \frac{\text{g}}{\text{mol}}}{1} = \boxed{80.912 \text{ g/mol}}$$

(b) $H_2SO_3 \longrightarrow 2H^+ + SO_3^{2-}$

$$(EW)_{H_2SO_3} = \frac{82.076 \frac{\text{g}}{\text{mol}}}{2} = \boxed{41.038 \text{ g/mol}}$$

(c) $H_3PO_4 \longrightarrow 3H^+ + PO_4^{3-}$

$$(EW)_{H_3PO_4} = \frac{98.0 \frac{\text{g}}{\text{mol}}}{3} = \boxed{32.67 \text{ g/mol}}$$

(d) $LiOH \longrightarrow Li^+ + OH^-$

$$(EW)_{LiOH} = \frac{23.95 \frac{\text{g}}{\text{mol}}}{1} = \boxed{23.95 \text{ g/mol}}$$

CHEMISTRY
Inorganic

11. $H_4P_2O_7 + 2Na^+ \longrightarrow Na_2H_2P_2O_7 + 2H^+$

For each mol of $Na_2H_2P_2O_7$ produced, 1 mol of $H_4P_2O_7$ is needed.

$$m_{H_4P_2O_7} = n_{Na_2H_2P_2O_7}(MW)_{H_4P_2O_7}$$

$$= \left(\frac{m_{Na_2H_2P_2O_7}}{(MW)_{Na_2H_2P_2O_7}} \right)(MW)_{H_4P_2O_7}$$

$$= \left(\frac{400 \text{ g}}{221.94 \frac{\text{g}}{\text{mol}}} \right)\left(177.98 \frac{\text{g}}{\text{mol}} \right)$$

$$= 320.8 \text{ g}$$

$$(EW)_{H_4P_2O_7} = \frac{(MW)_{H_4P_2O_7}}{\Delta \text{ oxidation number}}$$

$$= \frac{177.98 \frac{\text{g}}{\text{gmol}}}{2} = 88.99 \text{ g/mol}$$

$$\text{no. of } (EW)\text{s needed} = \frac{m_{H_4P_2O_7}}{(EW)_{H_4P_2O_7}}$$

$$= \frac{320.8 \text{ g}}{88.99 \text{ g}} = \boxed{3.6}$$

12. (a) $(EW)_{Na} = \dfrac{(MW)_{Na}}{\Delta \text{ oxidation number}}$

$$= \frac{22.99 \frac{\text{g}}{\text{mol}}}{1} = \boxed{22.99 \text{ g/mol}}$$

(b) $(EW)_{Ag} = \dfrac{(MW)_{Ag}}{\Delta \text{ oxidation number}}$

$$= \frac{107.868 \frac{\text{g}}{\text{mol}}}{1} = \boxed{107.868 \text{ g/mol}}$$

(c) $(EW)_{Cu} = \dfrac{(MW)_{Cu}}{\Delta \text{ oxidation number}}$

$$= \frac{63.546 \frac{\text{g}}{\text{mol}}}{2} = \boxed{31.773 \text{ g/mol}}$$

13. (a)

	% composition	AW (g/mol)	% comp./AW (mol/g)
Cu	79.9	63.6	1.26
O	20.1	16	1.26

$$\frac{Cu}{O} = \frac{1.26 \frac{\text{mol}}{\text{g}}}{1.26 \frac{\text{mol}}{\text{g}}} = 1$$

$$\boxed{CuO}$$

(b)

	% composition	AW (g/mol)	% comp./AW (mol/g)
Fe	46.56	55.9	0.833
S	53.44	32.1	1.665

$$\frac{Fe}{S} = \frac{0.833 \frac{\text{mol}}{\text{g}}}{1.665 \frac{\text{mol}}{\text{g}}} = 0.5$$

$$\boxed{FeS_2}$$

(c)

	% composition	AW (g/mol)	% comp./AW (mol/g)
Fe	63.53	55.9	1.136
S	36.47	32.1	1.136

$$\frac{Fe}{S} = \frac{1.136 \frac{\text{mol}}{\text{g}}}{1.136 \frac{\text{mol}}{\text{g}}} = 1$$

$$\boxed{FeS}$$

(d)

	% composition	AW (g/mol)	% comp./AW (mol/g)
C	85.62	12	7.14
H	14.38	1	14.38

$$\frac{C}{H} = \frac{7.14 \frac{\text{mol}}{\text{g}}}{14.38 \frac{\text{mol}}{\text{g}}} = 0.496$$

$$\boxed{CH_2}$$

(e)

	% composition	AW (g/mol)	% comp./AW (mol/g)
C	40.0	12	3.33
H	6.7	1	6.7
O	53.3	16	3.33

$$\frac{O}{H} = \frac{C}{H} = \frac{3.33 \frac{mol}{g}}{6.7 \frac{mol}{g}} = 0.497$$

$$\boxed{CH_2O}$$

(f) Notice the components add up to 100% without the H_2O. Some of the hydrogen and oxygen must be in the form of water vapor.

After substituting in atomic weights,

$$2H_2 + O_2 \longrightarrow 2H_2O$$
$$4 + 32 \longrightarrow 36$$
or $$\frac{1}{8} + 1 \longrightarrow \frac{9}{8}$$

Each % of oxygen in the form of water takes 1/8% hydrogen. Let x be the % of oxygen locked up in the form of water.

$$x + \frac{1}{8}x = 7.14\%$$

$$x = 6.35\%$$

The nonwater oxygen is $57.11\% - 6.35\% = 50.76\%$

The nonwater hydrogen is $2.4\% - 6.35\%/8 = 1.61\%$

	% composition	AW (g/mol)	% comp./AW (mol/g)
Ca	15.89	40.1	0.4
H	1.61	1.0	1.6
P	24.6	31.0	0.8
O	50.76	16	3.2
H_2O	7.14	18	0.4

$$\frac{H}{Ca} = \frac{1.6 \frac{g}{mol}}{0.4 \frac{g}{mol}} = 4$$

$$\frac{P}{Ca} = \frac{0.8 \frac{g}{mol}}{0.4 \frac{g}{mol}} = 2$$

$$\frac{O}{Ca} = \frac{3.2 \frac{g}{mol}}{0.4 \frac{g}{mol}} = 8$$

$$\boxed{CaH_4P_2O_8 \cdot H_2O}$$

14. (a) $$\boxed{\begin{array}{c} CH_4 + 2Cl_2 \longrightarrow C + 4HCl \\ \text{single replacement (redox)} \end{array}}$$

(b) $$\boxed{\begin{array}{c} AgNO_3 + HCl \longrightarrow HNO_3 + AgCl \\ \text{double replacement} \end{array}}$$

(c) $$\boxed{\begin{array}{c} 2AsCl_3 + 3H_2S \longrightarrow As_2S_3 + 6HCl \\ \text{double replacement} \end{array}}$$

(d) $$\boxed{\begin{array}{c} 2Cu_2O + Cu_2S \longrightarrow 6Cu + SO_2 \\ \text{redox} \end{array}}$$

(e) $$\boxed{\begin{array}{c} B_2O_3 + 3Mg \longrightarrow 3MgO + 2B \\ \text{redox} \end{array}}$$

(f) $$\boxed{\begin{array}{c} BaSO_4 + 4C \longrightarrow BaS + 4CO \\ \text{redox} \end{array}}$$

(g) $$\boxed{\begin{array}{c} 3Li_2O + P_2O_5 \longrightarrow 2Li_3PO_4 \\ \text{combination} \end{array}}$$

(h) $$\boxed{\begin{array}{c} H_2SO_4 + Ba(OH)_2 \longrightarrow 2H_2O + BaSO_4 \\ \text{double displacement} \end{array}}$$

(i) $$\boxed{\begin{array}{c} 2HNO_3 + CaO \longrightarrow Ca(NO_3)_2 + H_2O \\ \text{double displacement} \end{array}}$$

(j) $$\boxed{\begin{array}{c} 2H_3PO_4 + 3MgCO_3 \longrightarrow Mg_3(PO_4)_2 \\ + 3H_2O + 3CO_2 \\ \text{double displacement with decomposition} \end{array}}$$

15. Write the oxidation numbers for all elements.

(a) $$H_2 + Cl_2 \longrightarrow 2HCl$$
$$H = 0 \quad Cl = 0 \quad H = +1; \ Cl = -1$$

H has become less negative (oxidized).

Cl has become more negative (reduced).

$$\boxed{\begin{array}{l} \text{H oxidized, Cl reduced, 1 electron} \\ \text{transferred} \end{array}}$$

(b) $Zn + H_2SO_4 \longrightarrow H_2 + ZnSO_4$
$\quad Zn = 0 \quad H = +1 \quad H = 0 \quad Zn = +2$

H has become more negative.

Zn has become less negative.

> Zn oxidized, H reduced, 1 electron transferred

(c) $2KBr + Cl_2 \longrightarrow Br_2 + 2KCl$
$\quad K = +1; \quad Cl = 0 \quad K = +1; \quad Br = 0;$
$\quad Br = -1 \quad\quad\quad\quad\quad\quad Cl = -1$

Br has become less negative.

Cl has become more negative.

> Br oxidized, Cl reduced, 1 electron transferred

16. (a) $\boxed{3Zn + N_2 \longrightarrow Zn_3N_2}$

(b) $\boxed{2FeCl_3 + SnCl_2 \longrightarrow 2FeCl_2 + SnCl_4}$

17. (a) When $Cl_2(g)$ participates in a redox reaction, the oxidation numbers of its constituent chlorine atoms change to -1 (i.e., $Cl_2(g)$ is reduced). Since Fe has oxidation numbers of $+2$ or $+3$, $Cl_2(g)$ may be reduced if Fe is oxidized to $+3$.

$\quad\quad Fe^{+2} \; Cl^{-1} \; Cl^0 \quad\quad Fe^{+3} \; Cl^{-1}$

$\boxed{2FeCl_2(s) + Cl_2(g) \longrightarrow 2FeCl_3(s)}$

(b) Assume the reaction occurs in aqueous solution.

$$H_2SO_4 \rightleftharpoons H^+ + HSO_4^-$$
$$HSO_4^- \rightleftharpoons H^+ + SO_4^{2-}$$
$$KMnO_4 \rightleftharpoons K^+ + MnO_4^-$$

Given the following table of electronegativities, it would not be unreasonable to assume that during a reaction, (1) SO_4^{2-} remains intact because of the high electronegativity of S, and (2) H is responsible for the change in MnO_4^- because H has a higher electronegativity than Mn.

element	electronegativity
S	2.5
H	2.1
Mn	1.5
K	0.8

$\boxed{H_2SO_4 + 2KMnO_4 \longrightarrow K_2SO_4 + Mn_2O_7 + H_2O}$

(c) Al may attain an oxidation number of $+3$ and is less electronegative than H (1.5 versus 2.1), so Al is more likely to form a compound with SO_4^{2-}.

$\quad Al^0 \quad\quad H^{+1} \; SO_4^{-2} \quad\quad Al^{+3} \; SO_4^{2-} \quad\quad H^0$

$\boxed{2Al(s) + 3H_2SO_4(aq) \longrightarrow Al_2(SO_4)_3(aq) + 3H_2(g)}$

(d) Fe can attain an oxidation number of $+2$ or $+3$ and is less electronegative than Ag (1.8 versus 1.9), so Fe is more likely to form a compound with SO_4^{2-}.

$\quad\quad Fe^0 \quad Ag^{+1} \; SO_4^{-2} \quad\quad Fe^{+2} \; SO_4^{-2} \quad Ag^0$

$\boxed{Fe + Ag_2SO_4 \longrightarrow FeSO_4 + 2Ag}$

18. $m_{NaCl,pure} = (250 \text{ lbm})(0.945) = 236.25 \text{ lbm}$

From the reaction equation, 2 moles of NaCl produce 1 mole of Na_2SO_4.

$$\frac{m_{NaCl,pure}}{(MW)_{NaCl}} = 2\left(\frac{m_{Na_2SO_4,pure}}{(MW)_{Na_2SO_4}}\right)$$

$$m_{Na_2SO_4,pure} = \frac{1}{2}\left(\frac{(MW)_{Na_2SO_4}}{(MW)_{NaCl}}\right)(m_{NaCl,pure})$$

$$= \frac{1}{2}\left(\frac{142.1 \; \frac{g}{mol}}{58.5 \; \frac{g}{mol}}\right)(236.25 \text{ lbm})$$

$$= 286.9 \text{ lbm}$$

$$m_{Na_2SO_4,impure} = \frac{m_{Na_2SO_4,pure}}{purity}$$

$$= \frac{286.9 \text{ lbm}}{0.834} = \boxed{344 \text{ lbm}}$$

19. $\quad \rho_{sea} = \rho_{H_2O}(\text{specific gravity})_{sea}$

$$= \left(62.4 \; \frac{lbm}{ft^3}\right)(1) = 62.4 \text{ lbm/ft}^3$$

$$V = \frac{m_{Br}}{\rho_{sea(Br)}}$$

$$= \frac{1 \text{ lbm}}{\left(62.4 \; \frac{lbm}{ft^3}\right)(65 \times 10^{-6})}\left(\frac{7.48 \text{ gal}}{ft^3}\right)$$

$$= \boxed{1844 \text{ gal}}$$

20. (a) $n_{H_2O} = 2n_{H_2S} = (2)\left(\dfrac{m_{H_2S}}{(MW)_{H_2S}}\right)$

$$= (2)\left(\dfrac{9\,g}{34.1\,\frac{g}{mol}}\right) = \boxed{0.528\ mol}$$

(b) $n_{H_2S} = \dfrac{1}{2}n_{H_2O} = \dfrac{1}{2}\left(\dfrac{m_{H_2O}}{(MW)_{H_2O}}\right)$

$$= \dfrac{1}{2}\left(\dfrac{8\,g}{18\,\frac{g}{mol}}\right) = 0.222\ mol$$

Assume H_2S is an ideal gas at STP.

$$V = \dfrac{nR^*T}{p}$$

$$= \left[\dfrac{(0.222\ mol)\left(8.314\,\frac{J}{K\cdot mol}\right)(273.15K)}{101\,325\ Pa}\right]$$

$$\times\left(\dfrac{10^3\,l}{m^3}\right) = \boxed{4.98\ l}$$

21. (a) $\boxed{2\ mol}$

(b) $\dfrac{m_{Na_2SO_4}}{(MW)_{Na_2SO_4}} = \dfrac{m_{H_2SO_4}}{(MW)_{H_2SO_4}}$

$$m_{Na_2SO_4} = \left(\dfrac{(MW)_{Na_2SO_4}}{(MW)_{H_2SO_4}}\right)(m_{H_2SO_4})$$

$$= \left(\dfrac{142.1\,\frac{g}{mol}}{98.1\,\frac{g}{mol}}\right)(18\,g) = \boxed{26.07\ g}$$

(c) Assume the 100 g of reactants constitute one part H_2SO_4 for every two parts NaOH.

$$\dfrac{m_{H_2SO_4} + m_{NaOH}}{(MW)_{H_2SO_4} + (MW)_{2NaOH}} = \dfrac{1}{2}\left(\dfrac{m_{H_2O}}{(MW)_{H_2O}}\right)$$

$$m_{H_2O} = (2)\left(\dfrac{(MW)_{H_2O}}{(MW)_{H_2SO_4} + (MW)_{2NaOH}}\right)$$
$$\times (m_{H_2SO_4} + m_{NaOH})$$

$$= (2)\left(\dfrac{18\,\frac{g}{mol}}{98.1\,\frac{g}{mol} + 80\,\frac{g}{mol}}\right)(100\,g)$$

$$= \boxed{20.21\ g}$$

(d) Assume the 9 g of reactants constitute one part H_2SO_4 for every two parts NaOH.

$$n_{H_2O} = (2)\left(\dfrac{m_{H_2SO_4} + m_{NaOH}}{(MW)_{H_2SO_4} + (MW)_{2NaOH}}\right)$$

$$= (2)\left(\dfrac{9\,g}{98.1\,\frac{g}{mol} + 80\,\frac{g}{mol}}\right) = \boxed{0.1010\ mol}$$

$$n_{Na_2SO_4} = \dfrac{n_{H_2O}}{2} = \dfrac{0.1010\ mol}{2} = \boxed{0.0505\ mol}$$

22. step 1 :

$$Ag(s) + HNO_3(aq) \longrightarrow AgNO_3(aq) + \dfrac{1}{2}H_2(g)$$

step 2 :

$$AgNO_3(aq) + NaCl(aq) \longrightarrow NaNO_3(aq) + AgCl(s)$$

From step 2,

$$\dfrac{m_{AgNO_3}}{(MW)_{AgNO_3}} = \dfrac{m_{AgCl}}{(MW)_{AgCl}}$$

$$m_{AgNO_3} = \left(\dfrac{(MW)_{AgNO_3}}{(MW)_{AgCl}}\right)(m_{AgCl})$$

$$= \left(\dfrac{169.9\,\frac{g}{mol}}{143.4\,\frac{g}{mol}}\right)(7.2\,g) = 8.5305\ g$$

From step 1,

$$\dfrac{m_{Ag}}{(MW)_{Ag}} = \dfrac{m_{AgNO_3}}{(MW)_{AgNO_3}}$$

$$m_{Ag} = \left(\dfrac{(MW)_{Ag}}{(MW)_{AgNO_3}}\right)(m_{AgNO_3})$$

$$= \left(\dfrac{107.9\,\frac{g}{mol}}{169.9\,\frac{g}{mol}}\right)(8.5305\,g) = 5.418\ g$$

$$(m_{coin})(purity_{coin}) = m_{Ag}$$

$$purity_{coin} = \dfrac{m_{Ag}}{m_{coin}}$$

$$= \dfrac{5.418\,g}{5.82\,g}$$

$$= \boxed{0.931\ (93.1\%)}$$

23. Let $x_{i/j}$ = mass fraction of compound i in compound j.

$$x_{Cu/CuSO_4 \cdot 5H_2O} = \frac{(MW)_{Cu}}{(MW)_{CuSO_4 \cdot 5H_2O}}$$

$$= \frac{63.5 \frac{g}{mol}}{249.6 \frac{g}{mol}} = 0.2544$$

$$purity_{Cu} = \frac{m_{Cu}}{m_{sample}}$$

$$= \frac{m_{CuSO_4 \cdot 5H_2O} \, x_{Cu/CuSO_4 \cdot 5H_2O}}{m_{sample}}$$

$$= \frac{(30 \text{ g})(0.2544)}{10 \text{ g}} = \boxed{0.763 \ (76.3\%)}$$

24. Let x_i = mole fraction of component i in solution.

$$(MW)_{C_2H_4Br_2} = (2)(12.011) + (4)(1.0080)$$
$$+ (2)(79.904)$$
$$= 187.862 \text{ (say, 187.9)}$$

$$(MW)_{C_3H_6Br_2} = (3)(12.011) + (6)(1.0080)$$
$$+ (2)(79.904)$$
$$= 201.889 \text{ (say, 201.9)}$$

$$n_{C_2H_4Br_2} = \frac{m_{C_2H_4Br_2}}{(MW)_{C_2H_4Br_2}} = \frac{10 \text{ g}}{187.9 \frac{g}{mol}}$$

$$= 0.0532 \text{ mol}$$

$$n_{C_3H_6Br_2} = \frac{m_{C_3H_6Br_2}}{(MW)_{C_3H_6Br_2}} = \frac{80 \text{ g}}{201.9 \frac{g}{mol}}$$

$$= 0.3962 \text{ mol}$$

Applying Raoult's rule,

$$p_{C_2H_4Br_2} = (x_{C_2H_4Br_2})(p^0_{C_2H_4Br_2})$$

$$= \left(\frac{n_{C_2H_4Br_2}}{n_{C_2H_4Br_2} + n_{C_3H_6Br_2}} \right) (p^0_{C_2H_4Br_2})$$

$$= \left(\frac{0.0532 \text{ mol}}{0.0532 \text{ mol} + 0.3962 \text{ mol}} \right) (173 \text{ mm Hg})$$

$$= \boxed{20.48 \text{ mm Hg}}$$

$$p_{C_3H_6Br_2} = (x_{C_3H_6Br_2})(p^0_{C_3H_6Br_2})$$

$$= \left(\frac{n_{C_3H_6Br_2}}{n_{C_3H_6Br_2} + n_{C_2H_4Br_2}} \right) (p^0_{C_3H_6Br_2})$$

$$= \left(\frac{0.3962 \text{ mol}}{0.0532 \text{ mol} + 0.3962 \text{ mol}} \right) (127 \text{ mm Hg})$$

$$= \boxed{111.97 \ (112.0) \text{ mm Hg}}$$

25. Let x_{com} = mole fraction of the compound in solution.

$$\Delta p_{sol} = x_{com} p^0_{sol} = \left(\frac{n_{com}}{n_{com} + n_{H_2O}} \right) p^0_{sol}$$

$$n_{com} = \frac{n_{H_2O} \Delta p_{sol}}{p^0_{sol} - \Delta p_{sol}} = \frac{\left(\frac{m_{H_2O}}{(MW)_{H_2O}} \right) \Delta p_{sol}}{p^0_{sol} - \Delta p_{sol}}$$

$$= \frac{\left(\frac{500 \text{ g}}{18 \frac{g}{mol}} \right) (0.092 \text{ mm Hg})}{17.54 \text{ mm Hg} - 0.092 \text{ mm Hg}} = 0.146 \text{ mol}$$

$$(MW)_{com} = \frac{m_{com}}{n_{com}} = \frac{57 \text{ g}}{0.146 \text{ mol}} = \boxed{390.4 \text{ g/mol}}$$

26. Sucrose is $C_{12}H_{22}O_{11}$. Its molecular weight is

$$(MW)_{suc} = (12) \left(12 \frac{g}{mol} \right) + (22) \left(1 \frac{g}{mol} \right)$$
$$+ (11) \left(16 \frac{g}{mol} \right) = 342 \text{ g/mol}$$

$$n_{suc} = \frac{m_{suc}}{(MW)_{suc}} = \frac{13.5 \text{ g}}{342 \frac{g}{mol}} = 0.0395 \text{ mol}$$

$$M = \frac{n_{suc}}{V_{sol}} = \frac{0.0395 \text{ mol}}{0.100 \text{ l}} = \boxed{0.395 \text{ mol/l}}$$

The molality is

$$m = \frac{n_{suc}}{m_{H_2O}} = \frac{n_{suc}}{m_{sol} - m_{suc}} = \frac{n_{suc}}{V_{sol} \rho_{sol} - m_{suc}}$$

$$= \left[\frac{0.0395 \text{ mol}}{(100 \text{ ml}) \left(1.05 \frac{g}{ml} \right) - 13.5} \right] \left(\frac{10^3 \text{ g}}{kg} \right)$$

$$= \boxed{0.432 \text{ mol/kg}}$$

27. (a) Acetic acid is CH_3COOH.

$$n_{H_2O} = \frac{m_{H_2O}}{(MW)_{H_2O}} = \frac{125 \text{ g}}{18 \frac{g}{mol}} = 6.94 \text{ mol}$$

$$n_{CH_3COOH} = \frac{m_{CH_3COOH}}{(MW)_{CH_3COOH}}$$

$$= \frac{10 \text{ g}}{60 \frac{\text{g}}{\text{mol}}} = 0.167 \text{ mol}$$

$$m = \frac{n_{CH_3COOH}}{m_{H_2O}} = \left(\frac{0.167 \text{ mol}}{125 \text{ g}}\right)$$

$$= \boxed{1.33 \text{ mol/kg}}$$

(b) $$x_{CH_3COOH} = \frac{n_{CH_3COOH}}{n_{H_2O} + n_{CH_3COOH}}$$

$$= \frac{0.167 \text{ mol}}{6.94 \text{ mol} + 0.167 \text{ mol}}$$

$$= \boxed{0.023}$$

$$x_{H_2O} = 1 - x_{CH_3COOH} = \boxed{0.977}$$

28. Let GEW = gram equivalent weight.

Let x = gravimetric fraction.

normality = $0.5 \text{ N} = 0.5 \text{ GEW/l}$

$$x_{NaOH} = 1 - x_{H_2O} = 1 - 0.12 = 0.88$$
$$NaOH \rightleftharpoons Na^+ + OH^-$$

$$(EW)_{NaOH} = \frac{(MW)_{NaOH}}{\Delta \text{ oxidation number}}$$

$$= \left(\frac{40}{1}\right)\left(\frac{\text{g}}{\text{GEW}}\right) = 40 \text{ g/GEW}$$

$$n = \frac{\text{no. of GEW of NaOH}}{V_{sol}}$$

$$= \frac{\frac{m_{NaOH}}{(EW)_{NaOH}}}{V_{sol}} = \frac{\frac{m_{wet\ NaOH}x_{NaOH}}{(EW)_{NaOH}}}{V_{sol}}$$

$$m_{wet\ NaOH} = \frac{(\text{normality})V_{sol}(EW)_{NaOH}}{x_{NaOH}}$$

$$= \left[\frac{(0.5)\left(\frac{\text{GEW}}{\text{l}}\right)(60 \text{ l})\left(40 \frac{\text{g}}{\text{GEW}}\right)}{0.88}\right]$$

$$\times \left(\frac{\text{kg}}{10^3 \text{ g}}\right) = \boxed{1.364 \text{ kg}}$$

29. Let SG = specific gravity.

$$m_{sol} = \rho_{sol}V_{sol} = SG_{sol}\rho_{H_2O}V_{sol}$$

$$= (1.0835)\left(62.43 \frac{\text{lbm}}{\text{ft}^3}\right)(55 \text{ gal})\left(\frac{1 \text{ ft}^3}{7.48 \text{ gal}}\right)$$

$$= 497.4 \text{ lbm}$$

Let $x = \dfrac{\text{mass fraction}}{\text{gravimetric fraction}}$

$$m_{CaCl_2} = x_{CaCl_2}m_{sol} = (0.10)(497.4 \text{ lbm})$$

$$= \boxed{49.74 \text{ lbm}}$$

30. (a) $$m_{H_2SO_4} = (\text{molarity})V_{sol}(MW)$$

$$= \left(2 \frac{\text{mol}}{\text{l}}\right)(2.5 \text{ l})\left(98.1 \frac{\text{g}}{\text{mol}}\right)$$

$$= \boxed{490.5 \text{ g}}$$

(b) $$m_{Ba(OH)_2} = (\text{molarity})V_{sol}(MW)_{Ba(OH)_2}$$

$$= \left(0.525 \frac{\text{mol}}{\text{l}}\right)(5 \text{ l})\left(171.35 \frac{\text{g}}{\text{mol}}\right)$$

$$= \boxed{449.8 \text{ g}}$$

(c) $$Al_2(SO_4)_3 \rightleftharpoons 2Al^{3+} + 3SO_4^{2-}$$

$$(EW)_{Al_2(SO_4)_3} = \frac{(MW)_{Al_2(SO_4)_3}}{\Delta \text{ oxidation number}}$$

$$= \frac{342.3}{6}\left(\frac{\text{g}}{\text{GEW}}\right) = 57.05 \text{ g/GEW}$$

$$m_{Al_2(SO_4)_3} = N(EW)_{Al_2(SO_4)_3}V_{sol}$$

$$= \left(0.5 \frac{\text{GEW}}{\text{l}}\right)\left(57.05 \frac{\text{g}}{\text{GEW}}\right)(0.350 \text{ l})$$

$$= \boxed{9.98 \text{ g}}$$

(d) $$KMnO_4 \rightleftharpoons K^+ + MnO_4^-$$

$$(EW)_{KMnO_4} = \frac{(MW)_{KMnO_4}}{\Delta \text{ oxidation number}}$$

$$= \left(\frac{158.04}{1}\right)\left(\frac{\text{g}}{\text{GEW}}\right)$$

$$= 158.04 \text{ g/GEW}$$

$$m_{KMnO_4} = (N)(EW)_{KMnO_4} V_{sol}$$

$$= \left(1.9 \frac{GEW}{l}\right)\left(158.04 \frac{g}{GEW}\right)(0.025 \, l)$$

$$= \boxed{7.51 \, g}$$

31. $\Delta T_f = -mK_{f,H_2O} = \left(\frac{-n_{CH_3OH}}{m_{H_2O}}\right)(K_{f,\,H_2O})$

$$= \left[\frac{-\left(\frac{m_{CH_3OH}}{(MW)_{CH_3OH}}\right)}{V_{H_2O}\rho_{H_2O}}\right](K_{f,H_2O})$$

$$m_{CH_3OH} = \frac{-\Delta T_f V_{H_2O}\rho_{H_2O}(MW)_{CH_3OH}}{K_{f,H_2O}}$$

$$= \left[\frac{-(-12°C)(20 \, l)\left(\frac{kg}{l}\right)\left(32 \frac{g}{mol}\right)}{\left(1.86 \frac{kg·°C}{mol}\right)}\right]\left(\frac{kg}{10^3 \, g}\right)$$

$$= \boxed{4.129 \, kg}$$

32. $\Delta T_f = -mK_{f,H_2O}$

$$= -\left[\frac{\left(\frac{m_{(CH_2OH)_2}}{(MW)_{(CH_2OH)_2}}\right)}{V_{H_2O}\rho_{H_2O}}\right](K_{f,H_2O})$$

$$= -\left[\frac{\left(\frac{5000 \, g}{62 \frac{g}{mol}}\right)}{(15 \, l)\left(1\frac{kg}{l}\right)}\right]\left(1.86 \frac{kg·°C}{mol}\right)$$

$$= \boxed{-10°C}$$

33. $\Delta T_f = -mK_{f,ben}$

$$T_{f,sol} - T_{f,\text{pure solvent}} = \left[\frac{\frac{m_{nap}}{(MW)_{nap}}}{m_{ben}}\right]K_{f,ben}$$

$$(MW)_{nap} = \frac{-m_{nap}K_{f,ben}}{(T_{f,\text{solution}} - T_{f,\text{pure solvent}})m_{ben}}$$

$$= \frac{-(1.15 \, g)\left(5.12 \frac{kg·°C}{mol}\right)}{(4.95°C - 5.5°C)(0.100 \, kg)}$$

$$= \boxed{107.05 \, g/mol}$$

34. The boiling point constant for water is $0.512°C/m$. However, the complete ionization of NaCl doubles the ionic concentration since 2 ions, Na^+ and Cl^-, are produced for every molecule of NaCl.

$$\Delta T_b = mK_{b,H_2O} = \left(\frac{\frac{m_{NaCl}}{(MW)_{NaCl}}}{m_{H_2O}}\right)(K_{b,H_2O})$$

$$= \left[\frac{\frac{5 \, g}{58.5 \frac{g}{mol}}}{(2 \, qts)\left(\frac{gal}{4 \, qts}\right)\left(\frac{3.785 \, l}{gal}\right)\left(\frac{kg}{l}\right)}\right](2)\left(0.512\frac{kg·°C}{mol}\right)$$

$$= \boxed{0.046°C}$$

35. The molarity is the concentration of ions per liter.

$$pH = -\log[H_3O^+] = -\log\left(3 \times 10^{-3} \frac{ion}{l}\right)$$

$$= \boxed{2.52}$$

36. $pH = -\log[H_3O^+]$

$$[H_3O^+] = (10)(-pH) = 10^{-6.5} \frac{ion}{l}$$

$$= \boxed{3.16 \times 10^{-7} \, ion/l}$$

$$[OH^-] = \frac{10^{-14}\left(\frac{ion}{l}\right)^2}{[H_3O^+]} = \frac{10^{-14}\left(\frac{ion}{l}\right)^2}{3.16 \times 10^{-7}\frac{ion}{l}}$$

$$= \boxed{3.16 \times 10^{-8} \, ion/l}$$

CHEMISTRY
Inorganic

37. $HCl \rightleftharpoons H^+ + Cl^-$

    molarity = normality

$$\begin{aligned} pH &= -\log[H^+] \\ &= -\log[(\text{molarity})(\text{fraction ionized})] \\ &= -\log[(\text{normality})(\text{fraction ionized})] \\ &= -\log\left[\left(0.5\frac{\text{ion}}{l}\right)(0.93)\right] = \boxed{0.333} \end{aligned}$$

38. $pH = -\log[H_3O^+] = -\log\left(5 \times 10^{-6}\frac{\text{ion}}{l}\right)$

$$= 5.3$$

$$pOH = 14 - pH = 14 - 5.3 = \boxed{8.7}$$

39. $V_{Na_2CO_3} N_{Na_2CO_3} = V_{HCl} N_{HCl}$

$$\begin{aligned} N_{Na_2CO_3} &= \left(\frac{V_{HCl}}{V_{Na_2CO_3}}\right)(N_{HCl}) \\ &= \left(\frac{0.0658\,l}{0.050\,l}\right)\left(3\frac{GEW}{l}\right) = 3.948\ GEW/l \end{aligned}$$

$$Na_2CO_3 \longrightarrow 2Na^+ + CO_3^{2-}$$

$$\begin{aligned} (EW)_{Na_2CO_3} &= \frac{(MW)_{Na_2CO_3}}{\Delta\ \text{oxidation number}} \\ &= \left(\frac{106}{2}\right)\left(\frac{g}{GEW}\right) = 53\ g/GEW \end{aligned}$$

Consider 1 l of Na_2CO_3 solution.

$$\begin{aligned} m_{Na_2CO_3} &= N_{Na_2CO_3}\ V_{Na_2CO_3}\ (EW)_{Na_2CO_3} \\ &= \left(3.948\frac{GEW}{l}\right)(1\ l)\left(53\frac{g}{GEW}\right) \\ &= 209.2\ g \end{aligned}$$

$$\begin{aligned} m_{sol} &= V_{sol}(SG)_{sol}\ \rho_{H_2O} \\ &= (1\ l)(1.25)\left(\frac{1000\ g}{l}\right) = 1250\ g \end{aligned}$$

$$x_{Na_2CO_3} = \frac{m_{Na_2CO_3}}{m_{sol}} = \frac{209.2\ g}{1250\ g} = \boxed{0.167}$$

40. Note that molarity and formality are identical for NaOH and H_2SO_4.

$$V_{NaOH}\ N_{NaOH} = V_{H_2SO_4}\ N_{H_2SO_4}$$

$$V_{NaOH} F_{NaOH} \Delta_{\text{base charge}} = V_{H_2SO_4} F_{H_2SO_4} \Delta_{\text{acid charge}}$$

$$\begin{aligned} F_{H_2SO_4} &= \frac{V_{NaOH}\ F_{NaOH}\ \Delta_{\text{base charge}}}{V_{H_2SO_4}\ \Delta_{\text{acid charge}}} \\ &= \frac{(0.0383\ l)\left(0.103\frac{FW}{l}\right)\left(1\frac{\text{ion}}{FW}\right)}{(0.020\ l)\left(2\frac{\text{ion}}{FW}\right)} \\ &= \boxed{0.0986\ FW/l} \end{aligned}$$

41. $V_{HCl} F_{HCl} = V_{NaOH} F_{NaOH}$

$$\begin{aligned} F_{HCl} &= \frac{V_{NaOH}\ F_{NaOH}}{V_{HCl}} = \frac{(0.0243\ l)(0.1035)}{0.025\ l} \\ &= \boxed{0.1006} \end{aligned}$$

42. $NaC_2H_3O_2 \rightleftharpoons Na^+ + C_2H_3O_2^-$

       $HC_2H_3O_2 \rightleftharpoons H^+ + C_2H_3O_2^-$

$$K_a = \frac{[H^+][C_2H_3O_2^-]}{[HC_2H_3O_2]} \qquad \text{[Eq. 1]}$$

$[C_2H_3O_2^-]$ comes from two sources:

    1. from $HC_2H_3O_2$, in the same concentration as the H^+

    2. from the dissociation of $NaC_2H_3O_2$

$$\begin{aligned} [C_2H_3O_2^-] &= [H^+] + M_{NaC_2H_3O_2}(\%\ \text{ionization}) \\ &= [H^+] + \left(0.01\frac{mol}{l}\right)(1) \approx 0.01\ mol/l \end{aligned}$$

(since $[H^+]$ is much smaller than 0.01)

$$\begin{aligned} [HC_2H_3O_2] &= M_{HC_2H_3O_2} - [H^+] \\ &= 0.01\frac{mol}{l} - [H^+] \approx 0.01\ mol/l \end{aligned}$$

From Eq. 1,

$$1.8 \times 10^{-5}\frac{mol}{l} \approx \frac{[H^+]\left(0.01\frac{mol}{l}\right)}{0.01\frac{mol}{l}}$$

$$[H^+] = 1.8 \times 10^{-5}\ mol/l$$

$$pH = -\log[H^+] = -\log\left[1.8 \times 10^{-5}\frac{mol}{l}\right]$$

$$= \boxed{4.74}$$

43. $[H_3O^+] = 10^{-pH} = 1 \times 10^{-5} \text{ mol/l}$

$[\text{acid}] = 0.005 \text{ mol/l}$

$$\text{percent ionization} = \frac{[H_3O^+]}{[H_3O^+] + [\text{acid}]}$$

$$= \frac{1 \times 10^{-5} \frac{\text{mol}}{\text{l}}}{1 \times 10^{-5} \frac{\text{mol}}{\text{l}} + 0.005 \frac{\text{mol}}{\text{l}}} = \boxed{0.002 \ (0.2\%)}$$

44. $NH_3 + H_2O \rightleftharpoons NH_4^+ + OH^-$

$$K_b = \frac{[NH_4^+][OH^-]}{[NH_3]}$$

$$= \frac{x^2}{1-x} = \frac{\left(0.004 \frac{\text{mol}}{\text{l}}\right)^2}{1 - 0.004 \frac{\text{mol}}{\text{l}}}$$

$$= \boxed{1.606 \times 10^{-5} \text{ mol/l}}$$

45. $NH_3 + H_2O \rightleftharpoons NH_4^+ + OH^-$

$$K_b = \frac{[NH_4^+][OH^-]}{[NH_3]} = \frac{[OH^-]^2}{M_{\text{sol}} - [OH^-]}$$

$$\approx \frac{[OH^-]^2}{M_{\text{sol}}}$$

$$[OH^-] = \sqrt{K_b M_{\text{sol}}}$$

$$= \sqrt{\left(1.606 \times 10^{-5} \frac{\text{mol}}{\text{l}}\right)\left(0.05 \frac{\text{mol}}{\text{l}}\right)}$$

$$= \boxed{8.96 \times 10^{-4} \text{ mol/l}}$$

46. $BaCO_3 \rightleftharpoons Ba^+ + CO_3^-$

$$[Ba^+] = [CO_3^-] = \frac{\left(\frac{m_{BaCO_3}}{(MW)_{BaCO_3}}\right)}{V_{\text{sol}}}$$

$$= \frac{\left(\frac{1.4 \times 10^{-3} \text{ g}}{197.3 \frac{\text{g}}{\text{mol}}}\right)}{0.100 \text{ l}} = 7.096 \times 10^{-5} \text{ mol/l}$$

$$K_{sp} = [Ba^+][CO_3^-] = \left(7.096 \times 10^{-5} \frac{\text{mol}}{\text{l}}\right)^2$$

$$= \boxed{5.04 \times 10^{-9} \text{ mol}^2/\text{l}^2}$$

47. $Pb F_2 \rightleftharpoons Pb^{2+} + 2F^-$

$$K_{sp} = [Pb^{2+}][F^-]^2$$

$$[F^-] = 2[Pb^{2+}]$$

so $[F^-]^2 = 4[Pb^{+2}]^2$

$$[Pb^{2+}] = \left(\frac{K_{sp}}{4}\right)^{\frac{1}{3}}$$

$$= \left(\frac{3.2 \times 10^{-8} \frac{\text{mol}^3}{\text{l}^3}}{4}\right)^{\frac{1}{3}}$$

$$= \boxed{2 \times 10^{-3} \text{ mol/l}}$$

$$[F^-] = (2)[Pb^{2+}]$$

$$= (2)[Pb^{2+}] = (2)\left(2 \times 10^{-3} \frac{\text{mol}}{\text{l}}\right)$$

$$= \boxed{4 \times 10^{-3} \text{ mol/l}}$$

48. $K_{sp} = [Ag^+]^2 [CrO_4^{2-}]$

The vast majority of the Ag^+ is contributed by $AgNO_3$.

$$K_{sp} = [AgNO_3]^2 [CrO_4^{2-}]$$

$$[CrO_4^{2-}] = \frac{K_{sp}}{[AgNO_3]^2} = \frac{1.1 \times 10^{12} \frac{\text{mol}^3}{\text{l}^3}}{\left(0.5 \frac{\text{mol}}{\text{l}}\right)^2}$$

$$= \boxed{4.4 \times 10^{-12} \text{ mol/l}}$$

49. $n = \dfrac{m}{MW} = \dfrac{\text{no. of faradays}}{\text{change in charge}}$

$$\text{no. of faradays} = \frac{m(\text{change in charge})}{MW}$$

$$AgNO_2 \longrightarrow Ag^+ + NO_2^-$$

$$SnCl_2 \longrightarrow Sn^{2+} + 2Cl^-$$

The same number of faradays are used in each reaction.

$$\frac{m_{Ag}(\text{change in charge})_{Ag}}{(MW)_{Ag}} = \frac{m_{Sn}(\text{change in charge})_{Sn}}{(MW)_{Sn}}$$

$$m_{Sn} = \left(\frac{(MW)_{Sn}}{(MW)_{Ag}}\right)(m_{Ag})\left(\frac{(\text{change in charge})_{Ag}}{(\text{change in charge})_{Sn}}\right)$$

$$= \left(\frac{118.69 \frac{\text{g}}{\text{gmol}}}{107.868 \frac{\text{g}}{\text{gmol}}}\right)(2 \text{ g})\left(\frac{1}{2}\right) = \boxed{1.10 \text{ g}}$$

50. $t = \dfrac{m_{Al}(96\,500)(\text{change in oxidation state})}{I(MW)_{Al}}$

$$Al_2O_3 \longrightarrow 2Al^{+3} + 3O^{2-}$$

$$t = \frac{(100\,g)\left(96\,500\,\dfrac{C}{mol}\right)(3)}{(125\,A)\left(26.9815\,\dfrac{g}{mol}\right)} = \boxed{8584\,s}$$

51. $2H^+(aq)^+ SO_4^{2-}(aq) \longrightarrow H_2(g) + 2O_2(g) + S(s)$

$$n_{H_2} = \frac{1}{2}n_{H^+} = \left(\frac{1}{2}\right)\left[\frac{It}{\left(96\,500\,\dfrac{C}{mol}\right)(\text{change in charge})}\right]$$

$$= \left(\frac{1}{2}\right)\left[\frac{(2.5\,A)(3600\,s)}{\left(96\,500\,\dfrac{C}{mol}\right)(1)}\right] = 0.0466\,mol$$

$$n_{O_2} = \frac{1}{2}n_{O^{2-}} = \left(\frac{1}{2}\right)\left[\frac{It}{\left(96\,500\,\dfrac{C}{mol}\right)(\text{change in charge})}\right]$$

$$= \left(\frac{1}{2}\right)\left[\frac{(2.5\,A)(3600\,s)}{\left(96\,500\,\dfrac{C}{mol}\right)(2)}\right] = 0.0233\,mol$$

$$n_{total} = n_{H_2} + n_{O_2}$$

$$= 0.0466\,mol + 0.0233\,mol = 0.0699\,mol$$

$$V = \frac{nR^*T}{p}$$

$$= \frac{(0.0699\,mol)\left(0.08206\,\dfrac{atm \cdot l}{mol \cdot K}\right)(273+25)K}{(780\,mm\,Hg)\left(\dfrac{atm}{760\,mm\,Hg}\right)}$$

$$= \boxed{1.665\,l}$$

52. (a) Corrosion is a chemical or physical reaction of metallic materials with the environment and/or with each other.

(b) Stress corrosion is a form of self-corrosion. When a metallic material is subjected to stresses over time, certain grains will be more stressed than others; consequently, they become relatively anodic. Thus, galvanic cells are simulated near the grain boundaries. Cracks propagate across these boundaries by virtue of the oxidation mechanism.

(c) Corrosion fatigue is stress corrosion where the stresses are cyclic. Corrosion fatigue can lead to failure below normal working stresses.

(d) Fretting corrosion occurs at the interface between two highly loaded members where sliding and rubbing can result in both wear and corrosion. Metals relying on surface oxides are especially susceptible.

(e) Cavitation is a result of high velocities in liquids that reduce the fluid pressure to below the vapor pressure such that pockets of the fluid boil and subsequently collapse. The adjoining metallic surfaces are stressed by the sudden pressure increase; they are work hardened, becoming increasingly brittle with repetitions of cavitation to the point where the metal away.

(f) Electrode potential is a quantification of the relative tendency of an electrode material to be reduced (or, oxidized). Hydrogen ion reduction (oxidation) is used as the reference, with a potential of 0 V.

(g) A galvanic cell is a device that produces an electric current of its own accord via oxidation/reduction reactions. It consists of an anode (where oxidation occurs), a cathode (where reduction occurs), and an electrolytic medium.

53. (a) $Fe^{3+} + e^- \longrightarrow Fe^{2+}$

This is a reduction, so the standard reduction potential is

$$\boxed{\varepsilon^0 = 0.771\,V}$$

(b) $Al \longrightarrow Al^{3+} + 3e^-$

This is an oxidation, so negate standard reduction potential.

$$\boxed{\varepsilon^0 = 1.66\,V}$$

54. Since steel and zinc have negative reduction potentials, both are good candidates for oxidation and corrosion in a moist environment. Since zinc's reduction potential is lower than that of any steel, a piece of zinc, when electrically attached to a piece of steel, would be oxidized if both metals were exposed to a moist environment because electrons would flow in the direction of a positive potential difference: from the zinc to the steel. Since the steel functions as the cathode, it does not rust because corrosion always occurs at the anode. Consequently, the piece of zinc is called a sacrificial anode.

ORGANIC CHEMISTRY AND COMBUSTION

1. (a) An alcohol is a compound containing a hydroxyl (OH^-) group.

   (b) An alkane is the hydrocarbon C_nH_{2n+2}.

   (c) An alkene is the hydrocarbon C_nH_{2n}.

   (d) An alkyne is the hydrocarbon C_nH_{2n-2}.

   (e) An aromatic hydrocarbon is a hydrocarbon with the form C_nH_n.

   (f) A phenol is an aromatic hydrocarbon with a hydroxyl group replacing one hydrogen.

   (g) An ether is a compound in which two hydrocarbon groups are linked by an oxygen.

   (h) An aldehyde is a compound containing a carbinol group (COH).

   (i) A ketone is a compound containing the carbonyl group.

   (j) A carboxylic acid is a hydrocarbon containing the carboxyl group (COOH).

   (k) A halide is a compound of the type RX where X is a halogen (F, Cl, Br, I, or At only) and R is another element or organic group.

   (l) An anhydride is a compound formed from an acid by the removal of water.

   (m) An amide is a compound containing the $CONH_2$ radical.

   (n) An ester is a compound containing the functional group COO.

   (o) An amino acid is a compound containing a carbon bonded to a hydrogen, a primary amino group, a carboxyl group, and some other distinguishing group.

   (p) A carbohydrate is the compound $C_x(H_2O)_y$.

2. Assume STP. The molar volume of any gas is 359.2 ft^3.

$$m_{CH_4} = n_{CH_4}(MW)_{CH_4} = \left(\frac{V_{CH_4}}{359.2 \frac{\text{ft}^3}{\text{lbmole}}}\right)(MW)_{CH_4}$$

$$= \left(\frac{7 \text{ ft}^3}{359.2 \frac{\text{ft}^3}{\text{lbmole}}}\right)\left(16 \frac{\text{lbm}}{\text{lbmole}}\right) = 0.3118 \text{ lbm}$$

$$m_{H_2O}c_{p,H_2O}\Delta T = Q_{actual} = \eta Q_{ideal}$$
$$= \eta m_{CH_4}(HV)_{CH_4}$$

$$m_{H_2O} = \frac{\eta m_{CH_4}(HV)_{CH_4}}{c_{p,H_2O}\Delta T}$$

$$= \frac{(0.5)(0.3118 \text{ lbm})\left(24{,}000 \frac{\text{BTU}}{\text{lbm}}\right)}{\left(1 \frac{\text{BTU}}{\text{lbm-}^\circ\text{F}}\right)(200^\circ\text{F} - 60^\circ\text{F})}$$

$$= \boxed{26.726 \text{ lbm}}$$

3. $C_3H_8 + 5O_2 \longrightarrow 3CO_2 + 4H_2O$

$$\dot{n}_{CO_2} = 3\dot{n}_{C_3H_8} = \frac{3\dot{m}_{C_3H_8}}{(MW)_{C_3H_8}}$$

$$= \frac{(3)\left(15 \frac{\text{lbm}}{\text{hr}}\right)}{44 \frac{\text{lbm}}{\text{lbmole}}} = 1.023 \text{ lbmole/hr}$$

$$\dot{V}_{CO_2} = \frac{\dot{n}_{CO_2}R^*T}{p}$$

$$= \frac{\left(1.023 \frac{\text{lbmole}}{\text{hr}}\right)\left(1545 \frac{\text{ft}^3\text{-lbf}}{\text{ft}^2\text{-lbmole-}^\circ\text{R}}\right)(460+70)^\circ\text{R}}{\left(14.7 \frac{\text{lbf}}{\text{in}^2}\right)\left(144 \frac{\text{in}^2}{\text{ft}^2}\right)}$$

$$= \boxed{395.7 \text{ ft}^3/\text{hr}}$$

4. $\dot{m}_{N_2, \text{ actual}} = (\dot{m}_{\text{air, actual}})\left(\text{mass fraction } \frac{N_2}{\text{air}}\right)$

$$= \left(\frac{\dot{V}_{\text{air, actual}}}{v_{\text{air}}}\right)(0.7685)$$

With 30% excess air,

$$\dot{m}_{N_2, \text{ actual}} = \left(\frac{1.3\dot{V}_{\text{air, ideal}}}{v_{\text{air}}}\right)(0.7685)$$

$$= \left[\frac{(1.3)\left(\dfrac{\dot{V}_{O_2, \text{ ideal}}}{\text{volume fraction } \frac{O_2}{\text{air}}}\right)}{v_{\text{air}}}\right](0.7685)$$

Consider air, O_2, N_2, and CH_4 to be ideal gases.

$$\dot{m}_{N2,actual} = \left[\frac{(1.3)\left(\dfrac{\dot{V}_{O2,\,ideal}}{0.209}\right)}{\dfrac{R_{air}T}{p}} \right] (0.7685)$$

Combustion is $CH_4 + 2O_2 \longrightarrow CO_2 + 2H_2O$, and ideal gases have equal molar volumes.

$$\dot{m}_{N2,actual} = \left[\frac{(1.3)\left(\dfrac{2\dot{V}_{CH_4}}{0.209}\right)}{\dfrac{R_{air}T}{p}} \right] (0.7685)$$

$$= \left[\frac{(1.3)\left(\dfrac{(2)\left(4000\,\dfrac{ft^3}{hr}\right)}{0.209}\right)}{\dfrac{\left(53.35\,\dfrac{ft\text{-}lbf}{lbm\text{-}°R}\right)(100+460)°R}{\left(15\,\dfrac{lbf}{in^2}\right)\left(\dfrac{12\,in}{ft}\right)^2}} \right] (0.7685)$$

$$= \boxed{2764.8\ lbm/hr}$$

5. Sulfur (S) and nitrogen (N) do not burn. Carbon (C) and hydrogen (H) burn according to

$$C + O_2 \longrightarrow CO_2$$
$$4H_2 + O_2 \longrightarrow 2H_2O$$

$$n_{O2} = n_C + \tfrac{1}{4}n_H = \frac{m_C}{(MW)_C} + \tfrac{1}{4}\left(\frac{m_H}{(MW)_H}\right)$$
$$= \frac{m_{fuel}G_C}{(MW)_C} + \tfrac{1}{4}\left(\frac{m_{fuel}G_H}{(MW)_H}\right)$$
$$= \frac{(1\,lbm)(0.84)}{12\,\dfrac{lbm}{lbmole}} + \frac{(1\,lbm)(0.153)}{4\,\dfrac{lbm}{lbmole}}$$
$$= 0.10825\ lbmole$$

$$m_{air} = n_{air}(MW)_{air}$$
$$= \left(\frac{n_{O2}}{volume\ fraction\ \dfrac{O_2}{air}}\right)(MW)_{air}$$
$$= \left(\frac{0.10825\ lbmole}{0.209}\right)\left(28.967\,\frac{lbm}{lbmole}\right)$$
$$= \boxed{15.00\ lbm}$$

6. Assume ideal gas behavior. Ideal combustion is

$$C_3H_8 + 5O_2 \longrightarrow 3CO_2 + 4H_2O$$

With 20% excess air,

$$C_3H_8 + (1.2)(5)O_2 \longrightarrow 3CO_2 + 4H_2O + O_2$$

Including nitrogen,

$$C_3H_8 + 6O_2 + (3.78)(6)N_2 \longrightarrow 3CO_2 + 4H_2O$$
$$+ O_2 + (3.78)(6)N_2$$

$$C_3H_8 + 6O_2 + 22.68\,N_2 \longrightarrow 3CO_2 + 4H_2O$$
$$+ O_2 + 22.68\,N_2$$

The mass balance is

$$m_{C3H8} + m_{O2} + m_{N2} = m_{CO2} + m_{H2O} + m_{O2} + m_{N2}$$

But, $m = n(MW)$.

$$G_{CO2} =$$
$$\frac{n(MW)_{CO2}}{n(MW)_{CO2} + n(MW)_{H2O} + n(MW)_{O2} + n(MW)_{N2}}$$
$$= \frac{(3\,lbmole)\left(44\,\dfrac{lbm}{lbmole}\right)}{(3)(44) + (4)(18) + (1)(32) + (22.68)(28)}$$
$$= \boxed{0.152\ (15.2\%)\ wet}$$

Note: This is a wet analysis because the weight of the water vapor was included.

7. B stands for volume fraction
    x stands for mole fraction

N_2 and CO_2 do not burn, but CO and H_2 do.

$$2CO + O_2 \longrightarrow 2CO_2$$
$$2H_2 + O_2 \longrightarrow 2H_2O$$

$$B_{H2O} = B_{H2} = 0.12$$
$$B_{O2} = \tfrac{1}{2}B_{CO} + \tfrac{1}{2}B_{H2} = \left(\tfrac{1}{2}\right)(0.22) + \left(\tfrac{1}{2}\right)(0.12) = 0.17$$

Assuming every component behaves as an ideal gas, 1 mole of each component occupies the same volume: 22.4 l. Therefore, $x_i = B_i$, so

$$(0.06)CO_2 + (0.22)CO + (0.12)H_2 + (0.60)N_2 + (0.17)O_2$$
$$\longrightarrow (0.60)N_2 + (0.28)CO_2 + (0.12)H_2O$$

CHEMISTRY
Organic

$$\Delta H_r = \sum \Delta H_{f,\text{products}} - \sum \Delta H_{f,\text{reactants}}$$

$$\Delta H_r = (0.60)\left(0 \frac{\text{kcal}}{\text{mol}}\right) + (0.28)\left(-94.05 \frac{\text{kcal}}{\text{mol}}\right)$$

$$+ (0.12)\left(-57.80 \frac{\text{kcal}}{\text{mol}}\right) - \left[(0.06)\left(-94.05 \frac{\text{kcal}}{\text{mol}}\right)\right.$$

$$+ (0.22)\left(-26.42 \frac{\text{kcal}}{\text{mol}}\right) + (0.12)\left(0 \frac{\text{kcal}}{\text{mol}}\right)$$

$$\left. - (0.60)\left(0 \frac{\text{kcal}}{\text{mol}}\right) + (0.17)\left(0 \frac{\text{kcal}}{\text{mol}}\right)\right]$$

$$\Delta H_r = \left(-21.8 \frac{\text{kcal}}{\text{mol}}\right)\left(\frac{\text{mol}}{22.4 \text{ l}}\right) = \boxed{-0.973 \text{ kcal/l}}$$

ΔH_r is negative, so the reaction is exothermic.

8. Carbon dioxide burns as follows:

$$2CO + O_2 \longrightarrow 2CO_2$$

With 25% excess air,

$$2CO + 1.25\,O_2 \longrightarrow 2CO_2 + 0.25\,O_2$$

$$CO + 0.625\,O_2 \longrightarrow CO_2 + 0.125\,O_2$$

$$\Delta H_r = \sum H_{f,\text{products}} - \sum H_{f,\text{reactants}}$$

$$= (1)\left(-94.05 \frac{\text{kcal}}{\text{mol}}\right) + (0.125)\left(0 \frac{\text{kcal}}{\text{mol}}\right)$$

$$- \left[(1)\left(-26.42 \frac{\text{kcal}}{\text{mol}}\right) + (0.625)\left(0 \frac{\text{kcal}}{\text{mol}}\right)\right]$$

$$= -67.63 \text{ kcal/mol}$$

The masses of the products present are

$$m_{CO_2} = n_{CO_2}(\text{MW})_{CO_2} = (1 \text{ mol})\left(44 \frac{\text{g}}{\text{mol}}\right) = 44 \text{ g}$$

$$m_{O_2} = n_{O_2}(\text{MW})_{O_2} = (0.125 \text{ mol})\left(32 \frac{\text{g}}{\text{mol}}\right) = 4 \text{ g}$$

Air is approximately 79.1% nitrogen; the rest is oxygen.

$$m_{N_2} = n_{\text{air}} x_{N_2}(\text{MW})_{N_2} = \left(\frac{n_{O_2}}{x_{O_2}}\right) x_{N_2}(\text{MW})_{N_2}$$

$$= \left(\frac{0.625 \text{ mol}}{1 - 0.791}\right)(0.791)\left(28 \frac{\text{g}}{\text{mol}}\right) = 66.23 \text{ g}$$

Pressure is constant at 1 atm, so use c_p. Assume all combustion heat goes into flue gases.

$$T_{\text{final}} = T_{\text{initial}} + \frac{|\Delta H_r|}{\sum c_{p,i} m_i}$$

$$= T_{\text{initial}} + \frac{|\Delta H_r|}{c_{p,CO_2} m_{CO_2} + c_{p,O_2} m_{O_2} + c_{p,N_2} m_{N_2}}$$

Assume $T_{\text{initial}} = 20°C$ and use $T = 20°C$ when choosing c_p values (although this is not entirely accurate, as temperature rises during reaction).

$$T_{\text{final}} = 20°C +$$

$$\frac{(1 \text{ mol})\left|\left(-67.63 \frac{\text{kcal}}{\text{mol}}\right)\left(\frac{10^3 \text{ cal}}{\text{kcal}}\right)\right|}{\left(0.205 \frac{\text{cal}}{\text{g}\cdot°C}\right)(44 \text{ g}) + (0.217)(4) + (0.247)(66.23)}$$

$$= \boxed{2597°C}$$

HEAT TRANSFER

1. $q = \dfrac{kA(T_1 - T_2)}{L}$

$= \dfrac{\left(0.05 \dfrac{\text{BTU}}{\text{hr-ft-}^\circ\text{F}}\right)(150 \text{ in}^2)(340^\circ\text{F} - 120^\circ\text{F})}{(1.5 \text{ in})\left(\dfrac{12 \text{ in}}{\text{ft}}\right)}$

$= \boxed{91.67 \text{ BTU/hr}}$

2. (a) $\text{Re} \equiv \dfrac{vD}{\nu} = \dfrac{\rho v D}{\mu}$

$v \equiv$ velocity of fluid

$D \equiv$ characteristic dimension of system

$\nu \equiv \dfrac{\rho}{\mu} =$ kinematic viscosity of fluid

$\rho \equiv$ density of fluid

$\mu \equiv$ absolute viscosity of fluid

(b) $\text{Pr} \equiv \dfrac{c_p \mu}{k}$

$c_p \equiv$ specific heat at constant pressure of fluid

$\mu \equiv$ absolute viscosity of fluid

$k \equiv$ thermal conductivity of fluid

(c) $\text{Nu} \equiv \dfrac{hD}{k}$

$h \equiv$ film coefficient of fluid

$D \equiv$ characteristic dimension of system

$k \equiv$ thermal conductivity of fluid

(d) $\text{Gr} \equiv \dfrac{L^3 \rho^2 \beta \Delta T g}{\mu^2}$

$L \equiv$ characteristic length of body

$\rho \equiv$ density of fluid

$\beta \equiv$ coefficient of thermal volumetric expansion of fluid

$\Delta T \equiv$ temperature gradient between hot body and fluid at infinity

$g \equiv$ acceleration of gravity

$\mu \equiv$ absolute viscosity of fluid

---

3. Let the hotter plate be plate 1 and the cooler plate be plate 2.

$$A_1 F_{1-2} = A_2 F_{2-1}$$

The heat transfer per unit area is the emissive power, E.

$|E| = E_{1-2} = -E_{2-1}$

$= \dfrac{q_{1-2}}{A_1} = -\dfrac{q_{2-1}}{A_2}$

$= \sigma F_{1-2}(T_1^4 - T_2^4)$

$= \left(0.1713 \times 10^{-8} \dfrac{\text{BTU}}{\text{hr-ft}^2\text{-}^\circ\text{R}^4}\right)(1.0)$

$\quad \times [(1000^\circ\text{R})^4 - (530^\circ\text{R})^4]$

$= \boxed{1578 \text{ BTU/hr-ft}^2}$

4. Filament is body 1. Enclosure is body 2.

$T_2 = 460 + 70 = 530^\circ\text{R}$

$q_{1-2} = \sigma A_1 F_{1-2}(T_1^4 - T_2^4)$

$\quad = \sigma A_1 \epsilon_1 (T_1^4 - T_2^4)$

$\quad = \left(0.1713 \times 10^{-8} \dfrac{\text{BTU}}{\text{hr-ft}^2\text{-}^\circ\text{R}^4}\right)(0.5 \text{ in}^2)\left(\dfrac{\text{ft}}{12 \text{ in}}\right)^2$

$\qquad \times (0.35)[(5000^\circ\text{R})^4 - (530^\circ\text{R})^4]$

$\quad = \boxed{1301.1 \text{ BTU/hr}}$

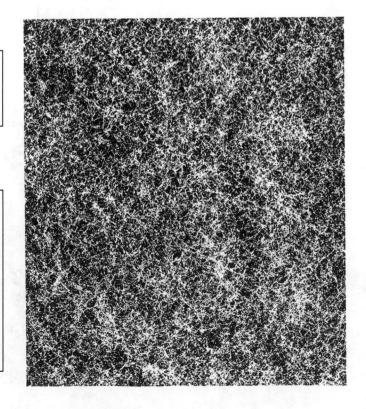

DETERMINATE STATICS

1. The block's freebody is

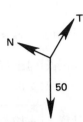

The 50 pounds is resolved into components parallel and perpendicular to the plane.

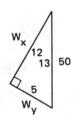

$$W_x = (50\,\text{lbf})\left(\frac{12}{\sqrt{(12)^2 + (5)^2}}\right) = (50\,\text{lbf})\left(\frac{12}{13}\right)$$

$$= \boxed{46.15\,\text{lbf}}$$

2. For the inclined load,

$$F_x = \left(\frac{3}{5}\right)(100\,\text{lbf}) = 60\,\text{lbf}$$

$$F_y = \left(\frac{4}{5}\right)(100\,\text{lbf}) = 80\,\text{lbf}$$

The moment about point A is

$$M_A = (600\,\text{lbf})(4\,\text{ft}) - (80\,\text{lbf})(9\,\text{ft})$$

$$= \boxed{1680\,\text{ft-lbf (clockwise)}}$$

$$\sum F_x = 200\,\text{lbf} + 60\,\text{lbf} = 260\,\text{lbf}$$

$$\sum F_y = -600\,\text{lbf} + 80\,\text{lbf} = -520\,\text{lbf}$$

$$F_R = \sqrt{(260)^2 + (-520)^2} = \boxed{581.4\,\text{lbf}}$$

$$\theta = \arctan\left(\frac{-520}{260}\right) = -63.4°$$

$$\text{location} = \frac{M_A}{\sum F_y} = \frac{1680\,\text{ft-lbf}}{520\,\text{lbf}}$$

$$= \boxed{3.23\,\text{ft from point A}}$$

3. $$\sum F_x = (100\,\text{lbf})\left(\frac{1}{\sqrt{2}}\right) - (200\,\text{lbf})(\cos\ 30°)$$

$$= \boxed{-102.5\,\text{lbf}}$$

$$\sum F_y = 300 + (100)\left(\frac{1}{\sqrt{2}}\right) + (200)(\sin\ 30°)$$

$$= \boxed{470.7\,\text{lbf}}$$

The moment about the center is

$$(300\,\text{lbf})(2\,\text{ft}) - (100\,\text{lbf})\left(\frac{1}{\sqrt{2}}\right)(2\,\text{ft}) - (200\,\text{lbf})(2\,\text{ft})$$

$$= \boxed{58.6\,\text{ft-lbf (counterclockwise)}}$$

4. (b) The centroid of the parabola is located at

$$y_c = \frac{3h}{5} = \frac{(3)(4\,\text{ft})}{5} = 2.4\,\text{ft up from } x\text{-axis}$$

The pressure at the centroid is the average pressure over the area.

$$\bar{p} = \gamma h_c = \gamma(4\,\text{ft} - y_c)$$

$$= \left(62.4\,\frac{\text{lbf}}{\text{ft}^3}\right)(4\,\text{ft} - 2.4\,\text{ft}) = \boxed{99.84\,\text{lbf/ft}^2}$$

(a) The area of the parabola is

$$A = \frac{4ah}{3} = \frac{(4)\left(\frac{4\,\text{ft}}{2}\right)(4\,\text{ft})}{3} = 10.667\,\text{ft}^2$$

The resultant force is

$$F_R = \bar{p}A = \left(99.8\,\frac{\text{lbf}}{\text{ft}^2}\right)(10.667\,\text{ft}^2) = \boxed{1064.6\,\text{lbf}}$$

The centroidal moment of inertia about a horizontal axis is

$$I_c = \frac{16ah^3}{175} = \frac{(16)(2\,\text{ft})(4\,\text{ft})^3}{175} = 11.70\,\text{ft}^4$$

The location of the resultant is

$$h_R = h_c + \frac{I_c}{Ah_c} = 1.6\,\text{ft} + \frac{11.70\,\text{ft}^4}{(10.667\,\text{ft}^2)(1.6\,\text{ft})}$$

$$= \boxed{2.286\,\text{ft from surface}}$$

(c) There are two hinges.

$$M = F_R h_c = \left(\frac{1064.6 \text{ lbf}}{2 \text{ hinges}}\right)(1.6 \text{ ft})$$

$$= \boxed{851.7 \text{ ft-lbf}}$$

5. $AB = \sqrt{(19)^2 + (24)^2} = 30.61 \text{ ft}$

Since the force in CB is 1000 lbf, the force in guy wire is

$$(1000 \text{ lbf})\left(\frac{30.61 \text{ ft}}{24 \text{ ft}}\right) = \boxed{1275.4 \text{ lbf}}$$

The force in the pole is

$$(1000 \text{ lbf})\left(\frac{19 \text{ ft}}{24 \text{ ft}}\right) = \boxed{791.7 \text{ lbf}}$$

6. Assume B_y is upward, A_y is downward, and A_x is to the left.

At point A:

$$\sum M_A = (B_y)(4) - (200)(6) + 500 = 0$$

$$B_y = \boxed{175 \text{ lbf upward}}$$

At point B:

$$\sum M_B = (A_y)(4) + 500 - (200)(6) = 0$$

$$A_y = \boxed{175 \text{ lbf downward}}$$

$$A_x = \boxed{200 \text{ lbf to the left}}$$

7. The total distributed load is

$$\left(\frac{1}{2}\right)(18 \text{ ft})\left(100 \frac{\text{lbf}}{\text{ft}}\right) = 900 \text{ lbf}$$

The resultant acts at $\left(\frac{2}{3}\right)(18) = 12 \text{ ft}$ up from point A.

At point A:

$$\sum M_A = (900 \text{ lbf})(12 \text{ ft}) - (T_1)(18 \text{ ft}) = 0$$

$$T_{\text{cord}} = \boxed{600 \text{ lbf to the left}}$$

At the tip:

$$\sum M_{\text{tip}} = (900 \text{ lbf})(6 \text{ ft}) - (F_A)(18 \text{ ft}) = 0$$

$$F_A = \boxed{300 \text{ lbf to the left}}$$

8. Assume F_B is directed towards A.

$$(F_B)_x = \frac{4}{5} F_B \text{ to the left}$$

$$(F_B)_y = \frac{3}{5} F_B \text{ down}$$

At point C, assuming counterclockwise is positive,

$$\sum M_C = (100)(8) + (F_B)_y(10) + (F_B)_x(8) - 500 = 0$$

$$F_B = -24.2 \text{ lbf}$$

$$(F_B)_x = (0.8)(-24.2 \text{ lbf}) = \boxed{-19.36 \text{ lbf to the left}}$$

$$= \boxed{19.36 \text{ lbf to the right}}$$

$$(F_B)_y = (0.6)(-24.2 \text{ lbf}) = \boxed{-14.52 \text{ lbf down}}$$

$$= \boxed{14.52 \text{ lbf up}}$$

Notice that distance BC is not the perpendicular moment arm.

9. $W = wL = \left(30 \frac{\text{lbf}}{\text{ft}}\right)(10 \text{ ft}) = 300 \text{ lbf}$

The resultant acts 5 ft from the hinge.

At the hinge:

$$\sum M_{\text{hinge}} = (T\sin 30°)(10) - (100)(10) - (300)(5) = 0$$

$$T = \boxed{500 \text{ lbf}}$$

10.

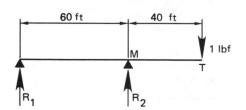

If the unit load is at the right end,

$$\sum F = 0: \quad R_1 + R_2 - 1 = 0$$

$$R_2 = 1 - R_1$$

$$\sum M_T = 0: \quad R_1(60 + 40) + R_2(40) = 0$$

$$100R_1 + (40)(1 - R_1) = 0$$

$$R_1 = -\frac{2}{3}$$

$$R_2 = 1 - R_1 = \frac{5}{3}$$

The right reaction at point M is a

$$\boxed{1.67 \text{ unit load up}}$$

11. $\sum M_C = (D_y)(6) - (8000)(6) + (1600)(16) = 0$

$D_y = 3733$ lbf up

Since DE is the only vertical member leaving point D,

$$DE = D_y = \boxed{3733 \text{ lbf up (compression)}}$$

12. By symmetry, $A_y = L_y$, assuming upward as positive,

$$2A_y = (60)(5) + (4)(5)$$
$$A_y = 160 \text{ kips}$$

For DE:

Assume DE is directed upward; cut as shown and sum vertical forces.

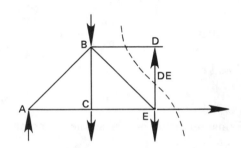

$$\sum F_y = 160 + DE - (60)(2) - 4 = 0$$

$$DE = \boxed{-36 \text{ kips up}}$$

$$\text{or } \boxed{36 \text{ kips down (compression)}}$$

For HJ:

Assume HJ is directed to the right; cut as shown, and sum moments about I.

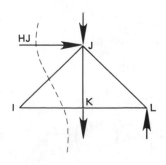

$\sum M_I = (HJ)(20) - (160)(60) + (60)(30) + (4)(30)$
$\qquad = 0$

$HJ = \boxed{384 \text{ kips to the right (compression)}}$

13. At pin F:

$$FG = 0 \text{ lbf}$$
$$FN = 2000 \text{ lbf}$$

At pin E:

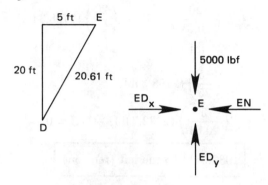

$$\sum F_y = 0 : ED_y - 5000 \text{ lbf} = 0$$
$$ED_y = 5000 \text{ lbf}$$
$$ED_x = \left(\frac{5}{20}\right)(5000 \text{ lbf}) = 1250 \text{ lbf}$$
$$\sum F_x = 0 : EN - 1250 \text{ lbf} = 0$$
$$EN = 1250 \text{ lbf}$$

At pin N:

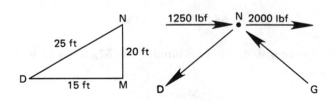

$$\sum F_y = 0 : ND_y = GN_y$$

Since the angles are the same, ND = GN.

$$\sum F_x = 0 : 2000 \text{ lbf} + 1250 \text{ lbf} - \frac{15}{25}ND - \frac{15}{25}GN$$
$$= 3250 \text{ lbf} - \frac{30}{25}ND$$

$$ND = 2708.3 \text{ lbf}$$

At pin G:

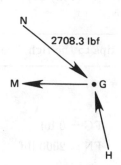

2708.3 lbf

M ← • G

N

H

$$\sum F_y = 0: \left(\frac{20}{20.61}\right)(GH) - \left(\frac{20}{25}\right)(2708.3 \text{ lbf}) = 0$$

$$GH = 2232.7 \text{ lbf}$$

$$\sum F_x = 0: \left(\frac{15}{25}\right)(2708.3 \text{ lbf})$$

$$- \left(\frac{5}{20.61}\right)(2232.7 \text{ lbf}) - MG = 0$$

$$MG = \boxed{1083.3 \text{ lbf to the left (tension)}}$$

14. Cut and sum moments about A.

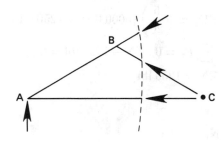

$$\mathbf{M_A} + \mathbf{M_{AB}} + \mathbf{M_{AC}} + \mathbf{M_{BC}} = 0$$

The moment arms associated with $\mathbf{M_A}$, $\mathbf{M_{AB}}$, and $\mathbf{M_{AC}}$ are zero, so

$$\mathbf{M_{BC}} = 0$$

Since the moment arm associated with $\mathbf{M_{BC}}$ is nonzero,

$$\boxed{\text{Force } \mathbf{BC} \text{ must be zero}}$$

15. By symmetry, $\mathbf{A_y} = \left(\frac{1}{2}\right)(100 + 100 + 100) = 150$.
Cut and sum moments about D.

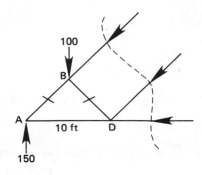

100

B

A ——— D
  10 ft

150

$$\sum M_D = (150)(10) - (100)(5) - (5)(BC_y) - (5)(BC_x)$$

$$= 1000 - (5)(BC)\cos 45° - (5)(BC)\sin 45°$$

$$= 1000 - (2)\left(\frac{1}{\sqrt{2}}\right)(5)(BC)$$

$$= 0$$

$$BC = \boxed{141.4 \text{ (compression)}}$$

16. $$S = c\left[\cosh\left(\frac{a}{c}\right) - 1\right]$$

$$10 \text{ ft} = c\left[\cosh\left(\frac{50 \text{ ft}}{c}\right) - 1\right]$$

By trial and error,

$$c = 126.6 \text{ ft}$$

The maximum tension occurs at the endpoint,

$$T = wy = w(c + S)$$

$$= \left(2\frac{\text{lbf}}{\text{ft}}\right)(126.6 \text{ ft} + 10 \text{ ft})$$

$$= \boxed{273.2 \text{ lbf}}$$

The midpoint tension is

$$H = wc = \left(2\frac{\text{lbf}}{\text{ft}}\right)(126.6 \text{ ft}) = \boxed{253.2 \text{ lbf}}$$

17. $$T = wy$$

$$y = \frac{T}{w} = \frac{500 \text{ lbf}}{2\frac{\text{lbf}}{\text{ft}}} = 250 \text{ ft}$$

$$y = c\left[\cosh\left(\frac{a}{c}\right)\right]$$

$$250 \text{ ft} = c\left[\cosh\left(\frac{50 \text{ ft}}{c}\right)\right]$$

By trial and error,

$$c = 245 \text{ ft}$$

$$S = y - c = 250 \text{ ft} - 245 \text{ ft}$$

$$= \boxed{5 \text{ ft}}$$

18. The horizontal distance between two towers is

$$\frac{120 \text{ ft}}{0.2} = 600 \text{ ft}$$

$$S_{\text{AC}} = c \left[\cosh \left(\frac{d}{c} \right) - 1 \right] = 30 \text{ ft}$$

$$\frac{30 \text{ ft} + c}{c} = \cosh \left(\frac{d}{c} \right)$$

$$c \left[\text{arccosh} \left(\frac{30 \text{ ft}}{c} + 1 \right) \right] = d$$

$$S_{\text{BC}} = c \left[\cosh \left(\frac{600 \text{ ft} - d}{c} \right) - 1 \right] = 150 \text{ ft}$$

$$150 \text{ ft} = c \left[\cosh \left(\frac{600 \text{ ft} - c \left[\text{arccosh} \left(\frac{30 \text{ ft}}{c} + 1 \right) \right]}{c} \right) - 1 \right]$$

By trial and error,

$$c = 591 \text{ ft}$$

(a) $H = wc = \left(10 \frac{\text{lbf}}{\text{ft}} \right) (591 \text{ ft}) = \boxed{5910 \text{ lbf}}$

$$T_{\text{A}} = w(S_{\text{AC}} + c) = \left(10 \frac{\text{lbf}}{\text{ft}} \right) (30 \text{ ft} + 591 \text{ ft})$$

$$= \boxed{6210 \text{ lbf}}$$

$$T_{\text{B}} = w(S_{\text{BC}} + c) = \left(10 \frac{\text{lbf}}{\text{ft}} \right) (150 \text{ ft} + 591 \text{ ft})$$

$$= \boxed{7410 \text{ lbf}}$$

(b) $d = c \left[\text{arccosh} \left(\frac{30 \text{ ft}}{c} + 1 \right) \right]$

$$= 591 \text{ ft} \left[\text{arccosh} \left(\frac{30}{591} + 1 \right) \right]$$

$$= \boxed{187.5 \text{ ft}}$$

19. $r = \sqrt{(7)^2 + (9)^2 + (4)^2} = 12.08$

$$F_x = F \left(\frac{x}{r} \right) = (190 \text{ lbf}) \left(\frac{7}{12.08} \right) = \boxed{110.10 \text{ lbf}}$$

$$F_y = F \left(\frac{y}{r} \right) = (190 \text{ lbf}) \left(\frac{9}{12.08} \right) = \boxed{141.56 \text{ lbf}}$$

$$F_z = F \left(\frac{z}{r} \right) = (190 \text{ lbf}) \left(\frac{4}{12.08} \right) = \boxed{62.91 \text{ lbf}}$$

20. Move the origin to the apex of the tripod so that the coordinates of the tripod base become

$$A = (5, 0, 0) - (0, 12, 0) = (5, -12, 0)$$

$$B = (0, 4, -8) - (0, 12, 0) = (0, -8, -8)$$

$$C = (-4, 5, 6) - (0, 12, 0) = (-4, -7, 6)$$

$$F_x = 1200 \quad F_y = 0 \quad F_z = 0$$

$$L_{\text{A}} = \sqrt{(5)^2 + (-12)^2 + (0)^2} = 13$$

$$\cos \theta_{\text{A},x} = \frac{5}{13} = 0.385$$

$$\cos \theta_{\text{A},y} = \frac{-12}{13} = -0.923$$

$$\cos \theta_{\text{A},z} = 0$$

$$L_{\text{B}} = \sqrt{(0)^2 + (-8)^2 + (-8)^2} = 11.31$$

$$\cos \theta_{\text{B},x} = \frac{0}{11.31} = 0$$

$$\cos \theta_{\text{B},y} = \frac{-8}{11.31} = -0.707$$

$$\cos \theta_{\text{B},z} = \frac{-8}{11.31} = -0.707$$

$$L_{\text{C}} = \sqrt{(-4)^2 + (-7)^2 + (6)^2} = 10.05$$

$$\cos \theta_{\text{C},x} = \frac{-4}{10.05} = -0.398$$

$$\cos \theta_{\text{C},y} = \frac{-7}{10.05} = -0.697$$

$$\cos \theta_{\text{C},z} = \frac{6}{10.05} = 0.597$$

The three equilibrium equations for the apex are

$$0.385\,F_A \qquad\qquad -0.398\,F_C = -1200$$
$$-0.923\,F_A - 0.707\,F_B - 0.697\,F_C = 0$$
$$-0.707\,F_B + 0.597\,F_C = 0$$

Solving these equations simultaneously yields

$$F_A = \boxed{-1793 \text{ (compression)}}$$

$$F_B = \boxed{1080 \text{ (tension)}}$$

$$F_C = \boxed{1279 \text{ (tension)}}$$

21. B is in the xy plane, and B_y is the only force in y-direction, so $B_y = 12,000$.

$$B_x = \left(\frac{12}{9}\right) B_y = 16,000$$

C is in the xz plane, and C_z is the only force in z-direction, so $C_z = 6000$.

$$C_x = \left(\frac{12}{12}\right)(6000) = 6000$$
$$A_x + B_x + C_x = 0$$
$$A_x = 22,000 \text{ to the left}$$

$A_x = 22,000$ left,	$A_y = 0$,	$A_z = 0$
$B_x = 16,000$,	$B_y = 12,000$,	$B_z = 0$
$C_x = 6000$,	$C_y = 0$,	$C_z = 6000$

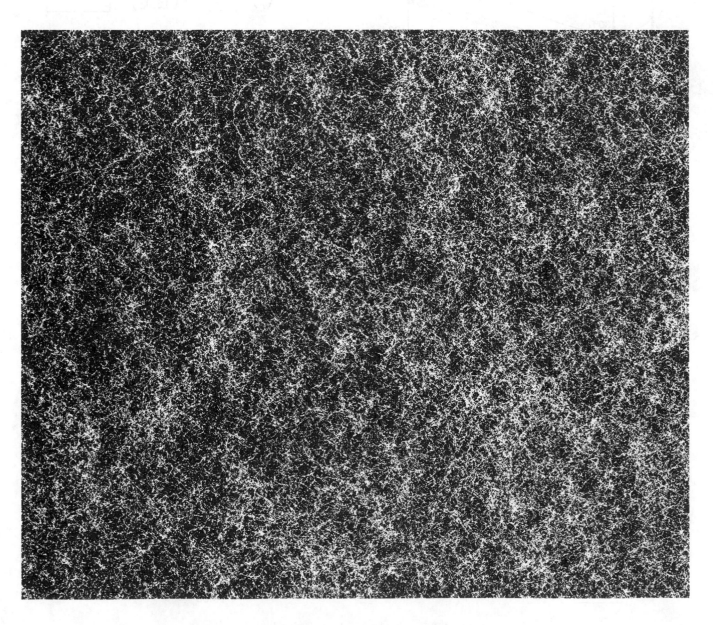

INDETERMINATE STATICS

1. The degree of indeterminacy is 3. Remove two re-actions (horizontal and vertical) from the pinned connection and one (vertical) from the mid-span support in order to make the structure statically determinate.

2. $\dfrac{\text{degree of}}{\text{indeterminacy}} = 3 + \dfrac{\text{no.}}{\text{members}} - (2)\left(\dfrac{\text{no.}}{\text{joints}}\right)$

$= 3 + 10 - (2)(6) = \boxed{1}$

3. The thermal strain is

$\varepsilon_{\text{th}} = \alpha \Delta T = \left(8.9 \times 10^{-6} \dfrac{1}{^\circ\text{F}}\right)(55^\circ\text{F}) = 0.000490$

The strain must be counteracted to maintain the rod in its original position.

$\sigma = E\varepsilon_{\text{th}} = \left(18 \times 10^6 \dfrac{\text{lbf}}{\text{in}^2}\right)(0.000490) = 8820\,\text{lbf/in}^2$

$F = \sigma A = \left(8820 \dfrac{\text{lbf}}{\text{in}^2}\right)(1\,\text{in})(2\,\text{in}) = \boxed{17{,}640\,\text{lbf}}$

4. (a) $\sigma = \dfrac{F}{A} = \dfrac{2250\,\text{lbf}}{\dfrac{(\pi)(2\,\text{in})^2}{4}} = \boxed{716.2\,\text{lbf/in}^2}$

(b) $\varepsilon = \dfrac{\sigma}{E} = \dfrac{\delta}{L_o}$

$\delta = L_o\dfrac{\sigma}{E} = (15\,\text{in})\left(\dfrac{716.2 \dfrac{\text{lbf}}{\text{in}^2}}{30 \times 10^6 \dfrac{\text{lbf}}{\text{in}^2}}\right)$

$= \boxed{3.58 \times 10^{-4}\,\text{in}}$

5. (a) Let F_c and F_{st} be the loads carried by the concrete and steel, respectively.

$F = F_c + F_{\text{st}} = 4500\,\text{lbf}$ [Eq. 1]

The deformation of the steel is

$\delta_{\text{st}} = \dfrac{F_{\text{st}} L}{A_{\text{st}} E_{\text{st}}}$

The deformation of the concrete is

$\delta_c = \dfrac{F_c L}{A_c E_c}$

The geometric constraint is $\delta_{\text{st}} = \delta_c$, or

$\dfrac{F_{\text{st}} L}{A_{\text{st}} E_{\text{st}}} = \dfrac{F_c L}{A_c E_c}$ [Eq. 2]

Solving Eqs. 1 and 2,

$F_c = \dfrac{F}{1 + \dfrac{A_{\text{st}} E_{\text{st}}}{A_c E_c}}$

$= \dfrac{4500\,\text{lbf}}{1 + \dfrac{\dfrac{(\pi)(0.4\,\text{in})^2}{4}\left(30 \times 10^6 \dfrac{\text{lbf}}{\text{in}^2}\right)}{\dfrac{(\pi)\left[(3\,\text{in})^2 - (0.4\,\text{in})^2\right]}{4}\left(2 \times 10^6 \dfrac{\text{lbf}}{\text{in}^2}\right)}}$

$= \boxed{3539\,\text{lbf}}$

$F_{\text{st}} = 4500\,\text{lbf} - F_c = 4500\,\text{lbf} - 3539\,\text{lbf} = \boxed{961\,\text{lbf}}$

$\sigma_{\text{st}} = \dfrac{F_{\text{st}}}{A_{\text{st}}} = \dfrac{961\,\text{lbf}}{\dfrac{(\pi)(0.4\,\text{in})^2}{4}} = \boxed{7647\,\text{lbf/in}^2}$

$\sigma_c = \dfrac{F_c}{A_c} = \dfrac{3539\,\text{lbf}}{\dfrac{(\pi)\left[(3\,\text{in})^2 - (0.4\,\text{in})^2\right]}{4}}$

$= \boxed{510\,\text{lbf/in}^2}$

(b) $\delta = \dfrac{F_{\text{st}} L}{A_{\text{st}} E_{\text{st}}} = \dfrac{(961\,\text{lbf})(12\,\text{in})}{\dfrac{(\pi)(0.4\,\text{in})^2}{4}\left(30 \times 10^6 \dfrac{\text{lbf}}{\text{in}^2}\right)}$

$= \boxed{3.06 \times 10^{-3}\,\text{in}}$

6. $A_{\text{st}} = \left(\dfrac{\pi}{4}\right)\left[(6\,\text{in})^2 - (5.4\,\text{in})^2\right] = 5.372\,\text{in}^2$

$A_b = \left(\dfrac{\pi}{4}\right)\left[(2\,\text{in})^2 - (1.6\,\text{in})^2\right] = 1.131\,\text{in}^2$

The thermal deformation that would occur if the pipes were free is

$\delta_b = \alpha_b L \Delta T = \left(10 \times 10^{-6} \dfrac{1}{^\circ\text{F}}\right)(60\,\text{in})(200^\circ\text{F})$

$= 0.120\,\text{in}$

$\delta_{\text{st}} = \alpha_{\text{st}} L \Delta T = \left(6.5 \times 10^{-6} \dfrac{1}{^\circ\text{F}}\right)(60\,\text{in})(200^\circ\text{F})$

$= 0.078\,\text{in}$

The two pipes expand by the same amount, δ, so that $\delta_{st} < \delta < \delta_b$.

$$\delta - \delta_{st} = \frac{F_{st}L}{A_{st}E_{st}} \qquad \text{[Eq. 1]}$$

$$\delta_b - \delta = \frac{F_bL}{A_bE_b} \qquad \text{[Eq. 2]}$$

F_{st} is a tensile force, F_b is a compressive force. Since there is no external force,

$$F_{st} = F_b = F$$

Subtracting Eq. 1 from Eq. 2,

$$\delta_b - \delta_{st} = \frac{FL}{A_{st}E_{st}} + \frac{FL}{A_bE_b}$$

$$= FL\left(\frac{1}{A_{st}E_{st}} + \frac{1}{A_bE_b}\right)$$

$$F = \frac{\dfrac{\delta_b - \delta_{st}}{L}}{\dfrac{1}{A_{st}E_{st}} + \dfrac{1}{A_bE_b}}$$

$$= \frac{\dfrac{0.120\text{ in} - 0.078\text{ in}}{60\text{ in}}}{\dfrac{1}{(5.372\text{ in}^2)\left(30\times10^6\frac{\text{lbf}}{\text{in}^2}\right)} + \dfrac{1}{(1.131\text{ in}^2)\left(15\times10^6\frac{\text{lbf}}{\text{in}^2}\right)}}$$

$$= 10{,}744\text{ lbf}$$

$$\sigma_{st} = \frac{F}{A_{st}} = \frac{10{,}744\text{ lbf}}{5.372\text{ in}^2} = \boxed{2000\text{ psi}}$$

$$\sigma_b = \frac{F}{A_b} = \frac{10{,}744\text{ lbf}}{1.131\text{ in}^2} = \boxed{9500\text{ psi}}$$

7. The deflection of the beam is $\delta_b = \dfrac{P_bL^3}{3EI}$ where P_b is the net load at the beam tip. If P_c is the tension in the cable,

$$\delta_c = \frac{P_cL}{AE}$$

$\delta_c = \delta_b$ is the constraint on the deformation. Therefore,

$$\frac{P_bL_b^3}{3EI} = \frac{P_cL_c}{AE} \qquad \text{[Eq. 1]}$$

Another equation is the equilibrium equation.

$$F - P_c = P_b \qquad \text{[Eq. 2]}$$

Solving Eqs. 1 and 2 simultaneously,

$$P_c = \frac{\dfrac{FL_b^3}{3I}}{\dfrac{L_c}{A} + \dfrac{L_b^3}{3I}}$$

$$= \frac{\dfrac{(270\text{ lbf})\left[(4\text{ ft})\left(\dfrac{12\text{ in}}{\text{ft}}\right)\right]^3}{(3)(10.0\text{ in}^4)}}{\dfrac{(2\text{ ft})\left(\dfrac{12\text{ in}}{\text{ft}}\right)}{0.0124\text{ in}^2} + \dfrac{\left[(4\text{ ft})\left(\dfrac{12\text{ in}}{\text{ft}}\right)\right]^3}{(3)(10.0\text{ in}^4)}}$$

$$= \boxed{177\text{ lbf}}$$

8. Let deflection down be positive. The deflection at the center of the beam is

$$\delta_b = \frac{5w(2L)^4}{384EI} - \frac{F(2L)^3}{48EI}$$

F is the force applied by the column at the beam center. The beam deflection must be equal to the shortening of the column.

$$\delta_c = \frac{Fh}{EA}$$

Since $\delta_b = \delta_c$,

$$\frac{Fh}{EA} = \frac{5w(2L)^4}{384EI} - \frac{F(2L)^3}{48EI}$$

$$F = \frac{5AwL^4}{24hI + 4AL^3} \qquad \text{[Eq. 1]}$$

Another equation is the equilibrium equation. Let R_1 and R_2 be the left and right support reactions, respectively, on the beam. By symmetry,

$$R_1 = R_2 = R$$

$$2R + F - 2wL = 0 \qquad \text{[Eq. 2]}$$

Solving Eqs. 1 and 2 simultaneously,

$$R = \frac{2wL - F}{2}$$

$$= \boxed{wL\left(\frac{48hI + 3AL^3}{48hI + 8AL^3}\right)}$$

9. (a) Let F_1 and F_2 and δ_1 and δ_2 be the tensions and the deformations in the cables, respectively. The moment equilibrium equation taken at the hinge of the rigid bar is

$$\sum M_o = aF_1 + (a + b)F_2 - (a + b + c)P = 0 \qquad \text{[Eq. 1]}$$

The relationship between the elongations is

$$\frac{\delta_1}{a} = \frac{\delta_2}{a+b}$$

This can be rewritten as

$$\frac{F_1 L_1}{AEa} = \frac{F_2 L_2}{AE(a+b)}$$

Since $\dfrac{L_1}{a} = \dfrac{L_2}{a+b}$, therefore,

$$F_1 = F_2 \qquad \text{[Eq. 2]}$$

Solving Eqs. 1 and 2,

$$aF + (a+b)F - (a+b+c)P = 0$$

$$F = F_1 = F_2$$

$$= \left(\frac{a+b+c}{2a+b}\right) P$$

$$= \left[\frac{6\,\text{ft} + 3\,\text{ft} + 3\,\text{ft}}{(2)(6\,\text{ft}) + 3\,\text{ft}}\right] (4500\,\text{lbf}) = \boxed{3600\,\text{lbf}}$$

(b) Take downward as a positive deflection. The slope of the rigid member is

$$\frac{d}{a+b+c}$$

$$\delta_1 = \left(\frac{F_1}{AE}\right) L_1 = \left(\frac{F_1}{AE}\right)\left(\frac{d}{a+b+c}\right) a$$

$$= \frac{3600\,\text{lbf}}{(0.124\,\text{in}^2)\left(10 \times 10^6\,\dfrac{\text{lbf}}{\text{in}^2}\right)}$$

$$\times \left(\frac{12\,\text{ft}}{6\,\text{ft} + 3\,\text{ft} + 3\,\text{ft}}\right)(6\,\text{ft})$$

$$= 1.74 \times 10^{-2}\,\text{ft} = \boxed{0.209\,\text{in}}$$

$$\delta_2 = \left(\frac{F_2}{AE}\right) L_2 = \left(\frac{F_2}{AE}\right)\left(\frac{d}{a+b+c}\right)(a+b)$$

$$= \frac{3600\,\text{lbf}}{(0.124\,\text{in}^2)\left(10 \times 10^6\,\dfrac{\text{lbf}}{\text{in}^2}\right)}$$

$$\times \left(\frac{12\,\text{ft}}{6\,\text{ft} + 3\,\text{ft} + 3\,\text{ft}}\right)(6\,\text{ft} + 3\,\text{ft})$$

$$= 2.61 \times 10^{-2}\,\text{ft} = \boxed{0.314\,\text{in}}$$

10. Let subscript s refer to the supported beam and subscript c refer to the cantilever beam. The deflections are equal.

$$\delta_s = \delta_c$$

$$\frac{P_s L^3}{48EI} = \frac{P_c L^3}{3EI}$$

$$P_s = 16 P_c \qquad \text{[Eq. 1]}$$

P_s and P_c are the net loads exerted on the supported beam center and the cantilever beam tip, respectively. The equilibrium equation for the supported beam is

$$\sum F_{s,y} = 2R - P_s = 0 \qquad \text{[Eq. 2]}$$

The equilibrium equation for the cantilever beam is

$$\sum F_{c,y} = P_c = F - P_s \qquad \text{[Eq. 3]}$$

Solving Eqs. 1 and 3 simultaneously,

$$P_c = \frac{F}{17}$$

$$P_s = \frac{16}{17}F$$

From Eq. 2,

$$R = \boxed{\frac{8}{17}F}$$

11. Assume deflection downward is positive.

step 1: Remove support 2 to make the structure statically determinate.

step 2: The deflection at the location of (removed) support 2 is the sum of the deflections induced by the discrete load and the distributed load.

From a beam deflection table,

$$\delta_{\text{discrete}} = \frac{Pb}{6EIL}\left[\left(\frac{L}{b}\right)(x-a)^3 + (L^2 - b^2)x - x^3\right]$$

$$(P = 5000,\ L = 15,\ a = 6,\ b = 9,\ x = 10)$$

From a beam deflection table,

$$\delta_{\text{distributed}} = -\frac{w}{24EI}(L^3 x - 2Lx^3 + x^4)$$

$$(w = 1000)$$

step 3: The deflection induced by R_2 alone considered as a load is

$$\delta_{R_2} = \frac{-R_2 a^2 b^2}{3EIL} \qquad (a = 10,\ b = 5)$$

step 4: The total deflection at the location of support 2 is zero.

$$\delta_{\text{discrete}} + \delta_{\text{distributed}} + \delta_{R_2} = 0$$

$$\left[\frac{(5000)(9)}{(6)(15)}\right]\left[\left(\frac{15}{9}\right)(10-6)^3 + [(15)^2 - (9)^2](10) - (10)^3\right]$$

$$+ \left(\frac{1000}{24}\right)[(15)^3(10) - (2)(15)(10)^3 + (10)^4]$$

$$- \frac{R_2(10)^2(5)^2}{(3)(15)} = 0$$

$$R_2 = \boxed{15,233}$$

12. First, remove the prop to make the structure statically determinate. The deflection induced by the distributed load at the tip is

$$\delta_{\text{distributed}} = \frac{-wL^4}{30EI} \quad \text{(down)}$$

The deflection caused by load R_1 is

$$\delta_R = \frac{RL^3}{3EI} \quad \text{(up)}$$

Since the deflection is actually zero at the tip,

$$\delta_{\text{distributed}} + \delta_R = 0$$

$$\frac{-wL^4}{30EI} + \frac{RL^3}{3EI} = 0$$

$$R = \frac{wL}{10} = \frac{(1000)(10)}{10} = \boxed{1000}$$

13. Use the three-moment method. The first moment of the area is

$$A_1a = \tfrac{1}{6}F(L^2 - c^2)$$

$$= (\tfrac{1}{6})(13,600)(22.5)\left((32.5)^2 - (22.5)^2\right)$$

$$= 28,050,000$$

Since there is no force between R_2 and R_3, $A_2b = 0$.

The left and right ends of the beam are simply supported; M_1 and M_3 are zero. Therefore, the three-moment equation becomes

$$2M_2(32.5 + 20) = (-6)\left(\frac{28,050,000}{32.5}\right)$$

$$M_2 = -49,318.7$$

M_2 can be written in terms of the load and reactions to the left of support 2.

$$M_2 = (-13,600)(10) + (32.5)(R_1) = -49,318.7$$

$$R_1 = \boxed{2667.1 \text{ lbf}}$$

Now that R_1 is known, moments can be taken about support 3 to the left.

$$\sum M_3 = (2667.1)(52.5) - (13,600)(30) + (R_2)(20) = 0$$

$$R_2 = \boxed{13,398.9 \text{ lbf}}$$

R_3 can be obtained by taking moments about support 1.

$$\sum M_1 = (22.5)(13,600) - (32.5)(13,398.9)$$

$$- (52.5)(R_3) = 0$$

$$R_3 = \boxed{-2466.0 \text{ lbf}}$$

14. Use the three-moment method. The first moments of the areas are

$$A_1a = \frac{1}{6}Fc(L_1^2 - c^2) = \left(\frac{1}{6}\right)(5000)(4)\left[(10)^2 - (4)^2\right]$$

$$= 280,000$$

$$A_2b = \frac{wL_2^4}{24} = \frac{(1000)(5)^4}{24} = 26,042$$

The two ends of the beam are simply supported; M_1 and M_3 are zero. Therefore, the three moment equation becomes

$$2M_2(10 + 5) = (-6)\left(\frac{280,000}{10} + \frac{26,042}{5}\right)$$

$$M_2 = -6642$$

M_2 can also be written in terms of the load and reactions to the left of support 2.

$$M_2 = (-5000)(6) + 10R_1 = -6642$$

$$R_1 = \boxed{2336}$$

Sum the moments around support 3.

$$\sum M_3 = (2336)(15) - (5000)(11) + 5R_2$$

$$- (5000)(2.5) = 0$$

$$R_2 = \boxed{6492}$$

Sum the moments around support 1.

$$\sum M_1 = (5000)(4) - (6492)(10) + (5000)(12.5)$$

$$- 15R_3 = 0$$

$$R_3 = \boxed{1172}$$

15. The equilibrium requirement is

$$R_1 + R_2 = 1000$$

The moment equation at support 1 is

$$\sum M_{R_1} = M_1 + M_2 + (1000)(30) - 40R_2 = 0$$

From a table of fixed-end moments,

$$M_1 = \frac{-Fb^2a}{L^2} = \frac{-(1000)(10)^2(30)}{(40)^2} = -1875$$

$$M_2 = \frac{Fa^2b}{L^2} = \frac{(1000)(30)^2(10)}{(40)^2} = 5625$$

$$R_2 = \boxed{843.75}$$

The moment equation at support 2 is

$$\sum M_{R_2} = M_1 + M_2 - (1000)(10) + 40R_1 = 0$$

$$R_1 = \boxed{156.25}$$

ENGINEERING MATERIALS

1. A polymer is a large molecule in the form of a long chain of repeating units called monomers, often abbreviated as *mers*.

2. Bifunctional monomers yield two reaction sites per monomer. Trifunctional monomers yield three reaction sites per monomer. Tetrafunctional monomers yield four reaction sites per monomer.

3. When heated, thermosetting polymers harden while forming complex, three-dimensional networks. Thermoplastic polymers, however, soften when heated; moreover, their chemical structures are not affected by heat.

4. An initiator is a substance used to start addition polymerization.

5. A saturated polymer cannot bond with a mer; an unsaturated polymer can.

6. The degree of polymerization is the average number of mers in a polymer.

7. In addition polymerization, mers combine sequentially into chains as their double bonds break, but no other products are produced.

Condensation polymerization is similar to addition polymerization, but additional products are produced. The repeating units formed in the condensation process are not similar to the mers from which they are formed.

8. Bifunctional mers form linear polymers, whereas trifunctional and tetrafunctional mers form network polymers.

9. The vinyl chloride mer (with double bond broken) is

$$
\begin{array}{c}
\text{H} \quad\; \text{H} \\
| \quad\;\; | \\
-\text{C}-\text{C}- \\
| \quad\;\; | \\
\text{Cl} \quad \text{H}
\end{array}
$$

With 20% efficiency, 5 molecules of HCl per PVC molecule are needed to supply the end H and Cl atoms. This is the same as 5 moles of HCl per mole of PVC.

$$
\frac{(5)\left(6.022 \times 10^{23}\; \dfrac{\text{molecules HCl}}{\text{mol}}\right)}{7000\; \dfrac{\text{g PVC}}{\text{mol}}}
$$

$$
= \boxed{4.3 \times 10^{20}\; \text{molecules HCl/g PVC}}
$$

10. Molecular weights:

$$
H_2O_2 = 34\; \text{g/mol}; \quad C_2H_4 = 28\; \text{g/mol}
$$

The number of H_2O_2 molecules in 10 ml is

$$
\frac{(10\; \text{ml})\left(1\; \dfrac{\text{g}}{\text{ml}}\right)\left(\dfrac{0.2}{100}\right)\left(6.022 \times 10^{23}\; \dfrac{\text{molecules}}{\text{mol}}\right)}{34\; \dfrac{\text{g}}{\text{mol}}}
$$

$$
= 3.54 \times 10^{20}\; \text{molecules}
$$

The number of ethylene molecules is

$$
\frac{(12\; \text{g})\left(6.022 \times 10^{23}\; \dfrac{\text{molecules}}{\text{mol}}\right)}{28\; \dfrac{\text{g}}{\text{mol}}}
$$

$$
= 2.58 \times 10^{23}\; \text{molecules}
$$

Since it takes 1 H_2O_2 molecule (that is, 2 OH$^-$ radicals) to stabilize a polyethylene molecule, there are 3.54×10^{20} polymers. The degree of polymerization is

$$
\boxed{\frac{2.58 \times 10^{23}\; \text{mers}}{3.54 \times 10^{20}\; \text{polymers}} = 729\; \text{mers/polymer}}
$$

11. Properties of typical ceramics:

Chemical
- high melting point
- low thermal conductivity
- high corrosion (acid) resistance
- low coefficient of thermal expansion

Mechanical
- high hardness
- high compressive strength
- low ductility (brittleness)
- high shear resistance (low slip)

Electrical
- low electrical conductivity

12. (a) kaolinite clay; $Al_2Si_2O_5(OH)_4$
(b) periclase; MgO
(c) quartz; SiO_2
(d) Al_2O_3
(e) CaF_2

13. (a)   (b)

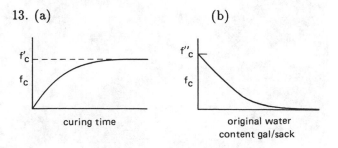

CRYSTALLINE STRUCTURES

1. (a) It follows directly from the Pythagorean theorem that the diagonal of the front face of the cube (whose length is the distance between atoms 1 and 2) has length $\sqrt{2}a$.

(b) It follows that the diagonal of the cube (whose length is the distance between atoms 1 and 3) has length $\sqrt{3}a$.

2.

direction	x-factor	y-factor	z-factor	crystallographic direction
a	1	0	1	$[101]$
b	1	$-\frac{1}{2}$	0	$[2\bar{1}0]$
c	0	1	0	$[010]$

3.

plane	x-intercept	y-intercept	z-intercept	Miller indices
a	1	$\frac{1}{2}$	∞	(120)
b	1	∞	∞	(100)
c	1	-1	1	$(1\bar{1}1)$

4. (a) 1 atom in the center and 1/8 atom is at each corner, so

$$1 \text{ atom} + (8)\left(\frac{1}{8} \text{ atom}\right) = \boxed{2 \text{ atoms}}$$

(b) 1/8 atom is at each corner and 1/2 atom is on each face, so

$$(8)\left(\frac{1}{8} \text{ atom}\right) + (6)\left(\frac{1}{2} \text{ atom}\right) = \boxed{4 \text{ atoms}}$$

(c) 1/12 atom is at each narrow corner and 1/6 atom is at each wide corner, so

$$(4)\left(\frac{1}{12} \text{ atom}\right) + (4)\left(\frac{1}{6} \text{ atom}\right) = \boxed{1 \text{ atom}}$$

5.

θ	$\sin^2\theta$	multiples of 0.0187	multiples rounded off
13.7°	0.056	2.99	3
16.0°	0.076	4.06	4
22.9°	0.151	8.07	8
27.1°	0.208	11.12	11
28.3°	0.225	12.03	12
32.6°	0.290	15.51	16

(0.0187 is determined by inspection or by trial and error.)

(a) $\boxed{\text{The material structure is compatible with that of FCC.}}$

(b) $d_{hkl} = \dfrac{\lambda}{2\sin\theta} = \dfrac{1.54 \text{ Å}}{2\sqrt{0.0187}}$

$\quad\quad = \boxed{5.63 \text{ Å } (5.63 \times 10^{-10} \text{ m})}$

6. $d_{hkl} = \dfrac{\lambda}{2\sin\theta} = \dfrac{0.58 \text{ Å}}{2\sin 9.5°}$

$\quad\quad = \boxed{1.76 \text{ Å } (1.76 \times 10^{-10} \text{ m})}$

7. (a) The smaller cation, of radius r, has a coordination number of 6 and lies at the geometric center of an octahedron with an anion of radius R lying at each corner so that the anions are touching.

Consider the following sectional view taken through the geometric center of the molecule

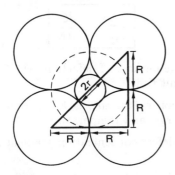

From the Pythagorean theorem,

$$4R^2 + 4R^2 = (2R + 2r)^2$$
$$r = (\sqrt{2} - 1)R = (\sqrt{2} - 1)(1.32 \text{ Å})$$
$$= \boxed{0.55 \text{ Å } (5.5 \times 10^{-11} \text{ m})}$$

(b) The smaller cation, of radius r, has a coordination number of 8 and lies at the geometric center of a cube with an anion of radius R lying at each corner so that the anions are touching.

Since the diagonal of the cube has length $2R + 2r$, it follows from the Pythagorean theorem that

$$4R^2 + 8R^2 = (2R + 2r)^2$$
$$r = (\sqrt{3} - 1)R = (\sqrt{3} - 1)(1.32 \text{ Å})$$
$$= \boxed{0.97 \text{ Å } (9.7 \times 10^{-11} \text{ m})}$$

8. packing factor $= PF = \dfrac{\text{volume of atoms}}{\text{volume of cell}}$

(a) For simple cubic (1 atom per cell),

$$PF = \dfrac{\frac{4}{3}\pi r^3}{(2r)^3} = \boxed{0.52}$$

(b) For FCC (4 atoms per cell),

$$PF = \dfrac{4\left(\frac{4}{3}\pi r^3\right)}{(2\sqrt{2}r)^3} = \boxed{0.74}$$

(c) For BCC (2 atoms per cell),

$$PF = \dfrac{2\left(\frac{4}{3}\pi r^3\right)}{\left(\frac{4}{\sqrt{3}}r\right)^3} = \boxed{0.68}$$

9. Fick's first law of diffusion is

$$\boxed{J = -D\dfrac{dC}{dx}}$$

J, a flux, is the number of defects moving across a unit surface per unit time.

D, the diffusion coefficient, is a function of material, activation energy, and temperature.

dC/dx is the defect concentration gradient in the direction of J.

Fick's second law of diffusion is

$$\boxed{\dfrac{dC}{dt} = D\dfrac{d^2C}{dx^2}}$$

dC/dt is the rate of change of the defect concentration with respect to time, and d^2C/dx^2 is the rate of change of the defect concentration gradient in the direction of J.

10. $J = -D\dfrac{dC}{dx}$

$$= -\left(0.4\times10^{-8}\,\dfrac{\text{cm}^2}{\text{s}}\right)\left(-\dfrac{0.5\,\frac{\text{atoms}}{\text{cm}^3}}{5\text{ cm}}\right)$$

$$= 4\times10^{-10}\,\dfrac{\text{atoms}}{\text{cm}^2\cdot\text{s}}$$

The number of silicon atoms is

$$(2\text{ cm}^2)(1\text{ s})\left(4\times10^{-10}\,\dfrac{\text{atoms}}{\text{cm}^2\cdot\text{s}}\right) = \boxed{8\times10^{-10}\text{ atoms}}$$

11. (a) A vacancy is a missing atom in a lattice.

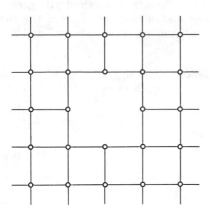

(b) An interstitiality is a foreign atom occupying an interstitial site that would be a void in a pure crystal.

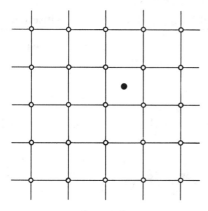

(c) A Schottky defect is an electrically neutral defect of ionic lattices, consisting of two vacancies: one cation and one anion.

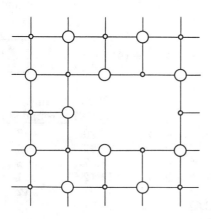

(d) A Frenkel defect is a vacant lattice site whose missing atom occupies an interstitial site that would be a void in a pure crystal.

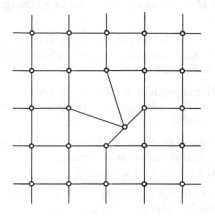

(e) An edge dislocation is an incomplete plane of atoms within a lattice.

(f) In a screw dislocation, the lattice planes are non-planar and spiral around the dislocation line.

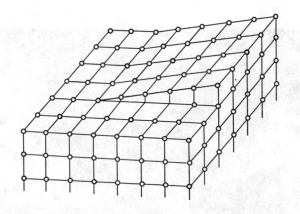

(g) A mixed dislocation constitutes an edge dislocation and a screw dislocation.

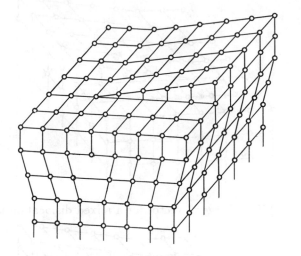

(h) A grain boundary is the interface between two or more crystals.

12. (a) The Burgers' vector for an edge dislocation is

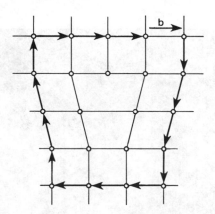

MATL SCI
Cryst Struct

(b) The Burgers' vector for a screw dislocation is

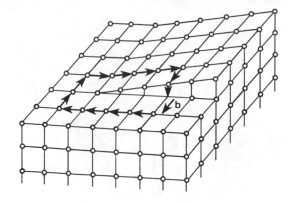

(c) The Burgers' vector for a mixed dislocation is

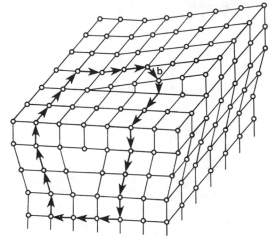

13. (a) Creep is a gradual, long-term increase in strain that occurs despite the absence of change in any other variable.

(b) Dislocation climb is the motion of an edge dislocation perpendicular to the slip plane.

(c) Hot working is a forming process that occurs above the recrystallization temperature.

(d) Cold working is a forming process that occurs below the recrystallization temperature.

14. (a) Ferromagnetism is exhibited by materials that experience a spontaneous and permanent alignment of electron spins when exposed to a magnetic field; the alignment of electron spins results in a strong magnetic field.

(b) Paramagnetism is the weak attractive effect in most alkali and transition metals when exposed to an external magnetic field.

(c) Diamagnetism is the weak repulsive effect in most nonmetals and organic materials when exposed to a magnetic field.

(d) Ferrimagnetism is the strong magnetism that occurs in crystals of certain ceramic compounds, such as ferrites, spinels, and garnets.

(e) Anti-ferromagnetism is exhibited by materials that are weakly attracted to a magnet.

15. $\boxed{1414^\circ \text{F}}$

16. $B = \dfrac{m}{A} = \dfrac{34 \times 10^{-3}\,\text{Wb}}{0.02\,\text{m}^2} = \boxed{1.7\,\text{T}}$

17. $\qquad B = \dfrac{m}{A}$

(a) $\quad m = BA = \left(1.2\,\dfrac{\text{Wb}}{\text{m}^2}\right)\pi(0.005\,\text{m})^2$

$\qquad = \boxed{9.4 \times 10^{-5}\,\text{Wb}}$

(b) $p_m = md = (9.4 \times 10^{-5}\,\text{Wb})(0.1\,\text{m})$

$\qquad = \boxed{9.4 \times 10^{-6}\,\text{Wb·m}}$

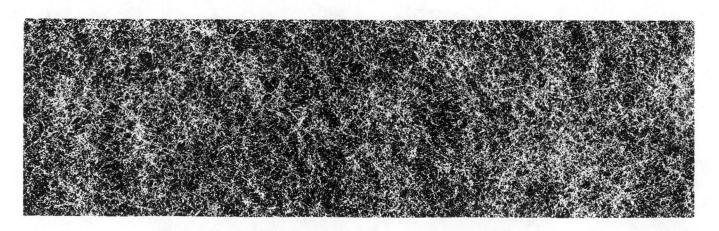

MATERIAL TESTING

1. $\sigma = s(1 + e) = \left(20{,}000 \, \dfrac{\text{lbf}}{\text{in}^2}\right)\left(1 + 0.0200 \, \dfrac{\text{in}}{\text{in}}\right)$

$= \boxed{20{,}400 \, \text{lbf/in}^2 \, \text{(psi)}}$

$\epsilon = \ln(1 + e) = \ln\left(1 + 0.0200 \, \dfrac{\text{in}}{\text{in}}\right)$

$= \boxed{0.0198 \, \text{in/in}}$

2. (a) 0.5% parallel offset yield strength: $\boxed{70 \, \text{ksi}}$

(b) Elastic modulus: $\dfrac{60 \, \text{ksi}}{0.02} = \boxed{3000 \, \text{ksi}}$

(c) Ultimate strength: $\boxed{80 \, \text{ksi}}$

(d) Fracture strength: $\boxed{70 \, \text{ksi}}$

(e) Percent elongation after fracture: $\boxed{5.7\%}$

3. $G = \dfrac{E}{2(1 + \nu)} = \dfrac{3000 \, \text{ksi}}{2(1 + 0.3)} = \boxed{1154 \, \text{ksi}}$

4. One measure of ductility is the reduction in area at failure.

$\dfrac{4.00 \, \text{in}^2 - 3.42 \, \text{in}^2}{4.00 \, \text{in}^2} = \boxed{0.145 \, (14.5\%)}$

5. The stress-strain curve may be approximated as follows:

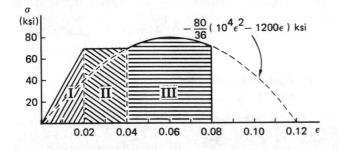

toughness = area under curve

= area of region I + area of region II

+ area of region III

$= \dfrac{1}{2}(0.02)(71 \, \text{ksi}) + (0.02)(71 \, \text{ksi})$

$+ \displaystyle\int_{0.04}^{0.08} -\dfrac{80}{36}(10^4\epsilon^2 - 1200 \, \epsilon)d\epsilon \, \text{ksi}$

$= 0.71 \, \text{ksi} + 1.42 \, \text{ksi} + 3.08 \, \text{ksi} = \boxed{5.21 \, \text{ksi}}$

6. The strain-time curve is

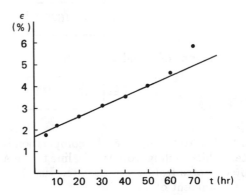

The steady-state creep rate, the slope of the line through points 2 through 7, is

$\dfrac{\Delta\epsilon}{\Delta t} = \dfrac{0.046 - 0.022}{60 \, \text{hr} - 10 \, \text{hr}} = \boxed{0.00048 \, (0.048 \, \%/\text{hr})}$

7. (a) Fatigue is a type of failure experienced by a material after repeated loadings, even if the stress level never exceeds its ultimate strength.

(b) An S-N curve is a curve used to describe the number of applications, N, of a stress with amplitude S repeated until failure.

(c) Fatigue life, denoted by N_p, is the number of cycles, N, associated with a particular stress level, S, on an S-N curve.

(d) The endurance limit, denoted by S'_e, is the stress amplitude level below which a material will withstand an infinite number of stress loadings without failing.

(e) Fatigue strength, denoted by S_p, is the stress associated with a particular number of loadings, N, on an S-N curve.

8. $\boxed{S_u \approx 500 \, (\text{BHN})}$

THERMAL TREATMENT OF METALS

1. $\left(\dfrac{0.40}{100}\right)(10 \text{ lbm}) = \boxed{0.04 \text{ lbm}}$

2. (a)   Composition of $\alpha = \boxed{0.4\% \text{ A; } 99.6\% \text{ B}}$

       Composition of $\beta = \boxed{4.7\% \text{ A; } 95.3\% \text{ B}}$

(b) Amount of α (solid) $= \dfrac{2-1}{2-0.4}$

$$= \boxed{0.62 \ (62\%)}$$

Amount of liquid $= 1 - 0.62$

$$= \boxed{0.38 \ (38\%)}$$

3. From 1650°F to the A_3 line, the composition is 100% austenite. While cooling from the A_3 line to the A_1 line, ferrite precipitates from the austenite. Just before the A_1 line, the amount of austenite is

$$\frac{0.5 - 0.025}{0.8 - 0.025} = 0.613 \ (61.3\%)$$

After cooling below the A_1 line, all the austenite turns into pearlite and remains as pearlite. The amount of pearlite formed after the alloy is cooled to room temperature is

$$(0.613)(50 \text{ lbm}) = \boxed{30.6 \text{ lbm}}$$

4. Recrystallization is the heating of any metal to induce strain-free grain growth within the grains already formed, thereby relieving stresses generated during cold-working.

Since recrystallization involves strain-free grain growth within the grains already formed, the larger the average initial grain size, the longer the recrystallization time.

The rate of recrystallization increases as temperature increases.

In general, complete recrystallization occurs in time spans of approximately an hour.

Grains grow from a nucleus and proceed from that nucleus outward.

5. (a) Normalizing is heating above the critical point, followed by air cooling.

(b) Tempering is holding an alloy for a certain amount of time below the critical temperature.

(c) Austempering is an interrupted quenching process resulting in an austenite-to-bainite transition.

(d) Ausforming is any forming process that is performed on semi-cooled alloy steels in the temperature range of 600–800°F, plastically deforming metastable austenite before it is converted into martensite. The two steps of quench and strain hardening produce an exceptionally strong product.

(e) Martempering is an interrupted quenching process resulting in an austenite-to-tempered martensite transition.

(f) Precipitation hardening is the primary method of hardening non-allotropic alloys, and constitutes at least two (sometimes three) steps: precipitation, quenching, and, optionally, artificial aging.

(g) Age hardening is precipitation hardening.

(h) Strain hardening is the increase in ultimate tensile strength (i.e., hardening) that results during strain working (i.e., any forming process that occurs below the annealing temperature). Also known as work hardening.

6. Precipitation hardening.

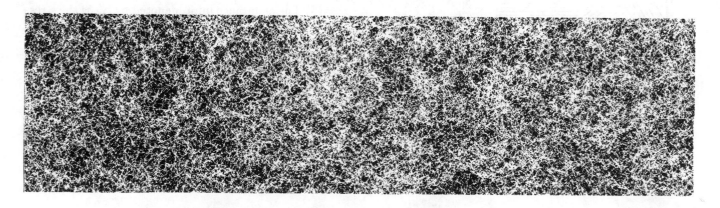

MANUFACTURING PROCESSES

1.   beam width $= 8$ in

$b = $ tool width $= 1$ in

Eight cutting passes are necessary. The total cutting length is

$$(8)(3\,\text{ft}) = 24\,\text{ft}$$

The cutting speed is

$$v = \frac{24\,\text{ft}}{10\,\text{min}} = \frac{24\,\text{ft}}{(10\,\text{min})\left(60\,\frac{\text{sec}}{\text{min}}\right)} = 0.04\,\text{ft/sec}$$

$$Z_w = bt_o v = \frac{\text{mass rate}}{\rho}$$

ρ is the density of wood, 30 lbm/ft^3.

The cutting depth is

$$t_o = \frac{\text{mass rate}}{\rho b v}$$

$$= \frac{\left(2\,\frac{\text{oz}}{\text{min}}\right)\left(\frac{\text{lbm}}{16\,\text{oz}}\right)\left(\frac{\text{min}}{60\,\text{sec}}\right)}{\left(30\,\frac{\text{lbm}}{\text{ft}^3}\right)(1\,\text{in})\left(\frac{\text{ft}}{12\,\text{in}}\right)\left(0.04\,\frac{\text{ft}}{\text{sec}}\right)}$$

$$= 0.0208\,\text{ft}$$

$$\phi = \tan^{-1}\left(\frac{r\cos\alpha}{1 - r\sin\alpha}\right)$$

$$= \tan^{-1}\left[\frac{(0.5)(\cos 45°)}{1 - (0.5)(\sin 45°)}\right] = 28.68°$$

$$F_s = F_h\cos\phi + F_v\sin\phi$$
$$= (10\,\text{lbf})(\cos 28.68°) + (15\,\text{lbf})(\sin 28.68°)$$
$$= 15.97\,\text{lbf}$$

$$\tau_{\text{ave}} = \frac{F_s\sin\phi}{bt_o} = \frac{(15.97\,\text{lbf})(\sin 28.68°)}{(1\,\text{in})(0.0208\,\text{ft})\left(12\,\frac{\text{in}}{\text{ft}}\right)}$$

$$= \boxed{30.7\,\text{lbf/in}^2}$$

2. (a) $P_{\text{heating}} = \dfrac{Q}{\Delta t} = \dfrac{mc_p\Delta T}{\Delta t} = \dfrac{\rho V c_p\Delta T}{\Delta t}$

$$= \frac{\left(172\,\frac{\text{lbm}}{\text{ft}^3}\right)\left[(4\,\text{in})\left(\frac{\text{ft}}{12\,\text{in}}\right)\right]^3\left(0.2\,\frac{\text{BTU}}{\text{lbm-°F}}\right)(35°\text{F})\left(778\,\frac{\text{ft-lbf}}{\text{BTU}}\right)}{(50\,\text{sec})\left(550\,\frac{\text{ft-lbf}}{\text{hp-sec}}\right)}$$

$$= 1.26\,\text{hp}$$

$P_{\text{cutting}} = P_{\text{heating}} = 1.26\,\text{hp}$

$$v = \pi D n$$
$$= (\pi)(0.2\,\text{in})\left(\frac{\text{ft}}{12\,\text{in}}\right)(1500\,\text{rpm})$$
$$= 78.5\,\text{ft/min}$$

$$F_h = \frac{P_{\text{cutting}}}{v} = \frac{(1.26\,\text{hp})\left(33{,}000\,\frac{\text{ft-lbf}}{\text{hp-min}}\right)}{78.5\,\frac{\text{ft}}{\text{min}}}$$

$$= \boxed{530\,\text{lbf}}$$

(b) $\eta = \dfrac{P_{\text{cutting}}}{P_{\text{electric}}} = \dfrac{1.26\,\text{hp}}{1.5\,\text{hp}} = \boxed{0.84}$

3. (a) $\alpha = 90° - \theta - \omega = 90° - 20° - 50° = 20°$

$$\phi = \tan^{-1}\left(\frac{r\cos\alpha}{1 - r\sin\alpha}\right)$$

$$= \tan^{-1}\left[\frac{(0.5)(\cos 20°)}{1 - (0.5)(\sin 20°)}\right] = 29.54°$$

$$\sigma = \frac{F_{ns}\sin\phi}{bt_o}$$

$$F_{ns} = F_h\sin\phi + F_v\cos\phi$$

Compute F_h from the cutting power.

$$F_h = \frac{\text{cutting power}}{v}$$

$$= \frac{(5.4\,\text{hp})\left(33{,}000\,\frac{\text{ft-lbf}}{\text{hp-min}}\right)}{\left(4\,\frac{\text{in}}{\text{sec}}\right)\left(\frac{\text{ft}}{12\,\text{in}}\right)\left(60\,\frac{\text{sec}}{\text{min}}\right)}$$

$$= 8910\,\text{lbf}$$

F_v can also be calculated from

$$F_s = F_h\cos\phi - F_v\sin\phi$$

$$\tau = \frac{F_s\sin\phi}{bt_o}$$

$$F_s = \frac{\tau bt_o}{\sin\phi} = \frac{\left(25{,}000\,\frac{\text{lbf}}{\text{in}^2}\right)(1\,\text{in})(0.1\,\text{in})}{\sin 29.54°}$$

$$= 5071\,\text{lbf}$$

$$F_v = \frac{F_h \cos\phi - F_s}{\sin\phi}$$

$$= \frac{(8910\,\text{lbf})(\cos 29.54°) - 5071\,\text{lbf}}{\sin 29.54°}$$

$$= 5437\,\text{lbf}$$

$$F_{ns} = F_h \sin\phi + F_v \cos\phi$$

$$= (8910\,\text{lbf})(\sin 29.54°) + (5437\,\text{lbf})(\cos 29.54°)$$

$$= 9123\,\text{lbf}$$

$$\sigma = \frac{F_{ns}\sin\phi}{bt_o} = \frac{(9123\,\text{lbf})(\sin 29.54°)}{(1\,\text{in})(0.1\,\text{in})}$$

$$= \boxed{44{,}979\,\text{lbf/in}^2}$$

(b) $$\beta = \tan^{-1}\left(\frac{F_t}{F_n}\right)$$

$$F_t = F_h \sin\alpha + F_v \cos\alpha$$

$$= (8910\,\text{lbf})(\sin 20°) + (5437\,\text{lbf})(\cos 20°)$$

$$= 8157\,\text{lbf}$$

$$F_n = F_h \cos\alpha - F_v \sin\alpha$$

$$= (8910\,\text{lbf})(\cos 20°) - (5437\,\text{lbf})(\sin 20°)$$

$$= 6513\,\text{lbf}$$

$$\beta = \tan^{-1}\left(\frac{8157\,\text{lbf}}{6513\,\text{lbf}}\right) = 51.39°$$

$$\phi_{\text{minimum energy}} = 45° + \frac{\alpha}{2} - \frac{\beta}{2}$$

$$= 45° + \frac{20°}{2} - \frac{51.39°}{2}$$

$$= \boxed{29.3°}$$

4. From Prob. 3,

$$F_s = 5071\,\text{lbf}$$

$$\phi = 29.54°$$

$$F_h = \frac{F_s + F_v \sin\phi}{\cos\phi}$$

$$= \frac{5071\,\text{lbf} + (4300\,\text{lbf})(\sin 29.54°)}{\cos 29.54°}$$

$$= 8265\,\text{lbf}$$

$$\text{cutting power} = F_h v$$

$$= \frac{(8265\,\text{lbf})\left(4\,\frac{\text{in}}{\text{sec}}\right)\left(\frac{\text{ft}}{12\,\text{in}}\right)\left(60\,\frac{\text{sec}}{\text{min}}\right)}{33{,}000\,\frac{\text{ft-lbf}}{\text{hp-min}}}$$

$$= \boxed{5.01\,\text{hp}}$$

If L is the length of the beam, the total energy (work) is

$$E = Pt = \frac{PL}{v}$$

$$= \frac{(5.01\,\text{hp})(6\,\text{ft})}{\left(4\,\frac{\text{in}}{\text{sec}}\right)\left(\frac{\text{ft}}{12\,\text{in}}\right)} = \boxed{90.2\,\text{hp-sec}}$$

5. $$U = \frac{P}{Z_w} = \frac{F_h}{bt_o}$$

$$= \frac{15\,\text{lbf}}{(1\,\text{in})\left(\frac{\text{ft}}{12\,\text{in}}\right)(0.021\,\text{ft})\left(778\,\frac{\text{ft-lbf}}{\text{BTU}}\right)}$$

$$= 11.02\,\text{BTU/ft}^3$$

$$T = T_o + \frac{U}{\rho c_p} = 68°\text{F} + \frac{11.02\,\frac{\text{BTU}}{\text{ft}^3}}{\left(30\,\frac{\text{lbm}}{\text{ft}^3}\right)\left(0.6\,\frac{\text{BTU}}{\text{lbm-}°\text{F}}\right)}$$

$$= \boxed{68.61°\text{F}}$$

6. Using Taylor's equation,

$$v_1 T_1^n = v_2 T_2^n \quad (T \text{ is time})$$

$$n = 0.1$$

$$T_2 = T_1\left(\frac{v_1}{v_2}\right)^{1/n} = (5\,\text{hr})\left(\frac{4.9\,\frac{\text{ft}}{\text{sec}}}{4.6\,\frac{\text{ft}}{\text{sec}}}\right)^{1/0.1}$$

$$= \boxed{9.4\,\text{hr}}$$

7. $$F = NS_{us}Lt$$

$$= (10)\left(70{,}000\,\frac{\text{lbf}}{\text{in}^2}\right)(\pi)(8\,\text{in})(0.2\,\text{in})$$

$$= \boxed{3.52 \times 10^6\,\text{lbf}}$$

8. $$E = mh\left(\frac{g}{g_c}\right)(\text{rate})\Delta t$$

$$= (4500\,\text{lbm})(5\,\text{ft})\frac{\left(32.2\,\frac{\text{ft}}{\text{sec}^2}\right)}{\left(32.2\,\frac{\text{lbm-ft}}{\text{lbf-sec}^2}\right)}$$

$$\times \left(10\,\frac{1}{\text{min}}\right)(2.5\,\text{min})$$

$$= \boxed{562{,}500\,\text{ft-lbf}}$$

PROPERTIES OF AREAS

1. Since the object is symmetrical about the x-axis, $y_c = 0$.

To find x_c, divide the object into 3 parts:

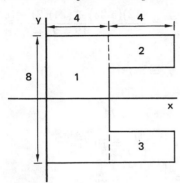

$$A_1 = (8)(4) = 32$$

$$x_{c1} = 2$$

$$A_2 = A_3 = (2)(4) = 8$$

$$x_{c2} = x_{c3} = 4 + 2 = 6$$

$$x_c = \frac{\Sigma A_i x_{ci}}{\Sigma A_i} = \frac{(32)(2) + (8)(6) + (8)(6)}{32 + 8 + 8}$$

$$= \boxed{3.33}$$

2. Divide the object into 2 parts:

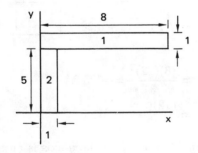

$$A_1 = (8)(1) = 8$$

$$x_{c1} = 4$$

$$A_2 = (5)(1) = 5$$

$$x_{c2} = \frac{1}{2}$$

$$x_c = \frac{\Sigma A_i x_{ci}}{\Sigma A_i} = \frac{(8)(4) + (5)\left(\frac{1}{2}\right)}{8 + 5} = 2.65$$

$$y_{c1} = -\frac{1}{2}$$

$$y_{c2} = -1 - \frac{5}{2} = -\frac{7}{2}$$

$$y_c = \frac{\Sigma A_i y_{ci}}{\Sigma A_i} = \frac{(8)\left(-\frac{1}{2}\right) + (5)\left(-\frac{7}{2}\right)}{8 + 5} = -1.65$$

$$\boxed{\text{The centroid is at } (2.65, -1.65)}$$

3. Divide into 3 parts:

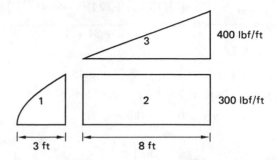

Part 1 is a parabola:

$$A_1 = \frac{2hb}{3} = \frac{(2)(3\text{ ft})\left(300\,\frac{\text{lbf}}{\text{ft}}\right)}{3} = 600\text{ lbf}$$

$$x_{c1} = \frac{3h}{5} = \frac{(3)(3\text{ ft})}{5} = 1.8\text{ ft from left end}$$

Part 2 is a rectangle:

$$A_2 = (8\text{ ft})\left(300\,\frac{\text{lbf}}{\text{ft}}\right) = 2400\text{ lbf}$$

$$x_{c2} = 4\text{ ft from right end}$$

Part 3 is a triangle:

$$A_3 = \left(\frac{1}{2}\right)(8\text{ ft})\left(400\,\frac{\text{lbf}}{\text{ft}}\right) = 1600\text{ lbf}$$

$$x_{c3} = \frac{h}{3} = \frac{8}{3} = 2.67\text{ ft from right end}$$

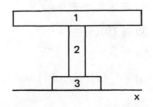

4. Divide into 3 parts:

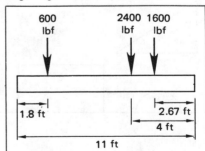

$$A_1 = \frac{1}{2}(6) = 3$$

$$y_{c1} = 12 + 1 + \frac{1}{4} = 13.25$$

$$A_2 = (12)(2) = 24$$

$$y_{c2} = 1 + 6 = 7$$

$$A_3 = (1)(4) = 4$$

$$y_{c3} = \frac{1}{2}$$

$$y_c = \frac{\Sigma A_i y_{ci}}{\Sigma A_i} = \frac{(3)(13.25) + (24)(7) + (4)\left(\frac{1}{2}\right)}{3 + 24 + 4}$$

$$= 6.77$$

$$d_1 = y_{c1} - y_c = 13.25 - 6.77 = 6.48$$

$$d_2 = y_{c2} - y_c = 7 - 6.77 = 0.23$$

$$d_3 = y_c - y_{c3} = 6.77 - 0.5 = 6.27$$

$$I_{c1} = \frac{bh^3}{12} = \frac{(6)\left(\frac{1}{2}\right)^3}{12} = 0.0625$$

$$I_{c2} = \frac{bh^3}{12} = \frac{(2)(12)^3}{12} = 288$$

$$I_{c3} = \frac{bh^3}{12} = \frac{(4)(1)^3}{12} = 0.333$$

Use the parallel axis theorem.

$$I_c = I_{c1} + A_1 d_1{}^2 + I_{c2} + A_2 d_2{}^2 + I_{c3} + A_3 d_3{}^2$$

$$= 0.0625 + (3)(6.48)^2 + 288 + (24)(0.23)^2$$

$$+ 0.33 + (4)(6.27)^2$$

$$= \boxed{572.88}$$

5. Divide the area into 3 parts:

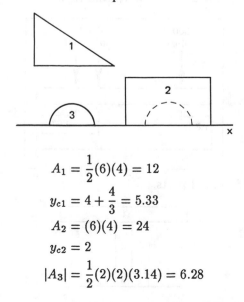

$$A_1 = \frac{1}{2}(6)(4) = 12$$

$$y_{c1} = 4 + \frac{4}{3} = 5.33$$

$$A_2 = (6)(4) = 24$$

$$y_{c2} = 2$$

$$|A_3| = \frac{1}{2}(2)(2)(3.14) = 6.28$$

Since part 3 is to be taken off,

$$A_3 = -6.28$$

$$y_{c3} = \frac{4r}{3\pi} = \frac{(4)(2)}{3\pi} = 0.85$$

$$y_c = \frac{\Sigma A_i y_{ci}}{\Sigma A_i} = \frac{(12)(5.33) + (24)(2) - (6.28)(0.85)}{12 + 24 - 6.28}$$

$$= 3.59$$

$$d_1 = y_{c1} - y_c = 5.33 - 3.59 = 1.74$$

$$d_2 = 3.59 - 2 = 1.59$$

$$d_3 = 3.59 - 0.85 = 2.74$$

$$I_{c1} = \frac{bh^3}{36} = \frac{(6)(4)^3}{36} = 10.67$$

$$I_{c2} = \frac{bh^3}{12} = \frac{(6)(4)^3}{12} = 32$$

$$I_{c3} = 0.11r^4 = (0.11)(2)^4 = 1.76$$

Since part 3 is to be taken off,

$$I_{c3} = -1.76$$

$$I_c = I_{c1} + A_1 d_1{}^2 + I_{c2} + A_2 d_2{}^2 + I_{c3} + A_3 d_3{}^2$$

$$= 10.67 + (12)(1.74)^2 + 32 + (24)(1.59)^2 - 1.76$$

$$- (6.28)(2.74)^2$$

$$= \boxed{90.77}$$

6. $$k = \sqrt{\frac{I}{A}} = \sqrt{\frac{90.77}{24 + 12 - 6.28}} = \boxed{1.748}$$

7. $$I_x = \frac{bh^3}{3} = \frac{(8)(12)^3}{3} = 4608 \text{ ft}^4$$

$$I_y = \frac{hb^3}{12} = \frac{(12)(8)^3}{12} = 512 \text{ ft}^4$$

$$P_{xy} = \int xy \, dA = 0 \quad \text{since the } y\text{-axis is an axis of symmetry}$$

$$I_{x'} = I_x \cos^2 \theta - 2P_{xy} \sin \theta \cos \theta + I_y \sin^2 \theta$$

$$= (4608 \text{ ft}^4)(\cos 60°)^2 - (2)(0)(\sin 60°)(\cos 60°)$$

$$+ (512 \text{ ft}^4)(\sin 60°)^2$$

$$= \boxed{1536 \text{ ft}^4}$$

$$I_{y'} = I_x \sin^2 \theta + 2P_{xy} \sin \theta \cos \theta + I_y \cos^2 \theta$$

$$= (4608 \text{ ft}^4)(\sin 60°)^2 + (2)(0)(\sin 60°)(\cos 60°)$$

$$+ (512 \text{ ft}^4)(\cos 60°)^2$$

$$= \boxed{3584 \text{ ft}^4}$$

STRENGTH OF MATERIALS

1. $\epsilon_{max} = \dfrac{\delta_{max}}{L} = \dfrac{0.02 \text{ in}}{200 \text{ in}} = 0.0001$

$\sigma_{max} = (E_{steel})(\epsilon_{max}) = \left(30 \times 10^6 \dfrac{\text{lbf}}{\text{in}^2}\right)(0.0001)$

$= 3000 \text{ lbf/in}^2 \ (< 10,000 \text{ lbf/in}^2)$

$A_{min} = \dfrac{F}{\sigma_{max}} = \dfrac{30,000 \text{ lbf}}{3000 \dfrac{\text{lbf}}{\text{in}^2}} = \boxed{10 \text{ in}^2}$

2. $\delta = L\epsilon = (16 \text{ in})(0.0012) = \boxed{0.0192 \text{ in}}$

$\sigma = E\epsilon = \left(2.5 \times 10^6 \dfrac{\text{lbf}}{\text{in}^2}\right)(0.0012)$

$= \boxed{3000 \text{ lbf/in}^2 \text{ (psi)}}$

3. $\sigma_{max} = \dfrac{\sigma_u}{\text{FS}} = \dfrac{60 \text{ ksi}}{5} = 12 \text{ ksi}$

$F = \sigma_{max} A = \sigma_{max}\left(\dfrac{\pi d^2}{4}\right)$

$d = \sqrt{\dfrac{4F}{\sigma_{max}\, \pi}} = \sqrt{\dfrac{(4)(7000 \text{ lbf})}{\left(12,000 \dfrac{\text{lbf}}{\text{in}^2}\right)\pi}} = \boxed{0.8618 \text{ in}}$

4. The maximum area in shear is

$A_\tau = \pi dt = \pi(0.75 \text{ in})(0.625 \text{ in}) = 1.4726 \text{ in}^2$

$F_{max} = \tau_{ultimate} A = \left(42 \dfrac{\text{kip}}{\text{in}^2}\right)(1.4726 \text{ in}^2)$

$= \boxed{61.85 \text{ kip}}$

5. $\epsilon_z = \dfrac{\sigma}{E} = \dfrac{F}{AE} = \dfrac{F}{\left(\dfrac{\pi d^2}{4}\right)E} = \dfrac{4F}{\pi d^2 E}$

$= \dfrac{(4)(40,000 \text{ lbf})}{\pi (0.75 \text{ in})^2 \left(30 \times 10^6 \dfrac{\text{lbf}}{\text{in}^2}\right)} = 3.018 \times 10^{-3}$

$\epsilon_x = \epsilon_z \nu = (3.018 \times 10^{-3})(0.3) = 9.054 \times 10^{-4}$

$d' = (\epsilon_x + 1)d = (9.054 \times 10^{-4} + 1)(0.75 \text{ in})$

$= \boxed{0.75068 \text{ in}}$

6. $E = \dfrac{\Delta\sigma}{\Delta\epsilon} = \dfrac{\dfrac{\Delta F}{A}}{\dfrac{\Delta L}{L_o}} = \dfrac{\dfrac{1500 \text{ lbf}}{\dfrac{\pi}{4}\left(\dfrac{5}{16} \text{ in}\right)^2}}{\dfrac{6 \text{ in}}{(1500 \text{ ft})\left(\dfrac{12 \text{ in}}{\text{ft}}\right)}}$

$= \boxed{5.87 \times 10^7 \text{ lbf/in}^2 \text{ (psi)}}$

7. The rod is initially in compression.

$\epsilon_{90°F} = \dfrac{\sigma_{90°F}}{E} = \dfrac{-2000 \dfrac{\text{lbf}}{\text{in}^2}}{30 \times 10^6 \dfrac{\text{lbf}}{\text{in}^2}} = -6.667 \times 10^{-5}$

$\epsilon_{th} = \alpha\Delta T = \left(6.5 \times 10^{-6} \dfrac{1}{°F}\right)(90°F - 0°F)$

$= 0.000585$

This strain is the fractional compression that an unrestrained bar would experience due to the given temperature change. The rod in question is restrained, so the net strain in the rod is

$\epsilon_{0°F} = \epsilon_{90°F} + \epsilon_{th} = -6.667 \times 10^{-5} + 0.000585$

$= 0.0005183$

The stress needed to maintain this strain is

$\sigma_{0°F} = E\epsilon_{0°F} = \left(30 \times 10^6 \dfrac{\text{lbf}}{\text{in}^2}\right)(0.0005183)$

$= \boxed{15,550 \text{ lbf/in}^2 \text{ (psi) tension}}$

8. $\epsilon_{th} = \alpha\Delta T = \left(6.5 \times 10^{-6} \dfrac{1}{°F}\right)(-80°F)$

$= -0.00052$

This strain must be counteracted to maintain the rod in its original position.

$\sigma = E\epsilon_{th} = \left(30 \times 10^6 \dfrac{\text{lbf}}{\text{in}^2}\right)(0.00052)$

$= 15,600 \text{ lbf/in}^2$

$F = \sigma A = \left(15,600 \dfrac{\text{lbf}}{\text{in}^2}\right)(0.5 \text{ in})(4 \text{ in})$

$= \boxed{31,200 \text{ lbf}}$

9. This can be reduced to a two-dimensional case.

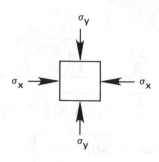

$$\sigma_x = \frac{F_x}{A_x} = \frac{-(2 \times 10^5 \text{ lbf})\left(\frac{\text{kip}}{10^3 \text{ lbf}}\right)}{(8 \text{ in})(4 \text{ in})} = -6.25 \text{ ksi}$$

$$\sigma_y = \frac{F_y}{A_y} = \frac{-(5 \times 10^4 \text{ lbf})\left(\frac{\text{kip}}{10^3 \text{ lbf}}\right)}{(4 \text{ in})(2 \text{ in})} = -6.25 \text{ ksi}$$

$$\tau = 0$$

$$\sigma_1, \sigma_2 = \frac{1}{2}(\sigma_x + \sigma_y) \pm \tau_{max}$$

$$= \left(\frac{1}{2}\right)(-6.25 \text{ ksi} - 6.25 \text{ ksi}) \pm 0 = \boxed{-6.25 \text{ ksi}}$$

10. Consider a rectangular element on the surface of the shaft:

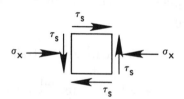

$$\sigma_x = \frac{F_x}{A_x} = \frac{F_x}{\frac{\pi d^2}{4}} = \frac{4F_x}{\pi d^2} = \frac{(4)(-15,000 \text{ lbf})}{\pi(2 \text{ in})^2}$$

$$= -4775 \text{ lbf/in}^2$$

$$\tau_s = \frac{Tr}{J} = \frac{T\left(\frac{d}{2}\right)}{\frac{\pi d^4}{32}} = \frac{16T}{\pi d^3} = \frac{(16)(25,000 \text{ in-lbf})}{\pi(2 \text{ in})^3}$$

$$= 15,915.5 \text{ lbf/in}^2$$

$$\tau_{1,2} = \pm\frac{1}{2}\sqrt{(\sigma_x - \sigma_y)^2 + (2\tau_s)^2}$$

$$= \pm\frac{1}{2}\sqrt{\left(4775 \frac{\text{lbf}}{\text{in}^2} - 0\right)^2 + \left[(2)\left(15,915.5 \frac{\text{lbf}}{\text{in}^2}\right)\right]^2}$$

$$\tau_{max} = \boxed{16,093.6 \text{ lbf/in}^2 \text{ (psi)}}$$

$$\sigma_{1,2} = \frac{1}{2}(\sigma_x + \sigma_y) \pm \tau_{max}$$

$$= \left(\frac{1}{2}\right)\left(-4775 \frac{\text{lbf}}{\text{in}^2} + 0\right) \pm 16,093 \frac{\text{lbf}}{\text{in}^2}$$

$$\boxed{\sigma_{max} = -18,481 \text{ lbf/in}^2 \text{ (psi)}}$$

11.

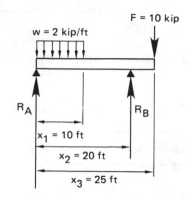

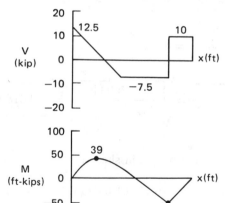

$$\Sigma M_A = \left(-\frac{x_1}{2}\right)(wx_1) + x_2 R_B - x_3 F = 0$$

$$R_B = \frac{x_3 F + \frac{wx_1^2}{2}}{x_2} = \frac{(25 \text{ ft})(10 \text{ kip}) + \frac{\left(2\frac{\text{kip}}{\text{ft}}\right)(10 \text{ ft})^2}{2}}{20 \text{ ft}}$$

$$= 17.5 \text{ kip}$$

$$\Sigma F_A = R_A - wx_1 + R_B - F = 0$$

$$R_A = wx_1 + F - R_B$$

$$= \left(2\frac{\text{kip}}{\text{ft}}\right)(10 \text{ ft}) + 10 \text{ kip} - 17.5 \text{ kip}$$

$$= 12.5 \text{ kip}$$

$$M_{\text{positive max}} = R_A x_0 - \frac{w x_0^2}{2}$$

$$= (12.5 \text{ kip})(6.25 \text{ ft}) - \frac{\left(2 \dfrac{\text{kip}}{\text{ft}}\right)(6.25 \text{ ft})^2}{2}$$

$$= \boxed{39.06 \text{ ft-kip}} \quad \left(\begin{array}{c}\text{used to graph}\\\text{moment diagram}\end{array}\right)$$

$$M_{\max} = \frac{-x_0^2 w}{2} + (x_0 - x_1)R_A$$

$$= \frac{-(10.1786 \text{ ft})^2 \left(1500 \dfrac{\text{lbf}}{\text{ft}}\right)}{2}$$

$$\quad + (10.1786 \text{ ft} - 3 \text{ ft})(15,267.86 \text{ lbf})$$

$$= \boxed{31,899 \text{ ft-lbf}}$$

12.

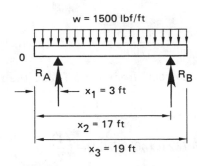

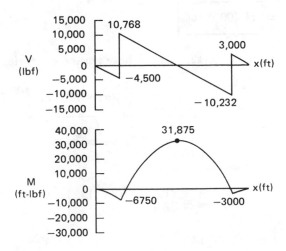

13.

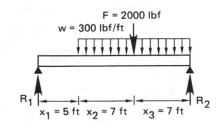

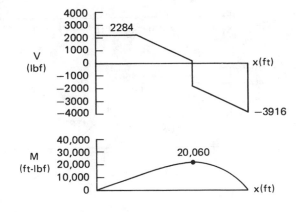

$$\Sigma M_A = -\left(\frac{x_3}{2} - x_1\right)(w x_3) + (x_2 - x_1)R_B = 0$$

$$R_B = \frac{\left(\dfrac{x_3}{2} - x_1\right)(w x_3)}{x_2 - x_1}$$

$$= \frac{\left(\dfrac{19 \text{ ft}}{2} - 3 \text{ ft}\right)\left(1500 \dfrac{\text{lbf}}{\text{ft}}\right)(19 \text{ ft})}{17 \text{ ft} - 3 \text{ ft}}$$

$$= 13,232.14 \text{ lbf}$$

$$\Sigma F = R_A + R_B - w x_3 = 0$$

$$R_A = w x_3 - R_B = \left(1500 \dfrac{\text{lbf}}{\text{ft}}\right)(19 \text{ ft}) - 13,232.14 \text{ ft}$$

$$= 15,267.86 \text{ lbf}$$

$M_{\max}$ occurs at

$$x_0 = \frac{15,267.86 \text{ lbf}}{1500 \dfrac{\text{lbf}}{\text{ft}}} = 10.1786 \text{ ft}$$

$$\Sigma M_1 = -\left(x_1 + \frac{x_2 + x_3}{2}\right)((x_2 + x_3)w) - (x_1 + x_2)F$$

$$\quad + (x_1 + x_2 + x_3)R_2 = 0$$

$$R_2 = \frac{\left(x_1 + \dfrac{x_2 + x_3}{2}\right)(x_2 + x_3)w + (x_1 + x_2)F}{x_1 + x_2 + x_3} =$$

$$\frac{\left(5 \text{ ft} + \dfrac{7 \text{ ft} + 7 \text{ ft}}{2}\right)(7 \text{ ft} + 7 \text{ ft})\left(300 \dfrac{\text{lbf}}{\text{ft}}\right) + (5 \text{ ft} + 7 \text{ ft})(2000 \text{ lbf})}{5 \text{ ft} + 7 \text{ ft} + 7 \text{ ft}}$$

$$= 3915.8 \text{ lbf}$$

$$\Sigma F = R_1 - (x_2 + x_3)w - F + R_2 = 0$$

$$R_1 = (x_2 + x_3)w + F - R_2$$

$$= (7 \text{ ft} + 7 \text{ ft})\left(300 \dfrac{\text{lbf}}{\text{ft}}\right) + 2000 \text{ lbf} - 3915.8 \text{ lbf}$$

$$= 2284.2 \text{ lbf}$$

$$M_{max} = (x_1 + x_2)R_1 - \frac{(x_2)^2 w}{2}$$

$$= (5 \text{ ft} + 7 \text{ ft})(2284.2 \text{ lbf}) - \frac{(7 \text{ ft})^2 \left(300 \frac{\text{lbf}}{\text{ft}}\right)}{2}$$

$$= \boxed{20{,}060.4 \text{ ft-lbf}}$$

$$\boxed{M_{max} \text{ located at } x = 12 \text{ ft}}$$

14.

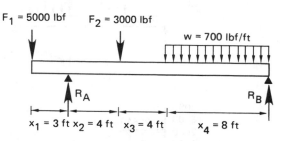

$F_1 = 5000 \text{ lbf}$ $F_2 = 3000 \text{ lbf}$ $w = 700 \text{ lbf/ft}$

R_A R_B

$x_1 = 3 \text{ ft}$ $x_2 = 4 \text{ ft}$ $x_3 = 4 \text{ ft}$ $x_4 = 8 \text{ ft}$

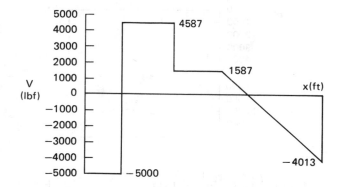

V (lbf) — 4587, 1587, −4013, −5000

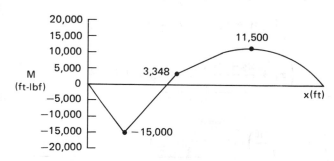

M (ft-lbf) — 11,500, 3,348, −15,000

$$\Sigma M_A = x_1 F_1 - x_2 F_2 - \left(x_2 + x_3 + \frac{x_4}{2}\right)(x_4 w)$$
$$+ (x_2 + x_3 + x_4)R_B = 0$$

$$R_B = \frac{-x_1 F_1 + x_2 F_2 + \left(x_2 + x_3 + \frac{x_4}{2}\right)(x_4 w)}{x_2 + x_3 + x_4}$$

$$= \frac{-(3 \text{ ft})(5000 \text{ lbf}) + (4)(3000) + \left(4+4+\frac{8}{2}\right)(8)(700)}{4 \text{ ft} + 4 \text{ ft} + 8 \text{ ft}}$$

$$= 4012.5 \text{ lbf}$$

$$\Sigma F = -F_1 + R_A - F_2 - x_4 w + R_B = 0$$
$$R_A = F_1 + F_2 + x_4 w - R_B$$

$$= 5000 \text{ lbf} + 3000 \text{ lbf} + (8 \text{ ft})\left(700 \frac{\text{lbf}}{\text{ft}}\right) - 4012.5 \text{ lbf}$$

$$= 9587.5 \text{ lbf}$$

Moment is maximum when shear $= 0$. Since the shear decreases linearly from the right end at the rate of 700 lbf/ft,

$$x = \frac{R_B}{w} = \frac{4012.5}{700} = 5.732 \text{ ft (from right end)}$$

Taking moments from $x = 5.732$ to the right,

$$M_{\text{positive max}} = xR_B - \frac{1}{2}wx^2$$

$$= (5.732 \text{ ft})(4012.5 \text{ lbf}) - \left(\frac{1}{2}\right)\left(700 \frac{\text{lbf}}{\text{ft}}\right)(5.732 \text{ ft})^2$$

$$= \boxed{11{,}500 \text{ ft-lbf}}$$

15. From Prob. 12, midway between the supports,

$$\sigma_{b,\text{mid}} = \frac{M_{\text{mid}}}{Z_{\text{rect}}} = \frac{6M_{\text{mid}}}{bh^2} = \frac{(6)(31{,}875 \text{ ft-lbf})\left(\frac{12 \text{ in}}{\text{ft}}\right)}{(10 \text{ in})(20 \text{ in})^2}$$

$$= \boxed{573.8 \text{ lbf/in}^2 \text{ (psi)}}$$

$$\tau_{\text{mid}} = \frac{3V_{\text{mid}}}{2A} = \frac{(3)(0)}{2A} = \boxed{0}$$

16. From Prob. 13,

$$M_{max} = 20{,}060 \text{ ft-lbf}$$
$$|V_{max}| = 3916 \text{ lbf}$$

$$\tau_{max} = \frac{3V}{2A} = \frac{(3)|V_{max}|}{2bh} = \frac{(3)(3916 \text{ lbf})}{(2)(5 \text{ in})(10 \text{ in})}$$

$$= \boxed{117.5 \text{ lbf/in}^2 \text{ (psi)}}$$

$$\sigma_{b,max} = \frac{M_{max}}{Z} = \frac{6M_{max}}{bh^2} = \frac{(6)(20{,}060 \text{ ft-lbf})\left(\frac{12 \text{ in}}{\text{ft}}\right)}{(5 \text{ in})(10 \text{ in})^2}$$

$$= \boxed{2888.6 \text{ lbf/in}^2 \text{ (psi)}}$$

17. From a beam elastic deflection table,

$$M_{max} = \frac{wL^2}{8}$$

$$\sigma_{b,\max} = \frac{M_{\max}}{Z} = \frac{\dfrac{wL^2}{8}}{\dfrac{bh^2}{6}} = \frac{3wL^2}{4bh^2}$$

$$= \frac{(3)\left(200\,\dfrac{\text{lbf}}{\text{ft}}\right)(14\text{ ft})^2\left(\dfrac{12\text{ in}}{\text{ft}}\right)}{(4)(3.625\text{ in})(7.625\text{ in})^2} = \boxed{1674\text{ lbf/in}^2\text{ (psi)}}$$

18. From Prob. 14, $M_{\max} = 15{,}044$ ft-lbf.

$$Z_{\text{req}} = \frac{M_{\max}}{\sigma_b} = \frac{(15{,}044\text{ ft-lbf})\left(\dfrac{12\text{ in}}{\text{ft}}\right)}{24{,}000\,\dfrac{\text{lbf}}{\text{in}^2}} = \boxed{7.52\text{ in}^3}$$

19.

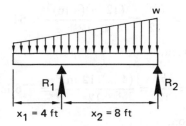

$$\Sigma M_2 = -x_2 R_1 + \left(\frac{x_1 + x_2}{3}\right)\left[\frac{w(x_1+x_2)}{2}\right] = 0$$

$$R_1 = \frac{(x_1+x_2)^2 w}{6x_2} = \left[\frac{(4\text{ ft}+8\text{ ft})^2}{(6)(8\text{ ft})}\right]w = (3\text{ ft})w$$

Place origin at the left end. The load between the left end and R_1 is

$$W_1 = \left(\frac{1}{2}\right)(4\text{ ft})\left(\frac{4w}{4+8}\right) = \frac{2w}{3}$$

The load's moment arm is 4 ft/3 from R_1. The moment at R_1 is

$$M_{R_1} = W_1 d_1 = \left(\frac{2w}{3}\right)\left(\frac{4\text{ ft}}{3}\right) \approx 0.889w$$

Between R_1 and R_2, the moment is caused by one reaction and the distributed load. Working to the left, the total load is

$$W_x = \left(\frac{1}{2}\right)(x_1+x)\left(\frac{w(x_1+x)}{4+8}\right) = \frac{w(x_1+x)^2}{24}$$

The load's moment arm is $\left(\dfrac{1}{3}\right)(x_1+x)$.

The moment between R_1 and R_2 is

$$M_x = \left(\frac{w(x_1+x)^2}{24}\right)\left[\left(\frac{1}{3}\right)(x_1+x)\right] - xR_1$$

$$= \frac{w(x_1+x)^3}{72} - 3xw$$

To find $M_{\max}$,

$$\frac{dM_x}{dx} = 0 = w\left[\frac{(3)(x_1+x)^2}{72} - 3\right]$$

Since $x_1 = 4$ ft, $x = 4.49$ ft (from R_1).

$$M_{\max} = \frac{w(4\text{ ft}+4.49\text{ ft})^3}{72\text{ ft}} - (3\text{ ft})(4.49\text{ ft})w$$

$$= (-4.97\text{ ft}^2)w$$

This value is the greater of the two.

$$\sigma_{b,\max} = \frac{M_{\max}c}{I_c} = \frac{(4.97\text{ ft}^2)w\left(\dfrac{h}{2}\right)}{I_c}$$

$$w \le \frac{2\sigma_{\text{all}}I_c}{(4.97\text{ ft}^2)h} \le \frac{(2)\left(22{,}000\,\dfrac{\text{lbf}}{\text{in}^2}\right)(78.5\text{ in}^4)}{(4.97\text{ ft}^2)\left(\dfrac{12\text{ in}}{\text{ft}}\right)(10\text{ in})}$$

$$= \boxed{5791.4\text{ lbf/ft}}$$

20. $\Sigma F = 4333\text{ lbf} - 4000\text{ lbf} - (5\text{ ft})w + 6067\text{ lbf} = 0$

$$w = \frac{4333\text{ lbf} - 4000\text{ lbf} + 6067\text{ lbf}}{5\text{ ft}} = 1280\text{ lbf/ft}$$

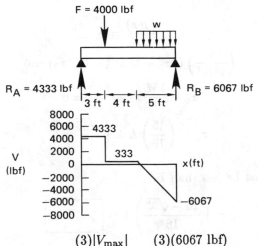

$$\tau_{\max} = \frac{(3)|V_{\max}|}{2A} = \frac{(3)(6067\text{ lbf})}{(2)(4\text{ in})(8\text{ in})}$$

$$= \boxed{284.4\text{ lbf/in}^2\text{ (psi)}}$$

21. $\sigma_{\max} = \sigma_{\text{axial}} + \sigma_{b,\max}$

$$= \frac{F}{A} + \frac{Mc}{I_c} = \frac{F}{bh} + \frac{Fe\left(\dfrac{h}{2}\right)}{\dfrac{bh^3}{12}} = \frac{F}{bh} + \frac{6Fe}{bh^2}$$

$$= \frac{5000\text{ lbf}}{(1\text{ in})(1\text{ in})} + \frac{(6)(5000\text{ lbf})(1.5\text{ in})}{(1\text{ in})(1\text{ in})^2}$$

$$= \boxed{50{,}000\text{ lbf/in}^2\text{ (psi)}}$$

22. $\sigma_{max} = \sigma_{axial} + \sigma_{b,max}$

$$= \frac{F}{A} + \frac{Mc}{I_c} = \frac{F}{bh} + \frac{Fe\left(\dfrac{h}{2}\right)}{\dfrac{bh^3}{12}} = \frac{F}{bh} + \frac{6\,Fe}{bh^2}$$

$$= \frac{4000 \text{ lbf}}{(0.75 \text{ in})(2 \text{ in})} + \frac{(6)(4000 \text{ lbf})(0.25 \text{ in} + 1 \text{ in})}{(0.75 \text{ in})(2 \text{ in})^2}$$

$$= \boxed{12{,}666.7 \text{ lbf/in}^2 \text{ (psi)}}$$

23. $\sigma_{max} = \sigma_{axial} + \sigma_{b,max}$

$$= \frac{F}{A} + \frac{Mc}{I_c} = \frac{F}{bh} + \frac{Fe\left(\dfrac{h}{2}\right)}{\dfrac{bh^3}{12}} = \frac{F}{bh} + \frac{6\,Fe}{bh^2}$$

$$= \frac{2000 \text{ lbf}}{(2.5 \text{ in})(4 \text{ in})} + \frac{(6)(2000 \text{ lbf})(4 \text{ in})}{(2.5 \text{ in})(4 \text{ in})^2}$$

$$= \boxed{1400 \text{ lbf/in}^2 \text{ (psi)}}$$

24. $y = \dfrac{w}{48EI}(2x^4 - 5Lx^3 + 3L^2x^2)$

Maximum deflection occurs at x_0, where

$$\frac{dy(x)}{dx} = 0$$

$$\left(\frac{w}{48EI}\right)(8x_0^3 - 15Lx_0^2 + 6L^2x_0) = 0$$

$$x_0^2 - \frac{15L}{8}x_0 + \frac{6L^2}{8} = 0$$

$$x_0 = \left(\frac{15}{16}\right)L \pm \left(\frac{\sqrt{33}}{16}\right)L$$

x_0 must be less than L.

$$x_0 = \left(\frac{15 - \sqrt{33}}{16}\right)L = \boxed{0.578\,L}$$

25. Consider the beam as the combination of two beams.

Beam A: Simply supported, length $L = 17$ ft, uniformly distributed load $w = 500$ lbf/ft across entire length.

Beam B: Simply supported, length $L = 17$ ft, concentrated load of $P = 2000$ lbf at $b = 5$ ft from right side.

$$y_{A,center} = \frac{-5wL^4}{384\,EI} = \frac{-(5)\left(500\,\dfrac{\text{lbf}}{\text{ft}}\right)(17 \text{ ft})^4\left(\dfrac{12 \text{ in}}{\text{ft}}\right)^3}{(384)\left(30 \times 10^6\,\dfrac{\text{lbf}}{\text{in}^2}\right)(200 \text{ in}^4)}$$

$$= -0.1566 \text{ in}$$

$$y_{B,center} = \left(\frac{-Pb}{6\,EI}\right)\left(\frac{3L^2}{8} - \frac{b^2}{2}\right)$$

$$= \frac{-(2000 \text{ lbf})(5 \text{ ft})}{(6)\left(30 \times 10^6\,\dfrac{\text{lbf}}{\text{in}^2}\right)(200 \text{ in}^4)}$$

$$\times \left[\frac{(3)(17 \text{ ft})^2}{8} - \frac{(5 \text{ ft})^2}{2}\right]\left(\frac{12 \text{ in}}{\text{ft}}\right)^3$$

$$= -0.04602 \text{ in}$$

$$y_{total} = y_A + y_B = -0.1566 \text{ in} - 0.04602 \text{ in}$$

$$= \boxed{-0.20262 \text{ in}}$$

26. $\qquad I_c = \dfrac{bh^3}{12} = \dfrac{(12 \text{ in})(4 \text{ in})^3}{12} = 64 \text{ in}^4$

Before rotation,

$$I_1 = \frac{bh^3}{12} = \frac{(4 \text{ in})(12 \text{ in})^3}{12} = 576 \text{ in}^4$$

After rotation,

$$I_2 = \frac{bh^3}{12} = \frac{(12 \text{ in})(4 \text{ in})^3}{12} = 64 \text{ in}^4$$

From the deflection equation in the beam table,

$$y_{max} = \frac{-PL^3}{48EI}$$

Deflection is inversely proportional to moment of inertia, so

$$y_{max,2} = \left(\frac{I_1}{I_2}\right)y_{max,1}$$

$$= \left(\frac{576 \text{ in}^4}{64 \text{ in}^4}\right)(-0.2 \text{ in}) = \boxed{-1.8 \text{ in}}$$

27. Break the beam cross section into 2 parts:

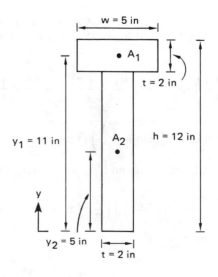

$$A_1 = b_1 h_1 = (5 \text{ in})(2 \text{ in}) = 10 \text{ in}^2$$
$$I_1 = \frac{b_1 h_1^3}{12} = \frac{(5 \text{ in})(2 \text{ in})^3}{12} = 3.33 \text{ in}^4$$

$$A_2 = b_2 h_2 = (2 \text{ in})(10 \text{ in}) = 20 \text{ in}^2$$
$$I_2 = \frac{b_2 h_2^3}{12} = \frac{(2 \text{ in})(10 \text{ in})^3}{12} = 166.67 \text{ in}^4$$

Let $\bar{y}_1, \bar{y}_2$, and $\bar{y}$ be the vertical distances to the centroids of areas 1 and 2, and the total area, respectively. Then,

$$\bar{y} = \frac{A_1 \bar{y}_1 + A_2 \bar{y}_2}{A_1 + A_2}$$
$$= \frac{(10 \text{ in}^2)(11 \text{ in}) + (20 \text{ in}^2)(5 \text{ in})}{10 \text{ in}^2 + 20 \text{ in}^2} = 7 \text{ in}$$

By the parallel axis theorem,

$$I = [I_1 + (\bar{y}_1 - \bar{y})^2 A_1] + [I_2 + (\bar{y}_2 - \bar{y})^2 A_2]$$
$$= [3.33 \text{ in}^4 + (11 \text{ in} - 7 \text{ in})^2 (10 \text{ in}^2)]$$
$$\quad + [166.67 \text{ in}^4 + (5 \text{ in} - 7 \text{ in})^2 (20 \text{ in}^2)]$$
$$= 410 \text{ in}^4$$

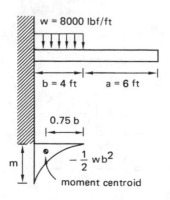

$$A_{\text{moment diagram}} = \frac{(\text{base})(\text{height})}{3}$$
$$= \frac{(b)\left(\frac{1}{2} w b^2\right)}{3} = \frac{w b^3}{6}$$

$$y_{\text{tip}} = \frac{-(A_{\text{moment diagram}})\left(\begin{array}{c}\text{distance from tip} \\ \text{to moment centroid}\end{array}\right)}{EI}$$

$$= \frac{-\left(\dfrac{w b^3}{6}\right)(a + 0.75\, b)}{EI}$$

$$= \frac{\left[\dfrac{-(8000\, \frac{\text{lbf}}{\text{ft}})(4 \text{ ft})^3}{6}\right](6 \text{ ft} + 0.75(4 \text{ ft}))\left(\dfrac{12 \text{ in}}{\text{ft}}\right)^3}{\left(30 \times 10^6\, \dfrac{\text{lbf}}{\text{in}^2}\right)(410 \text{ in}^4)}$$

$$= \boxed{-0.108 \text{ in}}$$

28. Use superposition.

Load 1: loaded with $P = +k y_{\max}$ at the center

Load 2: loaded with $-w$

From a beam elastic deflection table,

$$y_{\max,1} = \frac{P L^3}{48 EI} = \frac{(k y_{\max}) L^3}{48 EI}$$

$$y_{\max,2} = \frac{-5 w L^4}{384\, EI}$$

$$y_{\max} = y_{\max,1} + y_{\max,2} = \frac{8 k y_{\max} L^3 - 5 w L^4}{384\, EI}$$

$$w = \frac{8 k y_{\max} L^3 - 384\, EI\, y_{\max}}{5 L^4}$$

$$= \frac{(8)\left(30{,}000\, \dfrac{\text{lbf}}{\text{in}}\right)(-0.3 \text{ in})(10 \text{ ft})^3}{(5)(10 \text{ ft})^4}$$

$$\quad - \frac{(384)\left(30 \times 10^6\, \dfrac{\text{lbf}}{\text{in}^2}\right)(100 \text{ in}^4)(-0.3 \text{ in})\left(\dfrac{\text{ft}}{12 \text{ in}}\right)^3}{(5)(10 \text{ ft})^4}$$

$$= \boxed{2560 \text{ lbf/ft}}$$

ENGINEERING DESIGN

1.

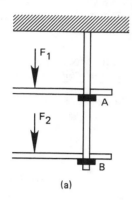 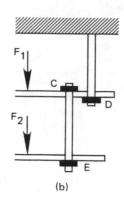

In (a), washers A and B each carry a single load: either F_1 or F_2.

In (b), washers C and E each carry a single load: F_2. However, washer D must carry the sum of both loads: $F_1 + F_2$. Furthermore, there are three washers (failure locations) compared to two washers in (a).

$$\boxed{\text{(a) is better}}$$

2. $\qquad L' = CL = (2)(24\,\text{ft}) = 48\,\text{ft}$

The cross section is symmetrical so it has only one radius of gyration.

$$I_c = k^2 A$$

$$k = \sqrt{\frac{I_c}{A}} = \frac{b}{\sqrt{12}} = \frac{(4\,\text{in})\left(\dfrac{\text{ft}}{12\,\text{in}}\right)}{\sqrt{12}} = 0.09623\,\text{ft}$$

The slenderness ratio is

$$(\text{SR}) = \frac{L'}{k} = \frac{48\,\text{ft}}{0.09623\,\text{ft}} = 498.8$$

This is larger than the typical maximum $(\text{SR})_{\text{crit}}$ (≈ 120), so slender column buckling theory may be used.

The critical buckling load is

$$F_e = \frac{\pi^2 E A}{\left(\dfrac{L'}{k}\right)^2} = \frac{(\pi^2)\left(1.5 \times 10^6\,\dfrac{\text{lbf}}{\text{in}^2}\right)(4\,\text{in})(4\,\text{in})}{(498.8)^2}$$

$$= \boxed{952.0\,\text{lbf}}$$

3. $\qquad L' = CL = (0.5)(8\,\text{ft}) = 4\,\text{ft}$

The cross section is symmetrical.

$$k^2 A = I_c$$

$$k = \sqrt{\frac{I_c}{A}} = \sqrt{\frac{\dfrac{\pi R^4}{4}}{\pi R^2}} = \frac{R}{2} = \frac{D}{4}$$

$$= \frac{1\,\text{in}}{4} = 0.25\,\text{in}$$

$$(\text{SR}) = \frac{L'}{k} = \frac{(4\,\text{ft})\left(\dfrac{12\,\text{in}}{\text{ft}}\right)}{0.25\,\text{in}} = 192 > 120$$

Slender column buckling theory applies.

If $\sigma_{\text{th}} \geq \sigma_e$, buckling will occur.

$$\sigma_{\text{th}} \geq \sigma_e$$

$$E\alpha\Delta T \geq \frac{\pi^2 E}{(\text{SR})^2}$$

$$\Delta T \geq \frac{\pi^2}{\alpha(\text{SR})^2} = \frac{\pi^2}{\left(6.5 \times 10^{-6}\,\dfrac{1}{\,^\circ\text{F}}\right)(192)^2}$$

$$\geq \boxed{41.2\,^\circ\text{F (condition for buckling)}}$$

4. $\sigma_a = \dfrac{S_y}{(\text{FS})} = \dfrac{44{,}000\,\dfrac{\text{lbf}}{\text{in}^2}}{2} = 22{,}000\,\text{lbf/in}^2$

To resist compressive failure,

$$\sigma = \frac{F}{A} = \frac{F}{\dfrac{\pi}{4}(D_o^2 - D_i^2)} \leq \sigma_a$$

$$D_i \leq \sqrt{D_o^2 - \frac{4F}{\pi\sigma_a}}$$

$$= \sqrt{(3\,\text{in})^2 - \frac{(4)(5000\,\text{lbf})}{(\pi)\left(22{,}000\,\dfrac{\text{lbf}}{\text{in}^2}\right)}} = 2.95138\,\text{in}$$

To resist buckling,

$$F \leq \frac{F_e}{(\text{FS})} = \frac{\pi^2 E I_c}{2L^2} = \frac{\pi^2 E\left(\dfrac{\pi}{4}\right)(R_o^4 - R_i^4)}{2L^2}$$

$$= \frac{\pi^3 E\left[\left(\dfrac{D_o}{2}\right)^4 - \left(\dfrac{D_i}{2}\right)^4\right]}{8L^2}$$

$$= \frac{\pi^3 E(D_o^4 - D_i^4)}{128L^2}$$

$$D_i \leq \left(D_o^4 - \frac{128\,L^2 F}{\pi^3 E}\right)^{\frac{1}{4}}$$

$$= \left[(3\text{ in})^4 - \frac{(128)(10\text{ ft})^2 \left(\frac{12\text{ in}}{\text{ft}}\right)^2 (5000\text{ lbf})}{(\pi^3)\left(30 \times 10^6\,\frac{\text{lbf}}{\text{in}^2}\right)}\right]^{\frac{1}{4}}$$

$$= 2.9037\text{ in}$$

The buckling restriction on D_i is the more restrictive.

wall thickness $= t = \dfrac{D_o - D_{i,\max}}{2}$

$$= \frac{3\text{ in} - 2.9037\text{ in}}{2} = \boxed{0.04815\text{ in}}$$

5. $F = k\delta$

$$\delta = \frac{F}{k} = \frac{18\text{ lbf}}{12\,\frac{\text{lbf}}{\text{ft}}} = \boxed{1.5\text{ ft}}$$

6. (a) $\sigma_h = \dfrac{pr}{t} = \dfrac{\left(1900\,\frac{\text{lbf}}{\text{in}^2}\right)\left(\frac{5\text{ in}}{2}\right)}{\left(\frac{1}{8}\text{ in}\right)}$

$$= \boxed{38{,}000\text{ lbf/in}^2}$$

(b) $\sigma_l = \dfrac{pr}{2t} = \dfrac{\sigma_h}{2}$

$$= \frac{38{,}000\,\frac{\text{lbf}}{\text{in}^2}}{2} = \boxed{19{,}000\text{ lbf/in}^2}$$

(c) | The principal stresses are the hoop stress and the long stress.

7. Use thin-wall equations.

$$p = \frac{F}{A} = \frac{F}{\left(\frac{\pi d^2}{4}\right)} = \frac{4F}{\pi d^2} = \frac{(4)(100{,}000.\text{lbf})}{(\pi)(10\text{ in}^2)}$$

$$= 1273.24\text{ lbf/in}^2$$

$$\sigma_h = \frac{pr}{t} = \frac{\left(1273.24\,\frac{\text{lbf}}{\text{in}^2}\right)\left(\frac{10\text{ in}}{2}\right)}{1\text{ in}}$$

$$= \boxed{6366.2\text{ lbf/in}^2}$$

8. To avoid failure in shear,

$$n_\tau \geq \frac{\tau_{\text{if one connector only}}}{\tau_a} = \frac{4F}{(\tau_a)(\pi d_{\text{bolt}}^2)}$$

$$= \frac{(4)(15{,}000\text{ lbf})}{\left(10{,}000\,\frac{\text{lbf}}{\text{in}^2}\right)(\pi)(0.75\text{ in})^2}$$

$$= 3.4; \quad n_{\tau,\min} = 4$$

To avoid failure in tension,

$$\sigma_{t,a} \geq \frac{F}{A_t} = \frac{F}{t(b - n_t d)}$$

$$n_t \leq \frac{(\sigma_{t,a})tb - F}{td\sigma_{t,a}} = \frac{b}{d} - \frac{F}{td\sigma_{t,a}}$$

$$= \frac{12\text{ in}}{0.75\text{ in}} - \frac{15{,}000\text{ lbf}}{(0.5\text{ in})(0.75\text{ in})\left(20{,}000\,\frac{\text{lbf}}{\text{in}^2}\right)}$$

$$= 14$$

$$n_{t,\max} = 14$$

To avoid failure in bearing,

$$n_p \geq \frac{\sigma_{p,\text{if only one connector}}}{\sigma_{p,a}}$$

$$= \frac{F}{dt\sigma_{p,a}} = \frac{15{,}000\text{ lbf}}{(0.75\text{ in})(0.5\text{ in})\left(65{,}000\,\frac{\text{lbf}}{\text{in}^2}\right)}$$

$$= 0.62; \quad n_{p,\min} = 1$$

Tensile stress and bearing stresses in the plate and shearing stresses in the bolts all exist simultaneously, so n_{design} must be chosen to satisfy all the three conditions.

(a) | $n_{\text{design}} = 4$

(b) | Failure in shear (minimum = 4) and tension (maximum = 14) considerations govern the design.

9. (a) ⊿ fillet

(b) ⊻ bevel

(c) OH overhead

(d) F flat

(e) H horizontal

10. $\tau_{max} = \dfrac{Tr}{J} = \dfrac{T\left(\dfrac{D}{2}\right)}{\dfrac{\pi D^4}{32}} = \dfrac{16T}{\pi D^3}$

$T = \dfrac{\pi D^3 \tau_{max}}{16}$

$\quad = \dfrac{(\pi)(2.5 \text{ in})^3 \left(10,000 \dfrac{\text{lbf}}{\text{in}^2}\right)\left(\dfrac{\text{ft}}{12 \text{ in}}\right)}{16}$

$\quad = \boxed{2556.63 \text{ ft-lbf}}$

$\theta = \dfrac{TL}{GJ} = \dfrac{TL}{G\left(\dfrac{\pi D^4}{32}\right)} = \dfrac{32TL}{G\pi D^4}$

$\quad = \left[\dfrac{(32)(2556.63 \text{ ft-lbf})(2 \text{ ft})\left(\dfrac{12 \text{ in}}{\text{ft}}\right)^2}{\left(11.5 \times 10^6 \dfrac{\text{lbf}}{\text{in}^2}\right)(\pi)(2.5 \text{ in})^4}\right]\left(\dfrac{360°}{2\pi \text{ rad}}\right)$

$\quad = \boxed{0.9566°}$

11. Each of the two restrictions (maximum allowable shear stress and maximum allowable angle of twist per unit length) has an associated lower limit on the required shaft diameter.

As limited by maximum shear,

$T = \dfrac{\left(63,025 \dfrac{\text{in-lbf}}{\text{hp-min}}\right)(200 \text{ hp})}{\left(1850 \dfrac{\text{rev}}{\text{min}}\right)} = 6813.5 \text{ in-lbf}$

$\tau_{max} \geq \dfrac{Tr}{J} = \dfrac{T\left(\dfrac{D}{2}\right)}{\dfrac{\pi D^4}{32}} = \dfrac{16T}{\pi D^3}$

$D \geq \left(\dfrac{16T}{\pi \tau_{max}}\right)^{\frac{1}{3}} = \left[\dfrac{(16)(6813.5 \text{ in-lbf})}{(\pi)\left(10,000 \dfrac{\text{lbf}}{\text{in}^2}\right)}\right]^{\frac{1}{3}}$

$\quad = 1.5140 \text{ in}$

As limited by maximum angle of twist per unit length,

$\theta = \dfrac{TL}{JG}$

$\dfrac{\theta}{L} = \dfrac{T}{JG}$

$\left(\dfrac{\theta}{L}\right)_{max} \geq \dfrac{T}{JG} = \dfrac{T}{\left(\dfrac{\pi D^4}{32}\right)G}$

$D \geq \left(\dfrac{32T}{\pi \left(\dfrac{\theta}{L}\right)_{max} G}\right)^{\frac{1}{4}}$

$\quad = \left[\dfrac{(32)(6813.5 \text{ in-lbf})}{(\pi)\left(\dfrac{1°}{\text{ft}}\right)\left(\dfrac{2\pi \text{ rad}}{360°}\right)\left(\dfrac{\text{ft}}{12 \text{ in}}\right)\left(11.5 \times 10^6 \dfrac{\text{lbf}}{\text{in}^2}\right)}\right]^{\frac{1}{4}}$

$\quad = 1.4272 \text{ in}$

The maximum shear criterion governs.

$D_{design} = \boxed{1.5140 \text{ in}}$

12. $T = \dfrac{(63,025)(\text{horsepower})}{n_{rpm}}$

$\quad = \dfrac{\left(63,025 \dfrac{\text{in-lbf}}{\text{hp-min}}\right)(200 \text{ hp})}{\left(875 \dfrac{\text{rev}}{\text{min}}\right)} = 14,405.7 \text{ in-lbf}$

$\tau_{max} = \dfrac{Tr}{J} = \dfrac{T\left(\dfrac{D}{2}\right)}{\dfrac{\pi D^4}{32}} = \dfrac{16T}{\pi D^3}$

$\quad = \dfrac{(16)(14,405.7 \text{ in-lbf})}{(\pi)(2 \text{ in})^3} = \boxed{9171 \text{ lbf/in}^2 \text{ (psi)}}$

13. $\tau_a = \dfrac{T_a r}{J}$

$T_a = \dfrac{J\tau_a}{r} = \dfrac{\left(\dfrac{\pi D^4}{32}\right)\tau_a}{\dfrac{D}{2}} = \dfrac{\pi D^3 \tau_a}{16}$

$\quad = \dfrac{(\pi)(2 \text{ in})^3 \left(12,000 \dfrac{\text{lbf}}{\text{in}^2}\right)}{16} = 18,849.6 \text{ in-lbf}$

$T = \dfrac{\left(63,025 \dfrac{\text{in-lbf}}{\text{hp-min}}\right)(\text{horsepower})}{n}$

$\text{horsepower} = \dfrac{nT}{63,025 \dfrac{\text{in-lbf}}{\text{hp-min}}}$

$\quad = \dfrac{(100 \text{ rpm})(18,849.6 \text{ in-lbf})}{63,025 \dfrac{\text{in-lbf}}{\text{hp-min}}}$

$\quad = \boxed{29.91 \text{ hp}}$

14. Assume both distances are measured flat. Let L' be the measured length and let L be the actual length.

$$L = L' + c_t + c_p$$

$$= L' + \alpha(T - T_{\text{std}})L' + \frac{(F - F_{\text{std}})L'}{AE}$$

$$= 100\,\text{ft} + \left(6.5 \times 10^{-6}\,\frac{1}{^\circ\text{F}}\right)(100^\circ\text{F} - 70^\circ\text{F})(100\,\text{ft})$$

$$+ \frac{(20\,\text{lbf} - 10\,\text{lbf})(100\,\text{ft})}{\left(\frac{3}{8}\,\text{in}\right)\left(\frac{1}{32}\,\text{in}\right)\left(30 \times 10^6\,\frac{\text{lbf}}{\text{in}^2}\right)}$$

$$= \boxed{100.022344\,\text{ft}}$$

15. Let L be the actual distance and let L' be the reading on the tape.

$$L = L' + c_t = L' + \alpha(T - T_{\text{std}})L$$

$$= 50\,\text{ft} + \left(6.5 \times 10^{-6}\,\frac{1}{^\circ\text{F}}\right)(90^\circ\text{F} - 68^\circ\text{F})(50\,\text{ft})$$

$$= \boxed{50.00715\,\text{ft}}$$

16. Let L be the actual distance and let L' be the reading on the tape.

$$L = L' + c_t + c_p$$

$$= L' + \alpha(T - T_{\text{std}})L' + \frac{(F - F_{\text{std}})L'}{AE}$$

$$L' = \frac{L}{1 + \alpha(T - T_{\text{std}}) + \frac{(F - F_{\text{std}})}{AE}}$$

$$= \frac{75\,\text{ft}}{1 + \left(6.5 \times 10^{-6}\,\frac{1}{^\circ\text{F}}\right)(90^\circ\text{F} - 68^\circ\text{F}) + \frac{15\,\text{lbf} - 5\,\text{lbf}}{(0.75)(0.001)(30 \times 10^6)}}$$

$$= \boxed{74.95597\,\text{ft}}$$

17. $n = \dfrac{E_{\text{steel}}}{E_{\text{copper}}} = \dfrac{3 \times 10^7\,\frac{\text{lbf}}{\text{in}^2}}{1.75 \times 10^7\,\frac{\text{lbf}}{\text{in}^2}} = 1.7143$

$$A_{\text{copper}} = A_{\text{total}} - A_{\text{steel}}$$

$$= (6\,\text{in})(0.75\,\text{in}) - (6\,\text{in})(0.5\,\text{in}) = 1.5\,\text{in}^2$$

$$A_t = A_{\text{copper}} + nA_{\text{steel}}$$

$$= 1.5\,\text{in}^2 + (1.7143)(6\,\text{in})(0.5\,\text{in}) = 6.6429\,\text{in}^2$$

$$\sigma_{\text{copper}} = \frac{F}{A_t} = \frac{100{,}000\,\text{lbf}}{6.6429\,\text{in}^2} = \boxed{15{,}054\,\text{lbf/in (psi)}}$$

$$\sigma_{\text{steel}} = \frac{nF}{A_t} = n\sigma_{\text{copper}} = (1.7143)\left(15{,}054\,\frac{\text{lbf}}{\text{in}^2}\right)$$

$$= \boxed{25{,}807\,\text{lbf/in}^2\,\text{(psi)}}$$

The strains are the same.

$$\epsilon_{\text{copper}} = \frac{\sigma_{\text{copper}}}{E_{\text{copper}}} = \frac{15{,}054\,\frac{\text{lbf}}{\text{in}^2}}{1.75 \times 10^7\,\frac{\text{lbf}}{\text{in}^2}} = \boxed{8.602 \times 10^{-4}}$$

$$\epsilon_{\text{steel}} = \frac{\sigma_{\text{steel}}}{E_{\text{steel}}} = \frac{25{,}807\,\frac{\text{lbf}}{\text{in}^2}}{3 \times 10^7\,\frac{\text{lbf}}{\text{in}^2}} = \boxed{8.602 \times 10^{-4}}$$

18. Using the transformation method,

$$n = \frac{E_{\text{steel}}}{E_{\text{copper}}} = \frac{30 \times 10^6}{17.5 \times 10^6} = 1.714$$

$$A_t = A_{\text{copper}} + 2nA_{\text{steel}}$$

$$= 0.6\,\text{in}^2 + (2)(1.714)(0.2\,\text{in}^2) = 1.2856\,\text{in}^2$$

$$\sigma_{\text{copper}} = \frac{F}{A_t} = \frac{(2)(10{,}000\,\text{lbf})}{1.2856\,\text{in}^2} = 15{,}557\,\text{lbf/in}^2$$

$$F_{\text{copper}} = \sigma_{\text{copper}}A_{\text{copper}} = \left(15{,}557\,\frac{\text{lbf}}{\text{in}^2}\right)(0.6\,\text{in}^2)$$

$$= \boxed{9334\,\text{lbf}}$$

$$\sigma_{\text{steel}} = n\sigma_{\text{copper}} = (1.714)\left(15{,}557\,\frac{\text{lbf}}{\text{in}^2}\right)$$

$$= 26{,}665\,\text{lbf/in}^2$$

$$F_{\text{steel}} = \sigma_{\text{steel}}A_{\text{steel}} = \left(26{,}665\,\frac{\text{lbf}}{\text{in}^2}\right)(0.2\,\text{in}^2)$$

$$= \boxed{5333\,\text{lbf}}$$

19. The elongation, δ_c, will be almost entirely due to compression of CD and elongation of BC; very little will be due to the bending of these members.

MECH MATL
Design

Cut bar AC at its ends. Using it as a free body,

$$\Sigma M_C = 0 : (5\text{ ft})(20,000\text{ lbf}) - (10\text{ ft})R_{A,y} = 0$$

$$R_{A,y} = 10,000\text{ lbf}$$

$$\Sigma F_y = 0 : R_{A,y} + R_{C,y} - F = 0$$

$$R_{C,y} = F - R_{A,y} = 20,000\text{ lbf} - 10,000\text{ lbf}$$

$$= 10,000\text{ lbf}$$

R_C will be applied to the rod's cross-sectional area at CD and BC, as $F_C = -R_C = -10,000$ lbf. Using the area transformation method,

$$n = \frac{E_{\text{steel}}}{E_{\text{copper}}} = \frac{30 \times 10^6}{17.5 \times 10^6} = 1.714$$

$$A_t = A_{CD} + nA_{BC}$$

$$= (1\text{ in}^2) + (1.714)(4\text{ in}^2) = 7.856\text{ in}^2$$

$$\delta_C = \delta_{CD} = \frac{\sigma_{CD}}{E_{CD}}L_{CD} = \frac{\frac{F_C}{A_{CD}}L_{CD}}{E_{CD}}$$

$$= \frac{\left(\frac{10,000\text{ lbf}}{1\text{ in}^2}\right)(4\text{ ft})\left(\frac{12\text{ in}}{\text{ft}}\right)}{17.5 \times 10^6 \frac{\text{lbf}}{\text{in}^2}} = \boxed{0.0274\text{ in}}$$

20.
$$A_2 = \frac{\pi d_2^2}{4} = \frac{\pi(1.5\text{ in})^2}{4} = 1.767\text{ in}^2$$

$$A_1 = \frac{\pi d_1^2}{4} = \frac{\pi(1\text{ in})^2}{4} = 0.7854\text{ in}^2$$

$$R_1 = T_1 = \sigma_1 A_1 = E_1 \epsilon_1 A_1 = E_1 \frac{\delta_1}{L_1} A_1$$

$$= \frac{\left(30 \times 10^6 \frac{\text{lbf}}{\text{in}^2}\right)(87.8 \times 10^{-6}\text{ in})(0.7854\text{ in}^2)}{20\text{ in}}$$

$$= \boxed{103.44\text{ lbf}}$$

$$\Sigma F_y = 0 : R_1 + R_2 - F = 0$$

$$R_2 = 200\text{ lbf} - 103.44\text{ lbf} = \boxed{96.56\text{ lbf}}$$

$$\delta_2 = \frac{R_2 L_2}{E_2 A_2} = \frac{(96.56\text{ lbf})(20\text{ in})}{\left(10.0 \times 10^6 \frac{\text{lbf}}{\text{in}^2}\right)(1.767\text{ in}^2)}$$

$$= \boxed{1.093 \times 10^{-4}\text{ in}}$$

PROPERTIES OF SOLID BODIES

1. Since the object is symmetric about the x-axis, $y_c = 0$ ft.

$$m_S = 64.4 \, \text{lbm}$$
$$x_{cS} = 1 \, \text{ft} + 6 \, \text{ft} + 1 \, \text{ft} = 8 \, \text{ft}$$
$$m_R = 32.2 \, \text{lbm}$$
$$x_{cR} = 1 \, \text{ft} + 3 \, \text{ft} = 4 \, \text{ft}$$
$$m_C = 64.4 \, \text{lbm}$$
$$x_{cC} = 0 \, \text{ft}$$

$$x_c = \frac{\Sigma m_i x_{ci}}{\Sigma m_i}$$
$$= [(64.4 \, \text{lbm}) \, (8 \, \text{ft}) + (32.2 \, \text{lbm}) \, (4 \, \text{ft})$$
$$+ \, (64.4 \, \text{lbm}) \, (0 \, \text{ft})]$$
$$\div (64.4 \, \text{lbm} + 32.2 \, \text{lbm} + 64.4 \, \text{lbm})$$
$$= \boxed{4 \, \text{ft}}$$

2.
$$m_A = (2 \, \text{ft}) \, (3 \, \text{ft}) \, (9 \, \text{ft}) \left(16 \, \frac{\text{slug}}{\text{ft}^3} \right)$$
$$= 864 \, \text{slugs}$$

$$m_B = (1 \, \text{ft}) \, (1 \, \text{ft}) \, (4 \, \text{ft}) \left(25 \, \frac{\text{slug}}{\text{ft}^3} \right)$$
$$= 100 \, \text{slugs}$$

$$x_{cA} = 1.5 \, \text{ft}$$
$$x_{cB} = 3 \, \text{ft} + 2 \, \text{ft} = 5 \, \text{ft}$$
$$x_c = \frac{\Sigma m_i x_{ci}}{\Sigma m_i} = \frac{(864 \, \text{slug}) \, (1.5 \, \text{ft}) + (100 \, \text{slug}) \, (5 \, \text{ft})}{864 \, \text{slug} + 100 \, \text{slug}}$$
$$= \boxed{1.863 \, \text{ft}}$$

$$y_{cA} = 4.5 \, \text{ft}$$
$$y_{cB} = 8 \, \text{ft} + 0.5 \, \text{ft} = 8.5 \, \text{ft}$$
$$y_c = \frac{(864 \, \text{slug}) \, (4.5 \, \text{ft}) + (100 \, \text{slug}) \, (8.5 \, \text{ft})}{864 \, \text{slug} + 100 \, \text{slug}}$$
$$= \boxed{4.915 \, \text{ft}}$$

$$z_{cA} = 1 \, \text{ft}$$
$$z_{cB} = 1 \, \text{ft} + 0.5 \, \text{ft} = 1.5 \, \text{ft}$$
$$z_c = \frac{(864 \, \text{slug}) \, (1 \, \text{ft}) + (100 \, \text{slug}) \, (1.5 \, \text{ft})}{864 \, \text{slug} + 100 \, \text{slug}}$$
$$= \boxed{1.052 \, \text{ft}}$$

3.
$$m_S = 1.5 \, \text{lbm}$$
$$y_{cS} = 3 \, \text{in} + 2.75 \, \text{in} - 1.5 \, \text{in} = 4.25 \, \text{in}$$
$$m_C = 0.125 \, \text{lbm}$$
$$y_{cC} = \frac{2h}{3} = \frac{(2) \, (3 \, \text{in})}{3} = 2 \, \text{in}$$
$$y_c = \frac{\Sigma m_i y_{ci}}{\Sigma m_i}$$
$$= \frac{(1.5 \, \text{lbm}) \, (4.25 \, \text{in}) + (0.125 \, \text{lbm}) \, (2 \, \text{in})}{1.5 \, \text{lbm} + 0.125 \, \text{lbm}}$$
$$= \boxed{4.08 \, \text{in from the } x\text{-axis}}$$

4. Since A is a rectangular parallelepiped,

$$I_{cA} = \frac{1}{12} m_A \, (a^2 + b^2)$$
$$= \left(\frac{1}{12} \right) (864 \, \text{slug}) \left[(2 \, \text{ft})^2 + (9 \, \text{ft})^2 \right]$$
$$= 6120 \, \text{slug-ft}^2$$

The distance, d, from the x-centroidal axis to the x-axis is found by

$$d = \sqrt{(y_c)^2 + (z_c)^2} = \sqrt{(4.5 \, \text{ft})^2 + (1 \, \text{ft})^2} = 4.61 \, \text{ft}$$
$$I_{xA} = I_{cA} + m_A d^2$$
$$= 6120 \, \text{slug-ft}^2 + (864 \, \text{slug}) \, (4.61 \, \text{ft})^2$$
$$= 24{,}482 \, \text{slug-ft}^2$$

For B,

$$I_{cB} = \frac{1}{12} m_B \, (a^2 + b^2)$$
$$= \left(\frac{1}{12} \right) (100 \, \text{slug}) \left[(1 \, \text{ft})^2 + (1 \, \text{ft})^2 \right]$$
$$= 16.67 \, \text{slug-ft}^2$$

$$d = \sqrt{(8.5 \, \text{ft})^2 + (1.5 \, \text{ft})^2} = 8.63 \, \text{ft}$$
$$I_{xB} = I_{cB} + m_B d^2$$
$$= 16.67 \, \text{slug-ft}^2 + (100 \, \text{slug}) \, (8.63 \, \text{ft})^2$$
$$= 7464 \, \text{slug-ft}^2$$

$$I_x = I_{xA} + I_{xB}$$
$$= 24{,}482 \, \text{slug-ft}^2 + 7464 \, \text{slug-ft}^2$$
$$= \boxed{31{,}946 \, \text{slug-ft}^2}$$

DYNAMICS
Masses

5. For the cone,

$$I_{xC} = \frac{3}{5} m_C \left(\frac{r^2}{4} + h^2 \right)$$

$$= \left(\frac{3}{5} \right) (0.125 \text{ lbm}) \left[\frac{(1.25 \text{ in})^2}{4} + (3 \text{ in})^2 \right]$$

$$= 0.704 \text{ lbm-in}^2$$

For the sphere,

$$I_{cS} = \frac{2}{5} m_S r^2 = \left(\frac{2}{5} \right) (1.5 \text{ lbm}) (1.5 \text{ in})^2$$

$$= 1.35 \text{ lbm-in}^2$$

$$I_{xS} = I_{cS} + m_S d^2$$

$$= 1.35 \text{ lbm-in}^2 + (1.5 \text{ lbm})(4.25 \text{ in})^2$$

$$= 28.44 \text{ lbm-in}^2$$

$$I_x = I_{xC} + I_{xS} = 0.704 \text{ lbm-in}^2 + 28.44 \text{ lbm-in}^2$$

$$= \boxed{29.14 \text{ lbm-in}^2}$$

6. $$m = (64.4 \text{ lbm}) \left(\frac{\text{slug}}{32.2 \text{ lbm}} \right) = 2 \text{ slug}$$

$$I_{CC} = I_{AA} - md^2 = 90 \text{ slug-ft}^2 - (2 \text{ slug}) (4 \text{ ft})^2$$
$$= 58 \text{ slug-ft}^2$$

$$I_{BB} = I_{CC} + md'^2 = 58 \text{ slug-ft}^2 + (2 \text{ slug}) (6 \text{ ft})^2$$

$$= \boxed{130 \text{ slug-ft}^2}$$

KINEMATICS

1. $x(t) = 2t^2 - 8t + 3$

$v(t) = \dfrac{dx}{dt} = 4t - 8$

$a(t) = \dfrac{dv}{dt} = 4$

At $t = 2$:

$x = (2)(2)^2 - (8)(2) + 3 = \boxed{-5}$

$v = (4)(2) - 8 = \boxed{0}$

$a = \boxed{4}$

At $t = 1$:

$x = (2)(1)^2 - (8)(1) + 3 = \boxed{-3}$

At $t = 3$:

$x = (2)(3)^2 - (8)(3) + 3 = \boxed{-3}$

displacement $= x_3 - x_1 = -3 - (-3) = 0$

total distance traveled from $t = 1$ to $t = 3$

$= \displaystyle\int_1^3 |v(t)|dt = \int_1^3 |4t - 8|dt$

$= \displaystyle\int_1^2 (8 - 4t)dt + \int_2^3 (4t - 8)dt$

$= 2 + 2 = \boxed{4}$

2. $a = \dfrac{\Delta v}{\Delta t} = \dfrac{\left(180 \; \dfrac{\text{mile}}{\text{hr}}\right)\left(\dfrac{5280 \text{ ft}}{\text{mile}}\right)\left(\dfrac{\text{hr}}{3600 \text{ sec}}\right)}{60 \text{ sec}}$

$= \boxed{4.4 \text{ ft/sec}^2}$

3. $a = \dfrac{v - v_0}{t} = \dfrac{20 \; \dfrac{\text{ft}}{\text{sec}} - 5 \; \dfrac{\text{ft}}{\text{sec}}}{(2 \text{ min})\left(\dfrac{60 \text{ sec}}{\text{min}}\right)}$

$= \boxed{0.125 \text{ ft/sec}^2}$

4. Assume uniform deceleration.

$v_0 = \left(60 \; \dfrac{\text{mile}}{\text{hr}}\right)\left(\dfrac{5280 \text{ ft}}{\text{mile}}\right)\left(\dfrac{\text{hr}}{3600 \text{ sec}}\right) = 88 \text{ ft/sec}$

$a = \dfrac{v - v_0}{t} = \dfrac{0 - 88 \; \dfrac{\text{ft}}{\text{sec}}}{5 \text{ sec}} = \boxed{-17.6 \text{ ft/sec}^2}$

$s = \dfrac{1}{2}t(v + v_0) = \dfrac{1}{2}tv_0$

$= \left(\dfrac{1}{2}\right)(5 \text{ sec})\left(88 \; \dfrac{\text{ft}}{\text{sec}}\right) = \boxed{220 \text{ ft}}$

5. $H = \dfrac{v_0^2 \sin^2 \phi}{2g}$

$= \dfrac{\left(2700 \; \dfrac{\text{ft}}{\text{sec}}\right)^2 (\sin^2 45°)}{(2)\left(32.2 \; \dfrac{\text{ft}}{\text{sec}^2}\right)} = \boxed{56{,}599 \text{ ft}}$

$R = \dfrac{v_0^2 \sin 2\phi}{g}$

$= \dfrac{\left(2700 \; \dfrac{\text{ft}}{\text{sec}}\right)^2 (\sin 90°)}{32.2 \; \dfrac{\text{ft}}{\text{sec}^2}} = \boxed{226{,}398 \text{ ft}}$

6. Neglect air friction.

$v_x = v_0 \cos \phi = \left(60 \; \dfrac{\text{ft}}{\text{sec}}\right)(\cos 36.87°)$

$= \boxed{48 \text{ ft/sec}}$

$t = \dfrac{s}{v_x} = \dfrac{72 \text{ ft}}{48 \; \dfrac{\text{ft}}{\text{sec}}} = 1.5 \text{ sec}$

$v_y = v_0 \sin \phi - gt$

$= \left(60 \; \dfrac{\text{ft}}{\text{sec}}\right)(\sin 36.87°) - \left(32.2 \; \dfrac{\text{ft}}{\text{sec}^2}\right)(1.5 \text{ sec})$

$= \boxed{-12.3 \text{ ft/sec}}$

DYNAMICS
Kinematics

$$y = v_0 \sin \phi \, t - \frac{1}{2}gt^2$$

$$= \left(60 \, \frac{\text{ft}}{\text{sec}}\right)(\sin 36.87°)(1.5 \text{ sec})$$

$$- \left(\frac{1}{2}\right)\left(32.2 \, \frac{\text{ft}}{\text{sec}^2}\right)(1.5 \text{ sec})^2$$

$$= \boxed{17.78 \text{ ft}}$$

7. $\quad H = z + \dfrac{v_0^2 \sin^2 \phi}{2g}$

$$= 12{,}000 \text{ ft} + \frac{\left(600 \, \dfrac{\text{ft}}{\text{sec}}\right)^2 (\sin^2 30°)}{(2)\left(32.2 \, \dfrac{\text{ft}}{\text{sec}^2}\right)}$$

$$= \boxed{13{,}398 \text{ ft}}$$

Let t_1 be the time the bomb takes to reach the maximum altitude.

$$t_1 = \frac{1}{2}\left(\frac{2v_0 \sin \phi}{g}\right) = \frac{\left(600 \, \dfrac{\text{ft}}{\text{sec}}\right)(\sin 30°)}{32.2 \, \dfrac{\text{ft}}{\text{sec}^2}} = 9.32 \text{ sec}$$

Let t_2 be the time the bomb takes to fall from H.

$$t_2 = \sqrt{\frac{2H}{g}} = \sqrt{\frac{(2)(13{,}398 \text{ ft})}{32.2 \, \dfrac{\text{ft}}{\text{sec}^2}}} = 28.85 \text{ sec}$$

$$t = t_1 + t_2 = 9.32 \text{ sec} + 28.85 \text{ sec}$$

$$= \boxed{38.17 \text{ sec}}$$

8. $\quad \omega(2) = (6)(2)^2 - (10)(2) = \boxed{4 \text{ clockwise}}$

$$\theta = \theta(3) - \theta(1) = \int_1^3 \omega(t)\,dt$$

$$= \int_1^3 (6t^2 - 10t)\,dt = \boxed{12}$$

To find the total distance traveled, check for sign reversals in $\omega(t)$ over the interval $t = 1$ to $t = 3$.

$$6t^2 - 10t = 0$$

$$\text{sign reversal at } t = \frac{5}{3}$$

The total angle turned is

$$\int_1^3 |\omega(t)|\,dt = \int_1^{\frac{5}{3}} (10t - 6t^2)\,dt + \int_{\frac{5}{3}}^3 (6t^2 - 10t)\,dt$$

$$= \boxed{15.26 \text{ rad}}$$

9. $\quad v = r\omega = r2\pi f = d\pi f$

$$= (14 \text{ in})\left(\frac{\text{ft}}{12 \text{ in}}\right)(\pi)\left(40 \, \frac{\text{rev}}{\text{min}}\right)\left(\frac{\text{min}}{60 \text{ sec}}\right)$$

$$= \boxed{2.44 \text{ ft/sec}}$$

10. $\quad \omega_1 = 2\pi f_1 = \left(2\pi \, \frac{\text{rad}}{\text{rev}}\right)\left(1200 \, \frac{\text{rev}}{\text{min}}\right)\left(\frac{\text{min}}{60 \text{ sec}}\right)$

$$= 125.66 \text{ rad/sec}$$

$$\omega_2 = \left(2\pi \, \frac{\text{rad}}{\text{rev}}\right)\left(3000 \, \frac{\text{rev}}{\text{min}}\right)\left(\frac{\text{min}}{60 \text{ sec}}\right)$$

$$= 314.16 \text{ rad/sec}$$

$$\alpha = \frac{\omega_2 - \omega_1}{\Delta t} = \frac{314.16 \, \dfrac{\text{rad}}{\text{sec}} - 125.66 \, \dfrac{\text{rad}}{\text{sec}}}{10 \text{ sec}}$$

$$= \boxed{18.85 \text{ rad/sec}^2}$$

11. $\quad f = \left(1750 \, \dfrac{\text{rev}}{\text{min}}\right)\left(\dfrac{\text{min}}{60 \text{ sec}}\right) = 29.167 \text{ rev/sec}$

$$\theta = (18°)\left(\frac{\text{rev}}{360°}\right) = 0.05 \text{ rev}$$

$$t = \frac{\theta}{f} = \frac{0.05 \text{ rev}}{29.167 \, \dfrac{\text{rev}}{\text{sec}}} = 0.001714 \text{ sec}$$

$$v = \frac{s}{t} = \frac{5 \text{ ft}}{0.001714 \text{ sec}} = \boxed{2916.7 \text{ ft/sec (fps)}}$$

12. $\quad \omega_B = \left(\dfrac{16}{24}\right)\omega_C = \left(\dfrac{16}{24}\right)\left(2 \, \dfrac{\text{rad}}{\text{sec}}\right) = 1.333 \text{ rad/sec}$

$$\alpha_B = \left(\frac{16}{24}\right)\alpha_C = \left(\frac{16}{24}\right)\left(6 \, \frac{\text{rad}}{\text{sec}^2}\right) = 4 \text{ rad/sec}^2$$

Since A and B are splined together,

$$\omega_A = \omega_B$$

$$\alpha_A = \alpha_B$$

$$v_D = r_A \omega_A = (0.5 \text{ ft}) \left(1.333 \frac{\text{rad}}{\text{sec}} \right)$$

$$= \boxed{0.6665 \text{ ft/sec (fps)}}$$

$$a_D = r_A \alpha_A = (0.5 \text{ ft}) \left(4 \frac{\text{rad}}{\text{sec}^2} \right)$$

$$= \boxed{2 \text{ ft/sec}^2}$$

13. $v_{\text{car}} = \left(45 \frac{\text{miles}}{\text{hr}} \right) \left(\frac{5280 \text{ ft}}{\text{mile}} \right) \left(\frac{\text{hr}}{3600 \text{ sec}} \right)$

$$= 66 \text{ ft/sec}$$

The separation distance after 1 second is

$$s(1) = \sqrt{\left[200 \text{ ft} + \left(15 \tfrac{\text{ft}}{\text{sec}} \right)(1 \text{ sec}) \right]^2 + \left[\left(66 \tfrac{\text{ft}}{\text{sec}} \right)(1 \text{ sec}) \right]^2}$$

$$= 224.9 \text{ ft}$$

The separation velocity is the difference in components of the car's and balloon's velocities along a mutually parallel line. Use the separation vector as this line.

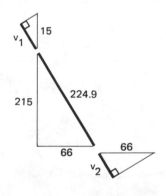

$$v_1 = \left(15 \frac{\text{ft}}{\text{sec}} \right) \left(\frac{215}{224.9} \right) = 14.34 \frac{\text{ft}}{\text{sec}}$$

$$v_2 = \left(66 \frac{\text{ft}}{\text{sec}} \right) \left(\frac{66}{224.9} \right) = 19.37 \frac{\text{ft}}{\text{sec}}$$

$$\Delta v = v_1 + v_2 = 14.34 \frac{\text{ft}}{\text{sec}} + 19.37 \frac{\text{ft}}{\text{sec}}$$

$$= \boxed{33.71 \text{ ft/sec}}$$

14. $\omega = \dfrac{v_0}{r_{\text{inner}}} = \dfrac{10 \frac{\text{ft}}{\text{sec}}}{2 \frac{\text{ft}}{\text{rad}}} = 5 \text{ rad/sec}$

$$v_{A/O} = r_{\text{outer}} \omega = \left(3 \frac{\text{ft}}{\text{rad}} \right) \left(5 \frac{\text{rad}}{\text{sec}} \right)$$

$$= \boxed{\begin{array}{l} 15 \text{ ft/sec, } 45° \text{ below the horizontal,} \\ \text{to the right} \end{array}}$$

$$v_{B/O} = r_{\text{outer}} \omega = \boxed{15 \text{ ft/sec, horizontal, to the left}}$$

15. $|AB|^2 = (3 \text{ ft})^2 + (3 \text{ ft})^2 - (2)(3 \text{ ft})^2 (\cos 135°)$

$$= 30.728 \text{ ft}^2$$

$$|AB| = 5.543 \text{ ft}$$

$$v_{B/A} = |AB| \omega = (5.543 \text{ ft}) \left(5 \frac{\text{rad}}{\text{sec}} \right)$$

$$= \boxed{\begin{array}{l} 27.72 \text{ ft/sec, } 22.5° \text{ above the horizontal,} \\ \text{to the left} \end{array}}$$

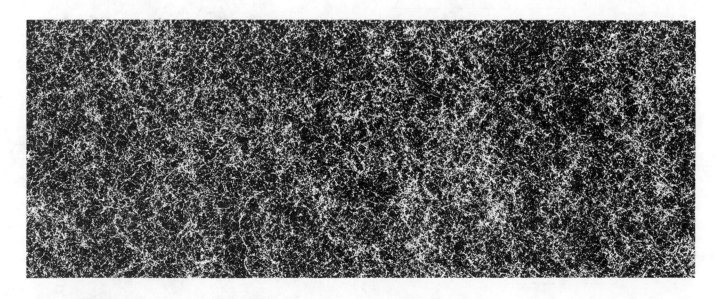

KINETICS

1. $v_t = \left(60\ \dfrac{\text{miles}}{\text{hr}}\right)\left(\dfrac{5280\ \text{ft}}{\text{mile}}\right)\left(\dfrac{\text{hr}}{3600\ \text{sec}}\right) = 88\ \text{ft/sec}$

$\tan\phi = \dfrac{v_t{}^2}{gr} = \dfrac{\left(88\ \dfrac{\text{ft}}{\text{sec}}\right)^2}{\left(32.2\ \dfrac{\text{ft}}{\text{sec}^2}\right)(6000\ \text{ft})} = \boxed{0.04\ (4\%)}$

2. The net acceleration vector is directed along the string.

$\phi = \arctan\left(\dfrac{a_c}{a_n}\right) = \arctan\left(\dfrac{v_t{}^2}{gr}\right)$

$= \arctan\left[\dfrac{\left(20\ \dfrac{\text{ft}}{\text{sec}}\right)^2}{\left(32.2\ \dfrac{\text{ft}}{\text{sec}^2}\right)(4\ \text{ft})}\right]$

$= \boxed{72.15°}\ \text{(independent of object's mass)}$

3. (a) $v_t = \omega r = 2\pi f r$

$= \left(2\pi\ \dfrac{\text{rad}}{\text{rev}}\right)\left(5\ \dfrac{\text{rev}}{\text{sec}}\right)(2\ \text{ft}) = 62.83\ \text{ft/sec}$

$a_n = \dfrac{v_t{}^2}{r} = \dfrac{\left(62.83\ \dfrac{\text{ft}}{\text{sec}}\right)^2}{2\ \text{ft}} = \boxed{1974\ \text{ft/sec}^2}$

(b) $F_{\text{centrifugal}} = \dfrac{ma_n}{g_c} = \dfrac{(10\ \text{lbm})\left(1974\ \dfrac{\text{ft}}{\text{sec}^2}\right)}{32.2\ \dfrac{\text{lbm-ft}}{\text{sec}^2\text{-lbf}}}$

$= \boxed{613.04\ \text{lbf (directed outward)}}$

(c) $|\mathbf{F}_{\text{centripetal}}| = |\mathbf{F}_{\text{centrifugal}}|$

$F_{\text{centripetal}} = \boxed{613.04\ \text{lbf (directed inward)}}$

(d) $\mathbf{H} = \dfrac{\mathbf{r} \times m\mathbf{v}_t}{g_c}$

$H = \dfrac{rm\mathbf{v}_t}{g_c}\quad \text{since } \mathbf{r} \perp \mathbf{v}_t$

$= \dfrac{(2\ \text{ft})(10\ \text{lbm})\left(62.83\ \dfrac{\text{ft}}{\text{sec}}\right)}{32.2\ \dfrac{\text{lbm-ft}}{\text{sec}^2\text{-lbf}}}$

$= \boxed{39.02\ \text{ft-lbf-sec}}$

4. $a = fg = (0.2)\left(32.2\ \dfrac{\text{ft}}{\text{sec}^2}\right) = 6.44\ \text{ft/sec}^2$

$s_{\text{skidding}} = \dfrac{v^2}{2a} = \dfrac{\left(12.88\ \dfrac{\text{ft}}{\text{sec}}\right)^2}{(2)\left(6.44\ \dfrac{\text{ft}}{\text{sec}^2}\right)} = \boxed{12.88\ \text{ft}}$

5. $a = fg = (0.333)\left(32.2\ \dfrac{\text{ft}}{\text{sec}^2}\right) = 10.72\ \text{ft/sec}^2$

$\Delta t = \dfrac{v}{a} = \dfrac{10\ \dfrac{\text{ft}}{\text{sec}}}{10.72\ \dfrac{\text{ft}}{\text{sec}^2}} = \boxed{0.933\ \text{sec}}$

6. $v_t = \left(40\ \dfrac{\text{mile}}{\text{hr}}\right)\left(\dfrac{5280\ \text{ft}}{\text{mile}}\right)\left(\dfrac{\text{hr}}{3600\ \text{sec}}\right)$

$= 58.67\ \text{ft/sec}$

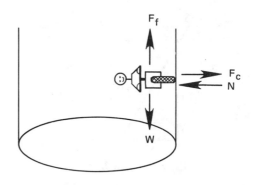

$N = F_c = \dfrac{ma_n}{g_c} = \left(\dfrac{m}{g_c}\right)\left(\dfrac{v_t{}^2}{r}\right)$

$W = F_f = fN$

$f = \dfrac{W}{N} = \dfrac{\dfrac{mg}{g_c}}{\dfrac{mv_t{}^2}{g_c r}} = \dfrac{gr}{v_t{}^2} = \dfrac{\left(32.2\ \dfrac{\text{ft}}{\text{sec}^2}\right)(50\ \text{ft})}{\left(58.67\ \dfrac{\text{ft}}{\text{sec}}\right)^2}$

$= \boxed{0.468}$

7. $a = \dfrac{v^2}{2s} = \dfrac{\left(12\ \dfrac{\text{ft}}{\text{sec}}\right)^2}{(2)(36\ \text{ft})} = 2\ \text{ft/sec}^2$

$F_f = fN = fW = (0.25)(10\ \text{lbf}) = 2.5\ \text{lbf}$

$$P = F_f + \frac{ma}{g_c} = 2.5 \text{ lbf} + \frac{(10 \text{ lbm})\left(2 \frac{\text{ft}}{\text{sec}^2}\right)}{32.2 \frac{\text{lbm-ft}}{\text{lbf-sec}^2}}$$

$$= \boxed{3.12 \text{ lbf}}$$

8. $a = g \sin\theta + fg \cos\theta$

$$= \left(32.2 \frac{\text{ft}}{\text{sec}^2}\right)\left[\sin(22.62°) + (0.1)(\cos(22.62°))\right]$$

$$= 15.36 \text{ ft/sec}^2$$

$$s = \frac{v_0^2}{2a} = \frac{\left(30 \frac{\text{ft}}{\text{sec}}\right)^2}{(2)\left(15.36 \frac{\text{ft}}{\text{sec}^2}\right)} = \boxed{29.30 \text{ ft}}$$

9.

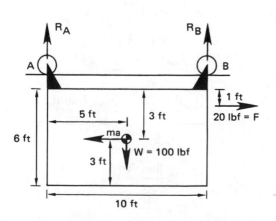

F = 100 lbf

block 2   50 lbm

block 1   100 lbm

$f_{1,G} = 0.2$

G (ground)

$$F_{f(1,G)} = (W_1 + W_2)f_{1,G} = (100 \text{ lbf} + 50 \text{ lbf})(0.2)$$

$$= 30 \text{ lbf } (< F; \text{ blocks move together}$$
$$\text{at acceleration} = a, \text{ assuming no slipping})$$

$$a = \frac{(F - F_{f(1,G)})g_c}{m_1 + m_2}$$

$$= \frac{(100 \text{ lbf} - 30 \text{ lbf})\left(32.2 \frac{\text{lbm-ft}}{\text{lbf-sec}^2}\right)}{100 \text{ lbm} + 50 \text{ lbm}} = 15.03 \text{ ft/sec}^2$$

If block 2 is not slipping,

$$F_{f(1,2)} = f_{1,2}W_2 = \frac{m_2 a}{g_c}$$

$$f_{1,2} = \frac{a}{g} = \frac{15.03 \frac{\text{ft}}{\text{sec}^2}}{32.2 \frac{\text{ft}}{\text{sec}^2}} = \boxed{0.467}$$

10.

R_A R_B

A   B

5 ft   3 ft   1 ft   20 lbf = F

6 ft   ma   W = 100 lbf

3 ft

10 ft

(a) $a_x = \dfrac{F_x g_c}{m} = \dfrac{(20 \text{ lbf})\left(32.2 \frac{\text{lbm-ft}}{\text{lbf-sec}^2}\right)}{100 \text{ lbm}}$

$$= \boxed{6.44 \text{ ft/sec}^2}$$

(b) $\sum M_A = -(5 \text{ ft})(100 \text{ lbf}) + (10 \text{ ft})R_B$
$$+ (1 \text{ ft})(20 \text{ lbf}) - (3 \text{ ft})(20) = 0$$

$$R_B = \boxed{54 \text{ lbf (upward)}}$$

$$\sum F_y = R_A + 54 \text{ lbf} - 100 \text{ lbf} = 0$$

$$R_A = \boxed{46 \text{ lbf (upward)}}$$

(c) Let y be the distance (positive upwards) from the center of gravity to F's line of action for $R_A = R_B = R$.

$$\sum M_{CG} = W(0) - (5 \text{ ft})R + (5 \text{ ft})R + y(20 \text{ lbf}) = 0$$

$$y = 0$$

$$\boxed{\text{Apply the force in line with the center of gravity.}}$$

11. $F_f = m\dfrac{g}{g_c}\sin\phi - \dfrac{ma}{g_c}$ [Eq. 1]

For constrained motion,

$$F_f r = I_0 \frac{\alpha}{g_c} = \left(\frac{2}{5}mr^2\right)\left(\frac{a}{r}\right)\left(\frac{1}{g_c}\right)$$

$$= \frac{2}{5}\frac{mar}{g_c}$$

$$F_f = \frac{2}{5}\frac{ma}{g_c}$$ [Eq. 2]

From Eq. 1 and Eq. 2, $mg\sin\phi - ma = \dfrac{2}{5}ma$.

$$a = \left(\frac{5}{7}\right)g\sin\phi = \left(\frac{5}{7}\right)\left(32.2 \frac{\text{ft}}{\text{sec}^2}\right)(\sin 30°)$$

$$= 11.5 \text{ ft/sec}^2$$

$$v = at = \left(11.5 \frac{\text{ft}}{\text{sec}^2}\right)(2 \text{ sec}) = \boxed{23 \text{ ft/sec (fps)}}$$

12. Let B's acceleration and velocity be a and v, positive upwards. Then A's acceleration and velocity are $-a$ and $-v$, respectively.

For B: $T - \dfrac{m_B g}{g_c} = \dfrac{m_B a}{g_c}$

$$T = \frac{m_B a + m_B g}{g_c}$$

For A: $T - \dfrac{m_A g \sin\phi}{g_c} = \dfrac{m_A(-a)}{g_c}$

$$T = \frac{m_A g \sin\phi - m_A a}{g_c}$$

DYNAMICS
Kinetics

$$a = \frac{m_A g \sin\phi - m_B g}{m_B + m_A}$$

$$= \frac{(10\ \text{lbm})\left(32.2\ \frac{\text{ft}}{\text{sec}^2}\right)\left[\sin(36.87°)\right] - (20\ \text{lbm})\left(32.2\ \frac{\text{ft}}{\text{sec}^2}\right)}{20\ \text{lbm} + 10\ \text{lbm}}$$

$$= -15.03\ \text{ft/sec}^2$$

$$v = at = \left(-15.03\ \frac{\text{ft}}{\text{sec}^2}\right)(3\ \text{sec})$$

$$= -45.1\ \text{ft/sec} = \boxed{45.1\ \text{ft/sec (downward)}}$$

13. $F = \dfrac{\dot{m}\Delta v}{g_c} = \dfrac{\left(560\ \dfrac{\text{lbm}}{\text{min}}\right)\left(\dfrac{\text{min}}{60\ \text{sec}}\right)\left(3.2\ \dfrac{\text{ft}}{\text{sec}}\right)}{32.2\ \dfrac{\text{lbm-ft}}{\text{lbf-sec}^2}}$

$$= \boxed{0.9275\ \text{lbf}}$$

14. $|\mathbf{Imp}| = \dfrac{|\Delta \mathbf{p}|}{g_c} = \dfrac{|\mathbf{p_2} - \mathbf{p_1}|}{g_c} = \dfrac{m v_2 - (-m v_1)}{g_c}$

$$= \frac{(0.4\ \text{lbm})\left(90\ \dfrac{\text{ft}}{\text{sec}} + 130\ \dfrac{\text{ft}}{\text{sec}}\right)}{32.2\ \dfrac{\text{lbm-ft}}{\text{lbf-sec}^2}} = \boxed{2.73\ \text{lbf-sec}}$$

15. $\Delta p_{\text{gun}} = -\Delta p_{\text{projectile}}$

$$\frac{m_{\text{gun}} v_{\text{gun}}}{g_c} = \frac{-m_{\text{projectile}} v_{\text{projectile}}}{g_c}$$

$$v_{\text{gun}} = -\frac{m_{\text{projectile}} v_{\text{projectile}}}{m_{\text{gun}}} = \frac{-(2.6\ \text{lbm})\left(2100\ \dfrac{\text{ft}}{\text{sec}}\right)}{1000\ \text{lbm}}$$

$$= \boxed{-5.46\ \text{ft/sec (opposite projectile)}}$$

16. In absence of external forces, $\Delta \mathbf{p} = 0$.

$$\frac{m_{\text{bullet}} v_{\text{bullet}}}{g_c} = \frac{m_{(\text{bullet} + \text{block})} v_{(\text{bullet} + \text{block})}}{g_c}$$

$$v_{(\text{bullet} + \text{block})} = \frac{m_{\text{bullet}} v_{\text{bullet}}}{m_{(\text{bullet} + \text{block})}}$$

$$= \frac{(0.15\ \text{lbm})\left(2300\ \dfrac{\text{ft}}{\text{sec}}\right)}{0.15\ \text{lbm} + 9\ \text{lbm}}$$

$$= \boxed{37.7\ \text{ft/sec}}$$

17.

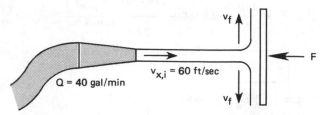

Neglecting gravity, water is turned through an angle of 90° in equal portions in all directions.

$$\dot{m} = \rho Q$$

$$= \left(62.4\ \frac{\text{lbm}}{\text{ft}^3}\right)\left(40\ \frac{\text{gal}}{\text{min}}\right)\left(\frac{\text{min}}{60\ \text{sec}}\right)\left(\frac{0.13368\ \text{ft}^3}{\text{gal}}\right)$$

$$= 5.56\ \text{lbm/sec}$$

For every direction other than along the x-axis, equal amounts of water are directed in opposite senses; no net force is applied in these directions.

$$F_x = \frac{\dot{m}\Delta v_x}{g_c} = \frac{-\dot{m} v_{x,i}}{g_c} = \frac{-\left(5.56\ \dfrac{\text{lbm}}{\text{sec}}\right)\left(60\ \dfrac{\text{ft}}{\text{sec}}\right)}{32.2\ \dfrac{\text{lbm-ft}}{\text{lbf-sec}^2}}$$

$$= \boxed{10.36\ \text{lbf (opposite water direction)}}$$

18. (a) $\dot{m} = \rho Q = \left(62.4\ \dfrac{\text{lbm}}{\text{ft}^3}\right)\left(100\ \dfrac{\text{gal}}{\text{sec}}\right)\left(\dfrac{0.13368\ \text{ft}^3}{\text{gal}}\right)$

$$= 834.16\ \text{lbm/sec}$$

Let inward directions be positive.

$$\dot{p}_{\text{out},x} = \dot{m} v_{\text{out}} \cos(160°)$$

$$= \left(834.16\ \frac{\text{lbm}}{\text{sec}}\right)\left(57\ \frac{\text{ft}}{\text{sec}}\right)(\cos(160°)) = -44{,}680\ \frac{\text{lbm-ft}}{\text{sec}^2}$$

$$\dot{p}_{\text{out},y} = \dot{m} v_{\text{out}} \sin(160°)$$

$$= \left(834.16\ \frac{\text{lbm}}{\text{sec}}\right)\left(57\ \frac{\text{ft}}{\text{sec}}\right)(\sin(160°)) = 16{,}262\ \text{lbm-ft/sec}^2$$

$$\dot{p}_{\text{in},x} = \dot{m} v_{\text{in}} = \left(834.16\ \frac{\text{lbm}}{\text{sec}}\right)\left(60\ \frac{\text{ft}}{\text{sec}}\right)$$

$$= 50{,}050\ \text{lbm-ft/sec}^2$$

$$F_x = \frac{\Delta \dot{p}_x}{g_c} = \frac{\dot{p}_{\text{out},x} - \dot{p}_{\text{in},x}}{g_c}$$

$$= \frac{-44{,}680\ \dfrac{\text{lbm-ft}}{\text{sec}^2} - 50{,}050\ \dfrac{\text{lbm-ft}}{\text{sec}^2}}{32.2\ \dfrac{\text{lbm-ft}}{\text{lbf-sec}^2}} = -2942\ \text{lbf}$$

$$F_y = \frac{\Delta \dot{p}_y}{g_c} = \frac{16{,}262\ \dfrac{\text{lbm-ft}}{\text{sec}^2} - 0}{32.2\ \dfrac{\text{lbm-ft}}{\text{lbf-sec}^2}} = 505.0\ \text{lbf}$$

DYNAMICS
Kinetics

$$F = \sqrt{F_x{}^2 + F_y{}^2} = \sqrt{(-2942\text{ lbf})^2 + (505\text{ lbf})^2}$$

$$= \boxed{2985\text{ lbf}}$$

(b) $\phi = \arctan\left(\dfrac{F_y}{F_x}\right) = \arctan\left(\dfrac{505\text{ lbf}}{-2942\text{ lbf}}\right)$

$$= \boxed{170.26°\text{ from the horizontal, counterclockwise}}$$

19. Since the electron is deflected from its original path, the collision is oblique. The ratio of the masses is

$$\frac{m_H}{m_e} = \frac{1.007277u}{0.0005486u} = 1836$$

Kinetic energy may or may not be conserved in this collision; there is insufficient information to make the determination. Momentum is always conserved, regardless of the axis along which it is evaluated. Consider the original path of the electron to be parallel to the x-axis. Then, by conservation of momentum in the x-direction,

$$m_e v_{e,x} + m_H v_{H,x} = m_e v'_{e,x} + m_H v'_{H,x}$$

Recognizing that $v_{H,x} = 0$ and substituting the ratio of masses,

$$v_{e,x} = v'_e \cos\theta_e + 1836\,v'_{H,x}$$

$$1500\,\frac{\text{ft}}{\text{sec}} = \left(65\,\frac{\text{ft}}{\text{sec}}\right)\cos 30° + (1836)(v'_{H,x})$$

$$v'_{H,x} = \boxed{0.786\text{ ft/sec}}$$

Notice that the x-component of v'_H was requested.

20. $m_{\text{left}}v_{\text{left}} + m_{\text{right}}v_{\text{right}} = m_{\text{couple}}v_{\text{couple}}$

$$v_{\text{couple}} = \frac{(5\text{ ton})\left(5\,\dfrac{\text{ft}}{\text{sec}}\right) + (5\text{ ton})\left(-4\,\dfrac{\text{ft}}{\text{sec}}\right)}{10\text{ ton}}$$

$$= \boxed{\frac{1}{2}\text{ ft/sec (to the right)}}$$

21. $v_{Ay} = 5\sin(45°) = 3.536\text{ ft/sec}$

$v_{Ax} = 5\cos(45°) = 3.536\text{ ft/sec}$

$v_{By} = \left(10\,\dfrac{\text{ft}}{\text{sec}}\right)\left(\dfrac{3}{5}\right) = 6\text{ ft/sec}$

$v_{Bx} = \left(10\,\dfrac{\text{ft}}{\text{sec}}\right)\left(-\dfrac{4}{5}\right) = -8\text{ ft/sec}$

The force of impact is in the x-direction only.

$$v'_{Ay} = v_{Ay}$$

$$v'_{By} = v_{By}$$

In the x-direction,

$$e = \frac{v'_{Ax} - v'_{Bx}}{v_{Bx} - v_{Ax}} = \frac{v'_{Ax} - v'_{Bx}}{-8\,\dfrac{\text{ft}}{\text{sec}} - 3.536\,\dfrac{\text{ft}}{\text{sec}}}$$

$$v'_{Ax} - v'_{Bx} = (0.8)\left(-8\,\frac{\text{ft}}{\text{sec}} - 3.536\,\frac{\text{ft}}{\text{sec}}\right)$$

$$= -9.229\text{ ft/sec} \qquad \text{[Eq. 1]}$$

$$m_A v_{Ax} + m_B v_{Bx} = m_A v'_{Ax} + m_B v'_{Bx}$$

Since $m_A = m_B$,

$$v'_{Ax} + v'_{Bx} = 3.536\,\frac{\text{ft}}{\text{sec}} - 8\,\frac{\text{ft}}{\text{sec}} = -4.464\text{ ft/sec}\ \text{[Eq. 2]}$$

Solving Eq. 1 and Eq. 2 simultaneously,

$v'_{Ax} = -6.846\text{ ft/sec}$

$v'_{Bx} = 2.382\text{ ft/sec}$

$v'_A = \sqrt{(v'_{Ax})^2 + (v'_{Ay})^2}$

$$= \sqrt{\left(3.536\,\frac{\text{ft}}{\text{sec}}\right)^2 + \left(-6.846\,\frac{\text{ft}}{\text{sec}}\right)^2} = \boxed{7.705\text{ ft/sec}}$$

$$v'_B = \sqrt{\left(6\,\frac{\text{ft}}{\text{sec}}\right)^2 + \left(2.382\,\frac{\text{ft}}{\text{sec}}\right)^2} = \boxed{6.456\text{ ft/sec}}$$

$$\phi_A = \arctan\left(\frac{3.536}{-6.846}\right) = \boxed{152.7°}$$

$$\phi_B = \arctan\left(\frac{6}{2.382}\right) = \boxed{68.3°}$$

22. (a) $mgh = \dfrac{1}{2}mv^2$

$$v = \sqrt{2gh} = \sqrt{(2)\left(32.2\,\frac{\text{ft}}{\text{sec}^2}\right)(3\text{ ft})} = \boxed{13.9\text{ ft/sec}}$$

(b) $F_c = \dfrac{mv^2}{g_c r} = \dfrac{(10\text{ lbm})\left(13.9\,\dfrac{\text{ft}}{\text{sec}}\right)^2}{\left(32.2\,\dfrac{\text{lbm-ft}}{\text{lbf-sec}^2}\right)(3\text{ ft})} = 20\text{ lbf}$

$$T = W + F_c = 10\text{ lbf} + 20\text{ lbf} = \boxed{30\text{ lbf}}$$

(c) $e = \dfrac{v_1' - v_2'}{v_2 - v_1} = \dfrac{v_1' - v_2'}{0 - 13.9\,\dfrac{\text{ft}}{\text{sec}}}$

$$v_1' - v_2' = (0.7)\left(-13.9\,\frac{\text{ft}}{\text{sec}}\right) = -9.73\text{ ft/sec}\ \ \text{[Eq. 1]}$$

$$m_1 v_1 + m_2 v_2 = m_1 v_1' + m_2 v_2'$$

$$(10\text{ lbm})v_1' + (50\text{ lbm})v_2' = 139\text{ ft/sec} \qquad \text{[Eq. 2]}$$

DYNAMICS
Kinetics

Solving Eq. 1 and Eq. 2 simultaneously,

$$v_1' = -5.79 \text{ ft/sec}$$

$$\boxed{v_2' = 3.94 \text{ ft/sec}}$$

(d) $\left(\dfrac{1}{2}\right)\left(\dfrac{m}{g_c}\right) v^2 = \dfrac{1}{2} k x^2$

$$k = \frac{m v^2}{g_c x^2} = \frac{(50 \text{ lbm})\left(3.94 \dfrac{\text{ft}}{\text{sec}}\right)^2}{\left(32.2 \dfrac{\text{lbm-ft}}{\text{lbf-sec}^2}\right)(0.5 \text{ ft})^2}$$

$$= \boxed{96.4 \text{ lbf/ft}}$$

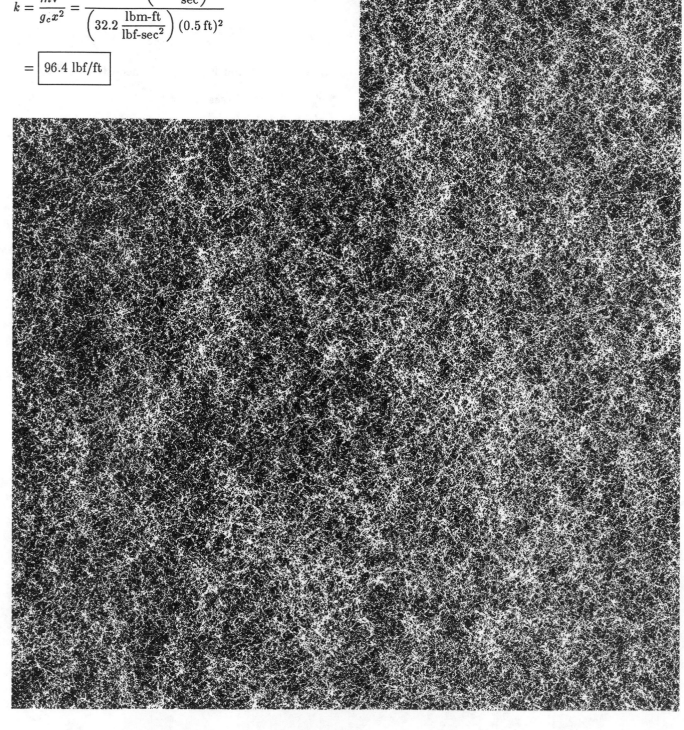

VIBRATING SYSTEMS

1. The force equilibrium equations for m_1 and m_2 are

$$\sum F_y = 0 = m_1 g - k_1 \delta_{1,\text{st}} + k_2(\delta_{2,\text{st}} - \delta_{1,\text{st}})$$
$$= 0 = m_2 g - k_2(\delta_{2,\text{st}} - \delta_{1,\text{st}})$$

Solving for $\delta_{1,\text{st}}$ and $\delta_{2,\text{st}}$,

$$\delta_{1,\text{st}} = \boxed{\dfrac{(m_1 + m_2)g}{k_1}}$$

$$\delta_{2,\text{st}} = \boxed{\left(\dfrac{m_2}{k_2} + \dfrac{m_1 + m_2}{k_1}\right)g}$$

2. The moment equilibrium equation is

$$\sum M = 0 = C\theta_{\text{st}} - mgl\cos\theta_{\text{st}}$$

θ_{st} satisfies the above equation. If θ_{st} is small, $\cos\theta_{\text{st}} \approx 1$.

$$\boxed{\theta_{\text{st}} \approx \dfrac{mgl}{C}}$$

3. (a) $\quad \omega = \sqrt{\dfrac{kg_c}{m}} = \sqrt{\dfrac{\left(10\,\dfrac{\text{lbf}}{\text{ft}}\right)\left(32.2\,\dfrac{\text{lbm-ft}}{\text{lbf-sec}^2}\right)}{1.9\,\text{lbm}}}$

$\qquad = 13.02\,\text{rad/sec}$

$f = \dfrac{\omega}{2\pi} = \dfrac{13.02\,\dfrac{\text{rad}}{\text{sec}}}{2\pi} = \boxed{2.08\,\text{Hz}}$

(b) $\quad T = \dfrac{2\pi}{\omega} = \dfrac{2\pi}{13.02\,\dfrac{\text{rad}}{\text{sec}}} = \boxed{0.4826\,\text{sec}}$

(c) $\quad F_0 = kx_0 = \left(10\,\dfrac{\text{lbf}}{\text{ft}}\right)(2\,\text{in})\left(\dfrac{\text{ft}}{12\,\text{in}}\right)$

$\qquad = \boxed{1.67\,\text{lbf}}$

(d) $\quad A = \sqrt{x_0^2 + \left(\dfrac{v_0}{\omega}\right)^2} = \sqrt{(2\,\text{in})^2 + 0} = 2\,\text{in}$

$$a_{\max} = A\omega^2 = (2\,\text{in})\left(13.02\,\dfrac{\text{rad}}{\text{sec}}\right)^2$$

$$= \boxed{339.04\,\text{in/sec}^2}$$

(e) $x_1 = A\cos(\omega t_1 - \phi)$

$$\phi = \arctan\left(\dfrac{v_0}{\omega x_0}\right) = \arctan(0) = 0^\circ$$

$$t_1 = \left(\dfrac{1}{\omega}\right)\left[\arccos\left(\dfrac{x_1}{A}\right)\right] + \dfrac{\phi}{\omega}$$

$$= \left(\dfrac{1}{13.02\,\dfrac{\text{rad}}{\text{sec}}}\right)\left[\arccos\left(\dfrac{1\,\text{in}}{2\,\text{in}}\right)\right] + 0$$

$$= 0.080\,\text{sec}$$

$$v(t) = \dot{x}(t) = -A\omega\sin(\omega t - \phi)$$

$$v_1 = -A\omega\sin(\omega t_1 - \phi)$$

$$= -(2\,\text{in})\left(13.02\,\dfrac{\text{rad}}{\text{sec}}\right)$$

$$\times\left[\sin\left(\left(13.02\,\dfrac{\text{rad}}{\text{sec}}\right)(0.080\,\text{sec}) - 0\right)\right]$$

$$= \boxed{-22.48\,\text{in/sec}}$$

(f) $v_{\max} = A\omega = (2\,\text{in})\left(13.02\,\dfrac{\text{rad}}{\text{sec}}\right)$

$$= \boxed{26.04\,\text{in/sec}}$$

4. $\quad \omega = \sqrt{\dfrac{g}{L}} = \sqrt{\dfrac{32.2\,\dfrac{\text{ft}}{\text{sec}^2}}{3\,\text{ft}}} = 3.276\,\text{rad/sec}$

$$A = \sqrt{\theta_0^2 + \left(\dfrac{\omega_0}{\omega}\right)^2} = \sqrt{\left[(10^\circ)\left(\dfrac{2\pi}{360^\circ}\right)\right]^2 + 0}$$

$$= 0.1745$$

$$\phi = \arctan\left(\dfrac{\omega_0}{\omega\theta_0}\right) = \arctan(0) = 0$$

$$\alpha_{\max} = A\omega^2 = (0.1745)\left(3.276\,\dfrac{\text{rad}}{\text{sec}}\right)^2$$

$$= 1.8728\,\text{rad/sec}^2$$

(a) $F_{max} = \dfrac{ma_{max}}{g_c} = \dfrac{mL\alpha_{max}}{g_c}$

$$= \dfrac{(0.4\text{ lbm})(3\text{ ft})\left(1.8728\,\dfrac{\text{rad}}{\text{sec}^2}\right)}{32.2\,\dfrac{\text{lbm-ft}}{\text{lbf-sec}^2}}$$

$$= \boxed{0.06979\text{ lbf}}$$

(b) $\omega_{max} = A\omega = (0.1745)\left(3.276\,\dfrac{\text{rad}}{\text{sec}}\right)$

$$= \boxed{0.5717\text{ rad/sec}}$$

(c) $v_{max} = L\omega_{max} = (3\text{ ft})\left(0.5717\,\dfrac{\text{rad}}{\text{sec}}\right)$

$$= \boxed{1.715\text{ ft/sec}}$$

5. (a) $I_0 = mr^2 = (0.05\text{ lbm})(0.3\text{ in})^2 \left(\dfrac{\text{ft}}{12\text{ in}}\right)^2$

$$= \boxed{3.125 \times 10^{-5}\text{ lbm-ft}^2}$$

(b) $T = 2\pi\sqrt{\dfrac{I_0}{kg_c}}$

$k = \left(\dfrac{2\pi}{T}\right)^2 \dfrac{I_0}{g_c}$

$$= \left(\dfrac{2\pi}{0.2\text{ sec}}\right)^2 \left(\dfrac{3.125 \times 10^{-5}\text{ lbm-ft}^2}{32.2\,\dfrac{\text{lbm-ft}}{\text{lbf-sec}^2}}\right)$$

$$= \boxed{0.0009578\text{ ft-lbf/rad}}$$

6. $I_c = \frac{1}{2}mr^2$

For rotation about an edge, by the parallel axis theorem,

$$I_0 = I_c + mr^2 = \tfrac{3}{2}mr^2$$

$$T = 2\pi\sqrt{\dfrac{I_0}{mgr}} = \boxed{2\pi\sqrt{\dfrac{3r}{2g}}}$$

7. $f = \left(\dfrac{1}{2\pi}\right)\sqrt{\dfrac{2g}{L}} = \left(\dfrac{1}{2\pi}\right)\sqrt{\dfrac{(2)\left(32.2\,\dfrac{\text{ft}}{\text{sec}^2}\right)}{(30\text{ in})\left(\dfrac{\text{ft}}{12\text{ in}}\right)}}$

$$= \boxed{0.81\text{ Hz}}$$

8. (a) $\omega = \boxed{11\pi\text{ rad/sec}}$

$$f = \dfrac{\omega}{2\pi} = \dfrac{11\pi\,\dfrac{\text{rad}}{\text{sec}}}{2\pi\,\dfrac{\text{rad}}{\text{cycle}}} = \boxed{\dfrac{11}{2}}\text{ Hz}$$

$$T = \dfrac{1}{f} = \boxed{\dfrac{2}{11}}\text{ sec}$$

(b) $x_{max} = A = \boxed{3}$

$$v_{max} = A\omega = \boxed{33\pi\text{ sec}^{-1}}$$

$$a_{max} = A\omega^2 = \boxed{363\pi^2\text{ sec}^{-2}}$$

At $t = 0.4$,

$$x(0.4) = 3\sin\left(11\pi(0.4) + \dfrac{\pi}{3}\right) = \boxed{2.23\text{ units}}$$

$$v(t) = \dfrac{dx(t)}{dt} = (11\pi)(3)\cos\left(11\pi t + \dfrac{\pi}{3}\right)$$

$$v(0.4) = 33\pi\cos\left(11\pi(0.4) + \dfrac{\pi}{3}\right)$$

$$= \boxed{-69.4\text{ units/sec}}$$

$$a(t) = \dfrac{dv(t)}{dt} = -(11\pi)(11\pi)(3)\sin\left(11\pi t + \dfrac{\pi}{3}\right)$$

$$a(0.4) = -363\pi^2\sin\left(11\pi(0.4) + \dfrac{\pi}{3}\right)$$

$$= \boxed{-2662\text{ units/sec}^2}$$

9.

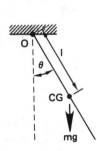

$$I_0\theta'' = -mgl\sin\theta$$

$$I_0\theta'' + mgl\,\theta = 0$$

For small values of θ,

$$\omega = \sqrt{\dfrac{mgl}{I_0}}$$

$$f = \dfrac{\omega}{2\pi} = \dfrac{1}{2\pi}\sqrt{\dfrac{mgl}{I_0}}$$

$$I_0 = \frac{mgl}{(2\pi f)^2}$$

$$= \frac{(9 \text{ lbm}) \left(32.2 \dfrac{\text{ft}}{\text{sec}^2}\right)(13 \text{ in})}{(2\pi)^2 \left(\dfrac{50}{60 \text{ sec}}\right)^2 \left(12 \dfrac{\text{in}}{\text{ft}}\right)}$$

$$= 1649 \text{ lbm-ft}^2$$

$$I_{CG} = I_0 - ml^2$$

$$= 1649 \text{ lbm-ft}^2 - (9 \text{ lbm}) \left(\frac{13 \text{ in}}{12 \dfrac{\text{in}}{\text{ft}}}\right)$$

$$= 1649 \text{ lbm-ft}^2 - 10.563 \text{ lbm-ft}^2$$

$$= \boxed{1638.46 \text{ lbm-ft}^2}$$

10. (b) Since $F = \delta A_0 E / L_0$, where A_0 is the cross-sectional area and L_0 is the length of the cable, the cable behaves like a spring of constant k.

$$k = \frac{F}{\delta} = \frac{A_0 E}{L_0} = \frac{(1.6 \text{ in}^2) \left(8.0 \times 10^6 \dfrac{\text{lbf}}{\text{in}^2}\right)}{65 \text{ ft}}$$

$$= 1.969 \times 10^5 \text{ lbf/ft}$$

$$\delta_{st} = \frac{mg}{kg_c} = \frac{(6500 \text{ lbm}) \left(32.2 \dfrac{\text{ft}}{\text{sec}^2}\right)}{\left(1.969 \times 10^5 \dfrac{\text{lbf}}{\text{ft}}\right) \left(32.2 \dfrac{\text{lbm-ft}}{\text{lbf-sec}^2}\right)}$$

$$= 0.033 \text{ ft}$$

$$\omega = \sqrt{\frac{kg_c}{m}}$$

$$= \sqrt{\frac{\left(1.969 \times 10^5 \dfrac{\text{lbf}}{\text{ft}}\right) \left(32.2 \dfrac{\text{lbm-ft}}{\text{lbf-sec}^2}\right)}{6500 \text{ lbm}}}$$

$$= \boxed{31.2 \text{ rad/sec}}$$

(a) $v_{max} = 4 \dfrac{\text{ft}}{\text{sec}} = A\omega$

$$A = \frac{v_{max}}{\omega} = \frac{4 \dfrac{\text{ft}}{\text{sec}}}{31.2 \dfrac{\text{rad}}{\text{sec}}} = \boxed{0.128 \text{ ft}}$$

(c) The maximum stress is obtained at the maximum amplitude.

$$\sigma = E\epsilon = E \frac{\delta}{L_0} = E \left(\frac{A + \delta_{st}}{L_0}\right)$$

$$= \left(8.0 \times 10^6 \frac{\text{lbf}}{\text{in}^2}\right) \left(\frac{0.128 \text{ ft} + 0.033 \text{ ft}}{65 \text{ ft}}\right)$$

$$= \boxed{19{,}815 \text{ lbf/in}^2}$$

11.

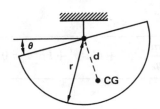

(a) $I_0 = \dfrac{mr^2}{4} = \dfrac{(1.0 \text{ lbm})(4 \text{ in})^2}{4} = 4 \text{ lbm-in}^2$

$d = \dfrac{4r}{3\pi} = \dfrac{(4)(4 \text{ in})}{3\pi} = 1.70 \text{ in}$

The equation of motion is

$$\boxed{\sum M_0 = I_0 \theta'' + mgd \sin\theta = 0}$$

(b) $\omega = \sqrt{\dfrac{mgd}{I_0}} = \sqrt{\dfrac{(1 \text{ lbm})\left(32.2 \dfrac{\text{ft}}{\text{sec}^2}\right)(1.70 \text{ in})}{(4 \text{ lbm-in}^2)\left(\dfrac{\text{ft}}{12 \text{ in}}\right)}}$

$$= \boxed{12.81 \text{ rad/sec}}$$

(c) The length, l, of the equivalent pendulum is

$$\omega_{eq} = \omega$$

$$\sqrt{\frac{g}{l}} = \sqrt{\frac{mgd}{I_0}}$$

$$l = \frac{I_0}{md} = \frac{4 \text{ lbm-in}^2}{(1 \text{ lbm})(1.70 \text{ in})} = \boxed{2.35 \text{ in}}$$

(d) $\omega_{max} = \theta'_{max} = \theta_{max}\omega$

$$= (5°)\left(\frac{\pi}{180} \text{ rad/}°\right)\left(12.81 \frac{\text{rad}}{\text{sec}}\right)$$

$$= \boxed{1.12 \text{ rad/sec}}$$

(e) $\alpha_{max} = \theta''_{max} = \theta_{max}\omega^2$

$$= (5°)\left(\frac{\pi}{180} \text{ rad/}°\right)\left(12.81 \frac{\text{rad}}{\text{sec}}\right)^2$$

$$= \boxed{14.3 \text{ rad/sec}^2}$$

12. (a) The equilibrium equation for moments taken about point O is

$$\frac{m}{g_c} L^2 \theta'' = -ka^2 \sin\theta - Ca^2 \frac{d}{dt}(\sin\theta)$$

$$\frac{m}{g_c} L^2 \theta'' + Ca^2 \theta' \cos\theta + ka^2 \sin\theta = 0$$

For small angles,

$$\frac{m}{g_c}L^2\theta'' + Ca^2\theta' + ka^2\theta = 0$$

(b) $\zeta = \dfrac{Ca^2}{2\sqrt{\dfrac{mL^2ka^2}{g_c}}} = \dfrac{Ca}{2L\sqrt{\dfrac{mk}{g_c}}}$

$$= \frac{\left(2.9\,\dfrac{\text{lbf-sec}}{\text{in}}\right)(20\text{ in})}{(2)(3.3\text{ ft})\left(\dfrac{12\text{ in}}{\text{ft}}\right)\sqrt{\dfrac{(2.2\text{ lbm})\left(115\,\dfrac{\text{lbf}}{\text{in}}\right)}{\left(32.2\,\dfrac{\text{lbm-ft}}{\text{lbf-sec}^2}\right)\left(12\,\dfrac{\text{in}}{\text{ft}}\right)}}}$$

$$= \boxed{0.91}$$

The system is slightly underdamped.

(c) $\omega_d = \omega\sqrt{1-\zeta^2}$

$$= \sqrt{\frac{ka^2g_c}{mL^2}}\sqrt{1-\frac{C^2a^4}{\dfrac{4mL^2ka^2}{g_c}}}$$

$$= \sqrt{\frac{ka^2g_c}{mL^2} - \left(\frac{Ca^2}{2mL^2}g_c\right)^2}$$

$\dfrac{ka^2g_c}{mL^2} = \dfrac{\left(115\,\dfrac{\text{lbf}}{\text{in}}\right)(20\text{ in})^2\left(32.2\,\dfrac{\text{lbm-ft}}{\text{lbf-sec}^2}\right)}{(2.2\text{ lbm})(3.3\text{ ft})^2\left(12\,\dfrac{\text{in}}{\text{ft}}\right)}$

$$= 5152\,\frac{1}{\text{sec}^2}$$

$\left(\dfrac{Ca^2g_c}{2mL^2}\right)^2 = \left(\dfrac{\left(2.9\,\dfrac{\text{lbf-sec}}{\text{in}}\right)(20\text{ in})^2\left(32.2\,\dfrac{\text{lbm-ft}}{\text{lbf-sec}^2}\right)}{(2)(2.2\text{ lbm})(3.3\text{ ft})^2\left(12\,\dfrac{\text{in}}{\text{ft}}\right)}\right)^2$

$$= 4220\,\frac{1}{\text{sec}^2}$$

$\omega_d = \sqrt{5152\,\dfrac{1}{\text{sec}^2} - 4220\,\dfrac{1}{\text{sec}^2}}$

$$= \boxed{30.53\text{ rad/sec}}$$

13. The equation of motion is

$$\sum F_y = mx'' + Cx' + kx = 0$$

$$\zeta = \frac{C}{2\sqrt{\dfrac{mk}{g_c}}}$$

$$C = 2\zeta\sqrt{\frac{mk}{g_c}}$$

$$\delta = \ln\frac{x_n}{x_{n+1}} = \frac{2\pi\zeta}{\sqrt{1-\zeta^2}}$$

$$3\delta = (3)\left(\frac{2\pi\zeta}{\sqrt{1-\zeta^2}}\right) = \ln\frac{x_1}{x_4} = \ln 2$$

$$\frac{2\pi\zeta}{\sqrt{1-\zeta^2}} = \frac{\ln 2}{3} = 0.231$$

$$4\pi^2\zeta^2 = (0.231)^2(1-\zeta^2)$$

$$\zeta = \sqrt{\frac{(0.231)^2}{4\pi^2 + (0.231)^2}} = 0.0367$$

$$C = (2)(0.0367)\sqrt{\frac{(44\text{ lbm})\left(86\,\dfrac{\text{lbf}}{\text{in}}\right)}{\left(32.2\,\dfrac{\text{lbm-ft}}{\text{lbf-sec}^2}\right)\left(12\,\dfrac{\text{in}}{\text{ft}}\right)}}$$

$$= \boxed{0.230\text{ lbf-sec/in}}$$

14. (a) $\omega = \sqrt{\dfrac{k_r g_c}{I}}$

$k_r = \dfrac{\pi d^4 G}{32L} = \dfrac{\pi(0.2\text{ in})^4\left(1.2\times10^7\,\dfrac{\text{lbf}}{\text{in}^2}\right)}{(32)(20\text{ in})}$

$$= 94.25\text{ lbf-in}$$

$I = \dfrac{mr^2}{2} = \dfrac{(6.5\text{ lbm})(12\text{ in})^2}{2} = 468\text{ lbm-in}^2$

$$\omega = \sqrt{\frac{(94.25\text{ lbf-in})\left(32.2\,\dfrac{\text{lbm-ft}}{\text{lbf-sec}^2}\right)}{(468\text{ lbm-in}^2)\left(\dfrac{\text{ft}}{12\text{ in}}\right)}}$$

$$= \boxed{8.82\text{ rad/sec}}$$

(b) $T = \dfrac{2\pi}{\omega} = \dfrac{2\pi}{8.82\,\dfrac{\text{rad}}{\text{sec}}} = \boxed{0.712\text{ sec}}$

(c) $A = \dfrac{\omega_0}{\omega} = \dfrac{2\,\dfrac{rad}{sec}}{8.82\,\dfrac{rad}{sec}} = 0.227\ rad = \boxed{13.0°}$

(d) $\alpha_{max} = A\omega^2 = (0.227\ rad)\left(8.82\,\dfrac{rad}{sec}\right)^2$

$\qquad = \boxed{17.7\ rad/sec^2}$

15. $\dfrac{1}{k_{eq}} = \dfrac{1}{k_{r1}} + \dfrac{1}{k_{r2}} = \left(\dfrac{32L}{\pi G}\right)\left(\dfrac{1}{d_1^4} + \dfrac{1}{d_2^4}\right)$

$= \left(\dfrac{(32)(10\ in)}{\pi\left(1.2 \times 10^7\,\dfrac{lbf}{in^2}\right)}\right)\left(\dfrac{1}{(0.16\ in)^4} + \dfrac{1}{(0.24\ in)^4}\right)$

$\quad k_{eq} = 64.47\ lbf\text{-}in$

For the two pendulums to be equivalent, their natural frequencies must be equal.

$\sqrt{\dfrac{k_{eq}g_c}{I_{eq}}} = \sqrt{\dfrac{k_r g_c}{I}}$

$\qquad I_{eq} = I\dfrac{k_{eq}}{k_r} = (468\ lbm\text{-}in^2)\left(\dfrac{64.47\ lbf\text{-}in}{94.25\ lbf\text{-}in}\right)$

$\qquad\quad = 320\ lbm\text{-}in^2$

$\qquad I_{eq} = \dfrac{m_{eq}r_{eq}^2}{2}$

$\qquad m_{eq} = \rho V_{eq} = \rho\pi r_{eq}^2 t$

ρ is the density of the disk material, and t is the thickness of the disk.

$\dfrac{m_{eq}}{m} = \dfrac{r_{eq}^2}{r^2}$

$\dfrac{I_{eq}}{I} = \dfrac{r_{eq}^4}{r^4}$

$r_{eq} = r\left(\dfrac{I_{eq}}{I}\right)^{1/4} = (12\ in)\left(\dfrac{320\ lbm\text{-}in^2}{468\ lbm\text{-}in^2}\right)^{1/4}$

$\qquad = \boxed{10.91\ in}$

16. The natural frequency of the system is

$$\omega = \sqrt{\dfrac{k_r g_c}{I}} = 8.82\ rad/sec$$

The magnification factor is

$$\beta = \dfrac{1}{1 - \left(\dfrac{\omega_f}{\omega}\right)^2} = \dfrac{1}{1 - \left(\dfrac{5\,\dfrac{rad}{sec}}{8.82\,\dfrac{rad}{sec}}\right)^2} = 1.474$$

$D = \beta\dfrac{T_0}{k}$

$T_0 = \dfrac{Dk}{\beta} = (5°)\left(\dfrac{\pi}{180°}\right)\left(\dfrac{94.25\ lbf\text{-}in}{1.474}\right)$

$\qquad = \boxed{5.58\ lbf\text{-}in}$

PROFESSIONAL PUBLICATIONS, INC. ● Belmont, CA

ELECTROSTATICS AND ELECTROMAGNETICS

1.

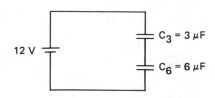

An equivalent circuit is

$$C_N = \frac{C_3 C_6}{C_3 + C_6} =$$

$$\frac{(3\,\mu F)(6\,\mu F)}{3\,\mu F + 6\,\mu F} = 2\,\mu F$$

$$Q_N = C_N V = (2 \times 10^{-6}\,F)(12\,V) = 24 \times 10^{-6}\,C$$

The charge across each capacitor is $\boxed{24\,\mu C.}$

Since $Q_N = C_3 V_3 = C_6 V_6$

$$V_3 = \frac{Q_N}{C_3} = \frac{C_N V}{C_3} = \frac{(2 \times 10^{-6}\,F)(12\,V)}{3 \times 10^{-6}\,F} = \boxed{8\,V}$$

$$V_6 = \frac{Q_N}{C_6} = \frac{C_N V}{C_6} = \frac{(2 \times 10^{-6}\,F)(12\,V)}{6 \times 10^{-6}\,F} = \boxed{4\,V}$$

2.

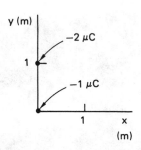

(a) The force between the point charges is

$$F = \left(\frac{1}{4\pi\epsilon_0}\right)\left(\frac{Q_1 Q_2}{r^2}\right)$$

$$= \left(8.987 \times 10^9\,\frac{N\cdot m^2}{C^2}\right)\left[\frac{(-1 \times 10^{-6}\,C)(-2 \times 10^{-6}\,C)}{(1\,m)^2}\right]$$

$$= \boxed{0.01797\,N}$$

The force is repulsive along the y-axis.

(b) The potential of the $-2\,\mu C$ charge is

$$V = \left(-\frac{1}{4\pi\epsilon_o}\right)\left(\frac{Q_1}{r}\right)$$

$$= -\left(8.987 \times 10^9\,\frac{N\cdot m^2}{C^2}\right)\left(\frac{-1 \times 10^{-6}\,C}{1\,m}\right)$$

$$= \boxed{8987\,J/C\ (8.987\,kV)}$$

3. (a)

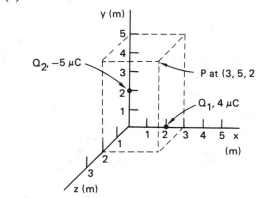

The force between the charges is

$$F = \left(\frac{1}{4\pi\epsilon_0}\right)\left(\frac{Q_1 Q_2}{r_{1-2}^2}\right)$$

$$= \left(8.987 \times 10^9\,\frac{N\cdot m^2}{C^2}\right)\left(\frac{(4 \times 10^{-6}\,C)(-5 \times 10^{-6}\,C)}{(2\,m)^2 + (2\,m)^2}\right)$$

$$= \boxed{-0.02247\,N\ (\text{attractive})}$$

(b) The electric potential at point P is

$$V_{Q_1} + V_{Q_2}$$

$$= \left(-\frac{1}{4\pi\epsilon_0}\right)\left(\frac{Q_1}{r_{1-P}}\right) - \left(\frac{1}{4\pi\epsilon_0}\right)\left(\frac{Q_2}{r_{2-P}}\right)$$

$$= -\left(8.987 \times 10^9\,\frac{N\cdot m^2}{C^2}\right)$$

$$\times \left[\frac{4 \times 10^{-6}\,C}{\sqrt{(1\,m)^2 + (5\,m)^2 + (2\,m)^2}} + \frac{-5 \times 10^{-6}\,C}{\sqrt{(3\,m)^2 + (3\,m)^2 + (2\,m)^2}}\right]$$

$$= \boxed{3017\,J/C\ (3.017\,kV)}$$

4. $E = \left(\dfrac{1}{4\pi\epsilon_0}\right)\left(\dfrac{Q}{r^2}\right)$

$= \left(8.987 \times 10^9\ \dfrac{\text{N}\cdot\text{m}^2}{\text{C}^2}\right)\left(\dfrac{0.5 \times 10^{-6}\ \text{C}}{(2\ \text{m})^2}\right)$

$= \boxed{1123\ \text{N/C} \ (1.123\ \text{kN/C})}$

5.

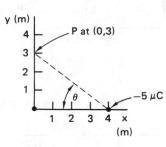

$E_x = \left(\dfrac{1}{4\pi\epsilon_0}\right)\left(\dfrac{|Q|}{r_{1-P}^2}\right)\cos\theta$

$= \left(8.987 \times 10^9\ \dfrac{\text{N}\cdot\text{m}^2}{\text{C}^2}\right)\left[\dfrac{5 \times 10^{-6}\ \text{C}}{(3\ \text{m})^2 + (4\ \text{m})^2}\right]$

$\times \left[\dfrac{4\ \text{m}}{\sqrt{(3\ \text{m})^2 + (4\ \text{m})^2}}\right]$

$= 1438\ \text{N/C}\ (1.438\ \text{kN/C})$

$E_y = \left(\dfrac{1}{4\pi\epsilon_o}\right)\left(\dfrac{|Q|}{r_{2-P}^2}\right) + \left(\dfrac{1}{4\pi\epsilon_o}\right)\left(\dfrac{|Q|}{r_{1-P}^2}\right)\sin\theta$

$= \left(8.987 \times 10^9\ \dfrac{\text{N}\cdot\text{m}^2}{\text{C}^2}\right)$

$\times \left[\dfrac{2 \times 10^{-6}\ \text{C}}{(3\ \text{m})^2} - \left(\dfrac{5 \times 10^{-6}\ \text{C}}{(3\ \text{m})^2 + (4\ \text{m})^2}\right)\left(\dfrac{3\ \text{m}}{\sqrt{(3\ \text{m})^2 + (4\ \text{m})^2}}\right)\right]$

$= 918.7\ \text{N/C}\ (0.9187\ \text{kN/C})$

The electric field strength magnitude at P is

$E = \sqrt{E_x^2 + E_y^2} = \sqrt{\left(1.438\ \dfrac{\text{kN}}{\text{C}}\right)^2 + \left(0.9187\ \dfrac{\text{kN}}{\text{C}}\right)^2}$

$= \boxed{1706\ \text{N/C}\ (1.706\ \text{kN/C})}$

6.

y (cm)
4 — • 2 nC at (3, 4)
3
2   θ
1
50 pC at (0, 0) — 1 2 3 4 x
-1   (cm)
-2
-3
-4 — • 2 nC at (3, -4)

$F_x = (2)\left(\dfrac{1}{4\pi\epsilon_0}\right)\left(\dfrac{|Q_1||Q_2|}{r_{2-50}^2}\right)\sin\theta$

$= (2)\left(8.987 \times 10^9\ \dfrac{\text{N}\cdot\text{m}^2}{\text{C}^2}\right)$

$\times \left[\dfrac{(2 \times 10^{-9}\ \text{C})(50 \times 10^{-2}\ \text{C})}{(0.05\ \text{m})^2}\right]$

$\times \left(\dfrac{3 \times 10^{-2}\ \text{m}}{5 \times 10^{-2}\ \text{m}}\right)$

$= -0.4314 \times 10^{-6}\ \text{N}\ (0.4314\ \mu\text{N})$

$F_y = 0$, by symmetry.

The force on the 5×10^{-11} C charge is $\boxed{0.4314\ \mu\text{N}}$.

7. (a) $E = \dfrac{V}{r} = \dfrac{100\ \text{V}}{0.005\ \text{m}} = 20\ \dfrac{\text{kV}}{\text{m}} = \boxed{2\ \text{kN/C}}$

(b) $U_v = \dfrac{1}{2}\epsilon_0 E^2$

$= \left(\dfrac{1}{2}\right)\left[\dfrac{1}{(4\pi)\left(8.987 \times 10^9\ \dfrac{\text{N}\cdot\text{m}^2}{\text{C}^2}\right)}\right]$

$\times \left(20 \times 10^3\ \dfrac{\text{N}}{\text{C}}\right)^2$

$= \boxed{1.77 \times 10^{-3}\ \text{J/m}^3\ (1.77\ \text{mJ/m}^3)}$

8. $V = Er = \left(10^6\ \dfrac{\text{V}}{\text{m}}\right)(0.005\ \text{m}) = \boxed{5000\ \text{V}\ (5\ \text{kV})}$

9.

$Q_{\mid}-$

r

z

$\mathbf{F} = Q\mathbf{E} = (Q)\left(-\dfrac{V}{r}\right)\mathbf{k}$

$= (-1.60 \times 10^{-19}\ \text{C})\left(-\dfrac{1200}{0.5}\ \dfrac{\text{N}}{\text{C}}\right)\mathbf{k}$

$= 0.384\ f\text{N}\mathbf{k}$

$U_e = \boxed{1200\ \text{J/C}\ (\text{V})}$

10.

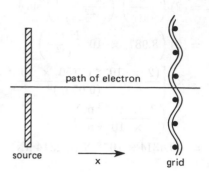

$$x = \frac{1}{2}at^2$$

$$t = \sqrt{\frac{2x}{a}}$$

$$v = at$$

Substituting for t,

$$v = \sqrt{2ax}$$

$$F = QE = ma$$

Substituting for F and rearranging, $a = \dfrac{QE}{m}$.

$$v = \sqrt{\frac{2\,QEx}{m}}$$

$$= \sqrt{\frac{(2)(1.60 \times 10^{-19}\,\text{C})\left(\dfrac{18}{0.003}\dfrac{\text{N}}{\text{C}}\right)(0.003\,\text{m})}{9.11 \times 10^{-31}\,\text{kg}}}$$

$$= \boxed{2.514 \times 10^6 \text{ m/s}}$$

11.

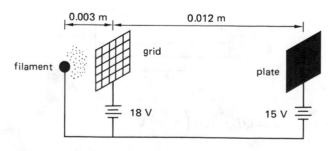

The electrical field at the filament (i.e., between the filament and the grid) due to the 2 batteries is

$$E = \frac{18\,\text{V}}{0.003\,\text{m}} + \frac{15\,\text{V}}{0.003\,\text{m} + 0.012\,\text{m}} = 7000\,\text{V/m}$$

An electron starting with $v = 0$ at the filament will achieve a velocity at the grid (which it passes through) of

$$v = \sqrt{\frac{2EQd}{m}}$$

$$= \sqrt{\frac{(2)\left(7000\,\dfrac{\text{V}}{\text{m}}\right)(1.602 \times 10^{-19}\,\text{C})(0.003\,\text{m})}{9.11 \times 10^{-31}\,\text{kg}}}$$

$$= 2.718 \times 10^6 \text{ m/s}$$

The electrical field between the grid and the plate is

$$E = \frac{15\,\text{V}}{0.003\,\text{m} + 0.012\,\text{m}} + \frac{15\,\text{V} - 18\,\text{V}}{0.012} = 750\,\text{V/m}$$

The velocity at the plate has reached

$$v = \sqrt{\frac{2EQd}{m} + v_0^2} =$$

$$\sqrt{\frac{(2)\left(750\,\dfrac{\text{V}}{\text{m}}\right)(1.602\times10^{-19}\,\text{C})(0.012\,\text{m})}{9.11 \times 10^{-31}\,\text{kg}} + \left(2.718\times10^6\,\dfrac{\text{m}}{\text{s}}\right)^2}$$

$$= \boxed{3.25 \times 10^6 \text{ m/s}}$$

12. (a) $\tan\theta = \dfrac{V_d|Q|L_2}{L_3 m v_0^2}$

$$\frac{V_d}{L_3} = E$$

$$\theta = \tan^{-1}\frac{E|Q|L_2}{m v_0^2}$$

$$= \tan^{-1}\left[\frac{\left(2\times10^4\,\dfrac{\text{N}}{\text{C}}\right)(1.60\times10^{-19}\,\text{C})(4\times10^{-2}\,\text{m})}{(9.11\times10^{-31}\,\text{kg})\left(2\times10^7\,\dfrac{\text{m}}{\text{s}}\right)^2}\right]$$

$$= \boxed{19.35°}$$

$$y = L_1 \tan\theta = (0.02\,\text{m} + 0.12\,\text{m})\tan(19.35°)$$

$$= \boxed{0.0492\,\text{m} \ (4.92\,\text{cm})}$$

13. $\quad B = \dfrac{\phi}{A} = \dfrac{34 \times 10^{-3}\,\text{Wb}}{0.02\,\text{m}^2} = \boxed{1.7\,\text{T}}$

14. (a) $\quad m = B_p A_p = \left(1.2\,\dfrac{\text{Wb}}{\text{m}^2}\right)(\pi)\left(\dfrac{0.01\,\text{m}}{2}\right)^2$

$$= \boxed{9.42 \times 10^{-5}\,\text{Wb} \ (942\,\mu\text{Wb})}$$

(b) Magnetic moment $= kL$

$$= (9.42 \times 10^{-5} \text{ Wb})(0.1 \text{ m})$$

$$= \boxed{9.42 \times 10^{-6} \text{ Wb·m}}$$

15. $F = mH$

$$H = \frac{F}{m} = \frac{3 \text{ N}}{4 \text{ Wb}} = \boxed{0.75 \text{ N/Wb}}$$

16. $F = \dfrac{m_1 m_2}{4\pi\mu r^2} = \dfrac{(8 \text{ Wb})(8 \text{ Wb})}{(4\pi)\left(4\pi \times 10^{-7} \dfrac{\text{Wb}}{\text{A·m}}\right)(0.01 \text{ m})^2}$

$$= \boxed{4.05 \times 10^{10} \text{ N (40.5 GN)}}$$

17.

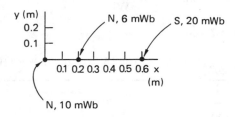

Force on the north pole at (0,0) is

$F_{10-6} + F_{10-20}$

$$= \frac{m_{10}m_6}{4\pi\mu r_{10-6}^2}\mathbf{i} + \frac{m_{10}m_{20}}{4\pi\mu r_{10-20}^2}\mathbf{i}$$

$$= \left[\frac{10 \times 10^{-3} \text{ Wb}}{(4\pi)\left(4\pi \times 10^{-7} \dfrac{\text{Wb}}{\text{A·m}}\right)}\right]$$

$$\times \left[\frac{20 \times 10^{-3} \text{ Wb}}{(0.6 \text{ m})^2} - \frac{6 \times 10^{-3} \text{ Wb}}{(0.2 \text{ m})^2}\right]\mathbf{i}$$

$$= \boxed{-59.8 \text{ N}\mathbf{i}}$$

18.

(a) $\mathbf{F}_{0-4} = \dfrac{m_0 m_4}{4\pi\mu r_{4-0}^2}$

$$= -\frac{(15 \text{ Wb})(15 \text{ Wb})}{(4\pi)\left(4\pi \times 10^{-7} \dfrac{\text{Wb}}{\text{A·m}}\right)(4 \times 10^{-2} \text{ m})^2}\mathbf{i}$$

$$= -8.905 \times 10^9 \text{ i N} = -8.905\mathbf{i} \text{ GN}$$

$\mathbf{F}_{0-3} = \dfrac{m_0 m_3}{4\pi\mu r_{3-0}^2}\mathbf{j}$

$$= \frac{(15 \text{ Wb})(15 \text{ Wb})}{(4\pi)\left(4\pi \times 10^{-7} \dfrac{\text{Wb}}{\text{A·m}}\right)(3 \times 10^{-2} \text{ m})^2}\mathbf{j}$$

$$= 1.583 \times 10^{10} \text{ j N} = 15.83\mathbf{j} \text{ GN}$$

$\mathbf{F}_0 = \mathbf{F}_{0-3} + \mathbf{F}_{0-4} = \boxed{-8.905\mathbf{i} \text{ GN} + 15.83\mathbf{j} \text{ GN}}$

(b) $\mathbf{F}_0 = m_0\mathbf{H}_0$

$$\mathbf{H}_0 = \frac{1}{m_0}\mathbf{F}_0$$

$$|\mathbf{H}_0| = \sqrt{\left(\frac{8.905 \times 10^9 \text{ N}}{15 \text{ Wb}}\right)^2 + \left(\frac{15.83 \times 10^9 \text{ N}}{15 \text{ Wb}}\right)^2}$$

$$= \boxed{1.211 \text{ GN/Wb}}$$

19. $F = NIBl = (1)(6.0 \text{ A})\left(1.1\dfrac{\text{Wb}}{\text{m}^2}\right)(0.3 \text{ m})$

$$= \boxed{1.98 \text{ N}}$$

20. $H = \dfrac{I}{2\pi r}$

$$B = \mu H = \frac{\mu I}{2\pi r}$$

$$\phi = BA = \frac{\mu I A}{2\pi r}$$

$$A = \frac{2\pi\phi r}{\mu I} = \frac{(2\pi)(30 \times 10^{-3} \text{ Wb})(3 \text{ m})}{\left(4\pi \times 10^{-7} \dfrac{\text{Wb}}{\text{A·m}}\right)(5 \text{ A})}$$

$$= 9.0 \times 10^4 \text{ m}^2$$

Substituting into the previous formula for ϕ while using the numerical value for A,

$$\phi = \frac{\left(4\pi \times 10^{-7} \dfrac{\text{Wb}}{\text{A·m}}\right)(4 \text{ A})(9.0 \times 10^4 \text{ m}^2)}{(2\pi)(3 \text{ m})}$$

$$= \boxed{0.024 \text{ Wb (24 mWb)}}$$

21. $V = n\dfrac{d\phi}{dt} = (100)\left(\dfrac{40\ \text{Wb}}{60\ \text{s}}\right) = \boxed{66.67\ \text{V}}$

22. $V = NBl\text{v} = (1)\left(4.0\ \dfrac{\text{Wb}}{\text{m}^2}\right)(0.10\ \text{m})\left(1.0\ \dfrac{\text{m}}{\text{s}}\right)$

$= \boxed{0.40\ \text{V}}$

23. $F = IN = (0.9\ \text{A})(400) = \boxed{360\ \text{A}}$

24. The analogous electrical circuit is

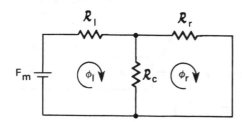

$\mathcal{R}_l = \sum \dfrac{l_l}{\mu_l A_l}$

$= \dfrac{18 \times 10^{-2}\ \text{m}}{\left(4\pi \times 10^{-7}\ \dfrac{\text{Wb}}{\text{A·m}}\right)(8400)(4 \times 10^{-4}\ \text{m}^2)}$

$= 4.263 \times 10^4\ \text{A/Wb}\ (42.63\ \text{kA/Wb})$

$\mathcal{R}_r = \mathcal{R}_l = 42.63\ \text{kA/Wb}$

$\mathcal{R}_c = \sum \dfrac{l_c}{\mu_c A_c}$

$= \dfrac{6 \times 10^{-2}\ \text{m}}{\left(4\pi \times 10^{-7}\ \dfrac{\text{Wb}}{\text{A·m}}\right)(8400)(4 \times 10^{-4}\ \text{m}^2)}$

$= 1.421 \times 10^4\ \text{A/Wb}\ (14.21\ \text{kA/Wb})$

$\mathcal{R}_c$ and $\mathcal{R}_r$ form a "flux divider" circuit.

$\mathcal{R}_c \phi_c = \mathcal{R}_r \phi_r$

$\phi_r = \left(\dfrac{\mathcal{R}_c}{\mathcal{R}_r}\right)\phi_c$

$\phi_l - \phi_r = \phi_c$

$\phi_l = \phi_r + \phi_c$

$F_m = IN = \mathcal{R}_l \phi_l + \mathcal{R}_c \phi_c$

$I = \dfrac{\mathcal{R}_l \phi_l + \mathcal{R}_c \phi_c}{N} = \dfrac{\phi_c\left[\mathcal{R}_l\left(1 + \dfrac{\mathcal{R}_c}{\mathcal{R}_r}\right) + \mathcal{R}_c\right]}{N} =$

$\dfrac{\left(3.6 \times 10^{-4}\ \text{Wb}\right)\left[\left(43 \times 10^3\ \dfrac{\text{A}}{\text{Wb}}\right)\left(1 + \dfrac{14 \times 10^3\ \frac{\text{A}}{\text{Wb}}}{43 \times 10^3\ \frac{\text{A}}{\text{Wb}}}\right) + \left(14 \times 10^3\ \dfrac{\text{A}}{\text{Wb}}\right)\right]}{300}$

$= \boxed{0.0852\ \text{A}\ (85.2\ \text{mA})}$

25. $\mathcal{R} = \sum \dfrac{l}{\mu A} = \dfrac{1}{\mu_o A}\sum \dfrac{l}{\mu_n}$

$= \dfrac{1}{\left(4\pi \times 10^{-7}\ \dfrac{\text{Wb}}{\text{A·m}}\right)(10^{-2}\ \text{m}^2)}$

$\times \left(\dfrac{1.59\ \text{m}}{800} + \dfrac{0.01}{1}\right)$

$= 0.9539 \times 10^6\ \dfrac{\text{A}}{\text{Wb}}$

$IN = \phi \mathcal{R}$

$N = \dfrac{\phi \mathcal{R}}{I} = \dfrac{(6.0 \times 10^{-3}\ \text{Wb})\left(0.9539 \times 10^6\ \dfrac{\text{A}}{\text{Wb}}\right)}{5.0\ \text{A}}$

$= \boxed{1144}$

26. $\mathbf{S} = \mathbf{E} \times \mathbf{H}.$

$\mathbf{E}$ is perpendicular to $\mathbf{H}$.

$\mathbf{S} = EH\mathbf{a}$

$\mathbf{a}$ is the unit vector in the direction of $\mathbf{E} \times \mathbf{H}$.

$|\mathbf{S}| = EH = \epsilon c E^2$

$H = \epsilon c E$

$= \left(8.854 \times 10^{-12}\ \dfrac{\text{C}^2}{\text{N·m}^2}\right)\left(3.00 \times 10^8\ \dfrac{\text{m}}{\text{s}}\right)$

$\times \left(10^{-3}\ \dfrac{\text{V}}{\text{cm}}\right)\left(100\ \dfrac{\text{cm}}{\text{m}}\right)$

$= \boxed{2.66 \times 10^{-4}\ \text{A/m}\ (0.266\ \text{mA/m})}$

$\mathbf{H}$ oscillates in an east-west direction.

DIRECT-CURRENT CIRCUITS

1. (a) $R = \dfrac{\rho L}{A} = \dfrac{\left(10.371 \, \frac{\Omega\text{-cmil}}{\text{ft}}\right)(500 \text{ ft})}{\left(\dfrac{0.064 \text{ in}}{0.001 \text{ in}}\right)^2}$

$= \boxed{1.266 \ \Omega}$

(b) $\rho = \rho_0(1 + \alpha \Delta T)$

$= \left(10.371 \, \dfrac{\Omega\text{-cmil}}{\text{ft}}\right)$

$\times \left[1 + \left(0.00402 \, \dfrac{1}{^\circ\text{C}}\right)(80^\circ\text{C} - 20^\circ\text{C})\right]$

$= 12.87 \ \Omega\text{-cmil/ft}$

$R = \dfrac{\rho L}{A} = \dfrac{\left(12.87 \, \frac{\Omega\text{-cmil}}{\text{ft}}\right)(500 \text{ ft})}{\left(\dfrac{0.064 \text{ in}}{0.001 \text{ in}}\right)^2}$

$= \boxed{1.57 \ \Omega}$

2. $\alpha = \dfrac{R_2 - R_1}{R_1(T_2 - T_1)} = \dfrac{400 \ \Omega - 30 \ \Omega}{30 \ \Omega(200^\circ\text{C} - 100^\circ\text{C})}$

$= \boxed{0.123^\circ\text{C}^{-1}}$

This uses 100°C as the reference temperature.

3. Find the current required to keep the motor running.

$P_m = IV$

$I = \dfrac{P_m}{V} = \dfrac{2000 \text{ W}}{240 \text{ V}} = 8.333 \text{ A}$

Find the resistance of the copper wire. The acceptable line loss is

$I^2 R = (0.05)(P_m)$

$R = (0.05)\left(\dfrac{P_m}{I^2}\right) = (0.05)\left[\dfrac{2000 \text{ W}}{(8.333 \text{ A})^2}\right] = 1.44 \ \Omega$

Solve for the minimum diameter of wire. Since there are two conductors, the total wire length is 2000 m.

$R = \dfrac{\rho L}{\left(\dfrac{D}{0.001 \text{ in}}\right)^2}$

$D = (0.001 \text{ in})\sqrt{\dfrac{\rho L}{R}} =$

$(0.001 \text{ in})\sqrt{\dfrac{\left(10.371 \, \frac{\Omega\text{-cmil}}{\text{ft}}\right)(2)(10^3 \text{ m})\left(\dfrac{\text{ft}}{0.3048 \text{ m}}\right)}{1.44 \ \Omega}}$

$= \boxed{0.217 \text{ in}}$

4. $R = \left(\dfrac{L}{A}\right)\rho_0[1 + \alpha(T - T_0)]$

$T = \left(\dfrac{A}{L\rho_0\alpha}\right)R + \left(T_0 - \dfrac{1}{\alpha}\right)$

The bracketed expressions in the previous equation are the slope and intercept of a line in the T-R plane. Therefore, given the coordinates $(11 \ \Omega, 5^\circ\text{C})$ and $(12 \ \Omega, 31^\circ\text{C})$,

$\text{slope} = \dfrac{31^\circ\text{C} - 5^\circ\text{C}}{12 \ \Omega - 11 \ \Omega} = 26^\circ\text{C}/\Omega$

$\text{intercept} = T - (\text{slope})(R) = 5^\circ\text{C} - \left(26 \, \dfrac{^\circ\text{C}}{\Omega}\right)(11 \ \Omega)$

$= -281^\circ\text{C}$

$T(R) = \left(26 \, \dfrac{^\circ\text{C}}{\Omega}\right)R - 281^\circ\text{C}$

$T(0) = \boxed{-281^\circ\text{C}}$

5. $E = \dfrac{P_m t_m}{\eta_m} + P_l t_l$

$= \left[\dfrac{1}{0.80}\right](5 \text{ hp})\left(0.7457 \, \dfrac{\text{kW}}{\text{hp}}\right)(15 \text{ day})\left(\dfrac{1 \text{ hr}}{\text{day}}\right)$

$+ (500 \text{ W})\left(\dfrac{\text{kW}}{10^3 \text{ W}}\right)(30 \text{ day})\left(\dfrac{4 \text{ hr}}{\text{day}}\right)$

$= \boxed{129.9 \text{ kW-hr}}$

6. $p = \rho g h$

$= \left(1 \, \dfrac{\text{kg}}{\text{l}}\right)\left(\dfrac{\text{l}}{10^3 \text{cm}^3}\right)\left(\dfrac{\text{cm}}{10^{-2} \text{ m}}\right)^3\left(9.8 \, \dfrac{\text{m}}{\text{s}^2}\right)$

$\times (120 \text{ ft})\left(\dfrac{0.3048 \text{ m}}{\text{ft}}\right)$

$= 3.584 \times 10^5 \text{ N/m}^2 \ (360 \text{ kPa})$

$$\frac{dV}{dt} = \left(2500 \,\frac{\text{gal}}{\text{min}}\right)\left(\frac{3.785\,\text{l}}{\text{gal}}\right)\left(\frac{10^3\text{cm}^3}{\text{l}}\right)$$

$$\times \left(\frac{10^{-2}\,\text{m}}{\text{cm}}\right)^3 \left(\frac{\text{min}}{60\,\text{s}}\right)$$

$$= 0.1577 \,\text{m}^3/\text{s}$$

$$p\frac{dV}{dt} = \eta P_r$$

$$P_r = \frac{p\dfrac{dV}{dt}}{\eta} = \frac{\left(3.584 \times 10^5 \,\dfrac{\text{N}}{\text{m}^2}\right)\left(0.1577\,\dfrac{\text{m}^3}{\text{s}}\right)}{0.70}$$

$$= \boxed{80.7 \,\text{kW}}$$

7. $P_m = \dfrac{Fv}{\eta} = IV$

$$I = \frac{Fv}{\eta V}$$

$$= \frac{(200\,\text{lbf})\left(\dfrac{4.4482\,\text{N}}{\text{lbf}}\right)\left(7\,\dfrac{\text{mi}}{\text{hr}}\right)\left(\dfrac{1.6093\,\text{km}}{\text{mi}}\right)\left(\dfrac{10^3\,\text{m}}{\text{km}}\right)\left(\dfrac{\text{hr}}{3600}\right)}{(0.75)(110\,\text{V})}$$

$$= \boxed{33.74 \,\text{A}}$$

8. $N = \dfrac{I}{Q_e} = \dfrac{P}{VQ_e} = \dfrac{800 \,\text{W}}{(110\,\text{V})(1.60 \times 10^{-19}\,\text{C})}$

$$= \boxed{4.55 \times 10^{19}}$$

9. $C = \dfrac{rP_p t}{\eta}$

$$= \frac{\left(\dfrac{\$0.05}{\text{kW-hr}}\right)(20\,\text{hp})\left(\dfrac{0.7457\,\text{kW}}{\text{hp}}\right)(20\,\text{day})\left(\dfrac{8\,\text{hr}}{\text{day}}\right)}{0.70}$$

$$= \boxed{\$\,170.45}$$

10. The circuit can be simplified to

From the current loop method,

$$V_l = R_l I_l + R_c(I_l - I_r) + V_c$$
$$V_c + R_c(I_l - I_r) = R_r I_r$$

$$I_r = \frac{V_c + \dfrac{R_c}{R_l}V_l}{\dfrac{R_c R_r}{R_l} + R_c + R_r}$$

$$= \frac{3.3\,\text{V} + \left(\dfrac{12\,\Omega}{12\,\Omega}\right)(3.3\,\text{V})}{\dfrac{(12\,\Omega)(5\,\Omega)}{12\,\Omega} + 12\,\Omega + 5\,\Omega} = \boxed{0.30 \,\text{A}}$$

11. $P = I^2 R = (0.5\,\text{A})^2(5\,\Omega) = \boxed{1.25 \,\text{W}}$

12. The circuit can be simplified to

From Kirchhoff's voltage law,

$$V_l + V_r = RI$$

$$I = \frac{V_l + V_r}{R} = \frac{24\,\text{V} + 6\,\text{V}}{14.1\,\Omega} = 2.13 \,\text{A}$$

$$P_l = V_l I = (24\,\text{V})(2.13\,\text{A}) = \boxed{51.1 \,\text{W}}$$

$$P_r = V_r I = (6\,\text{V})(2.13\,\text{A}) = \boxed{12.8 \,\text{W}}$$

13. (a) The circuit can be simplified to

From Kirchhoff's voltage law,

$$V_g = RI + V_b$$

$$I = \frac{V_g - V_b}{R} = \frac{100\,\text{V} - 66\,\text{V}}{8\,\Omega} = 4.25 \,\text{A}$$

$$P = I^2 R = (4.25\,\text{A})^2(8\,\Omega) = \boxed{144.5 \,\text{W}}$$

(b) $P_t = V_g I = (100 \text{ V})(4.25 \text{ A}) = 425 \text{ W}$

$E_s = P_s t_s = (P_t - P)t_s$

$= (425 \text{ W} - 144.5 \text{ W}) \left(\dfrac{\text{kW}}{10^3 \text{ W}} \right) (1 \text{ hr})$

$= \boxed{0.2805 \text{ kW-hr}}$

14. $\text{PL} = 10 \log_{10} \left(\dfrac{P_2}{P_1} \right) = 10 \log_{10} \left(\dfrac{100 \times 10^{-3} \text{ W}}{400 \text{ W}} \right)$

$= \boxed{-36.02 \text{ dB} \ (36.02 \text{ dB loss})}$

15. $A = 100 \text{ dB} + 25 \text{ dB} + 9 \text{ dB} = \boxed{134 \text{ dB}}$

16. $R_1 = \dfrac{R_a R_c}{R_a + R_b + R_c} = \dfrac{(30 \ \Omega)(20 \ \Omega)}{30 \ \Omega + 10 \ \Omega + 20 \ \Omega}$

$= \boxed{10 \ \Omega}$

$R_2 = \dfrac{R_a R_b}{R_a + R_b + R_c} = \dfrac{(30 \ \Omega)(10 \ \Omega)}{30 \ \Omega + 10 \ \Omega + 20 \ \Omega}$

$= \boxed{5 \ \Omega}$

$R_3 = \dfrac{R_b R_c}{R_a + R_b + R_c} = \dfrac{(10 \ \Omega)(20 \ \Omega)}{30 \ \Omega + 10 \ \Omega + 20 \ \Omega}$

$= \boxed{3.33 \ \Omega}$

17. The Thevenin equivalent circuit is

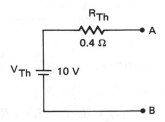

The Norton equivalent circuit is

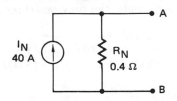

18. Given the coordinates (5 A, 8 V) and (10 A, 6 V) in the I-V plane, the line through the coordinates is

$V = (-0.4 \ \Omega)I + (10 \text{ V})$

For Thevenin, set $I = 0$ to get the open-circuit voltage.

$V_{\text{Th}} = 10 \text{ V}$

For Norton, set $V = 0$ to get the short-circuit current.

$I_N = \dfrac{10 \text{ V}}{0.4 \ \Omega} = 25 \text{ A}$

$R_{\text{Th}} = R_N = \dfrac{10 \text{ V}}{25 \text{ A}} = 0.4 \ \Omega$

The Thevenin equivalent circuit is

The Norton equivalent circuit is

19. The circuit can be represented as

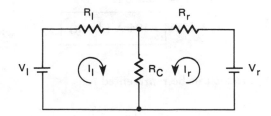

From the loop current method,

$V_l = R_l I_l + R_c (I_l + I_r)$

$V_r = R_r I_r + R_c (I_l + I_r)$

$I_l + I_r = \dfrac{R_r V_l + R_l V_r}{R_r R_l + R_r R_c + R_l R_c}$

$= \dfrac{(4 \ \Omega)(32 \text{ V}) + (2 \ \Omega)(20 \text{ V})}{(4 \ \Omega)(2 \ \Omega) + (4 \ \Omega)(8 \ \Omega) + (2 \ \Omega)(8 \ \Omega)}$

$= \boxed{3 \text{ A (down)}}$

If the first two equations are solved simultaneously,

$I_l = 4 \text{ A}$

$I_r = -1 \text{ A}$

20. The circuit can be represented as

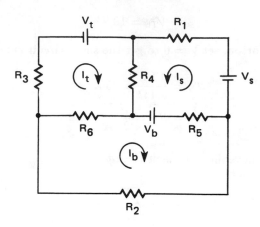

Using the current loop method,

$$V_t = R_4(I_t + I_s) + R_6(I_t - I_b) + R_3 I_t$$
$$V_s + V_b = R_1 I_s + R_4(I_t + I_s) + R_5(I_s + I_b)$$
$$V_b + R_6(I_t - I_b) = R_5(I_b + I_s) + R_2 I_b$$

After rearranging, substituting in the constants, and evaluating the coefficients, the previous three equations are

$$24\,\text{V} = (12\,\Omega)I_t + (6\,\Omega)I_s - (2\,\Omega)I_b$$
$$14\,\text{V} = (6\,\Omega)I_t + (10\,\Omega)I_s + (1\,\Omega)I_b$$
$$12\,\text{V} = -(2\,\Omega)I_t + (1\,\Omega)I_s + (8\,\Omega)I_b$$

Solving simultaneously,

$$I_t = 2.53\,\text{A}$$
$$I_s = -0.336\,\text{A}$$
$$I_t + I_s = \boxed{2.194\,\text{A (down)}}$$

21. The circuit can be represented as

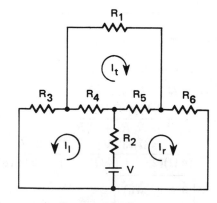

Using the current loop method,

$$V = R_2(I_l + I_r) + R_4(I_l + I_t) + R_3 I_l$$
$$R_s(I_r - I_t) = R_4(I_l + I_t) + R_1 I_t$$
$$V = R_2(I_l + I_r) + R_5(I_r - I_t) + R_b I_r$$

After rearranging, substituting in the constants, and evaluating the coefficients, the previous three equations are

$$48\,\text{V} = (9\,\Omega)I_l + (5\,\Omega)I_t + (2\,\Omega)I_r$$
$$0 = (5\,\Omega)I_l + (12\,\Omega)I_t - (3\,\Omega)I_r$$
$$48\,\text{V} = (2\,\Omega)I_l - (3\,\Omega)I_t + (10\,\Omega)I_r$$

Solving simultaneously,

$$I_t = \boxed{-1.42\,\text{A (1.42 A counterclockwise)}}$$

22. $R_s = \dfrac{R_c i_{max}}{i_t - i_{max}} = \dfrac{(500\,\Omega)(10^{-3}\,\text{A})}{10^{-1}\,\text{A} - 10^{-3}\,\text{A}} = \boxed{5.05\,\Omega}$

23. $R_m = \dfrac{V_t - i_{max} R_c}{i_{max}} = \dfrac{100\,\text{V} - (10^{-3}\,\text{A})(500\,\Omega)}{10^{-3}\,\text{A}}$

$$= \boxed{99{,}500\,\Omega\ (99.5\,\text{k}\Omega)}$$

24. (a)

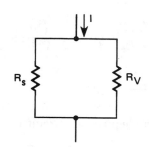

$$V = \left(\frac{R_s R_V}{R_s + R_V}\right) I$$

$$R_s = \frac{V R_V}{R_V I - V} = \frac{(50 \times 10^{-3}\,\text{V})(4.17\,\Omega)}{(4.17\,\Omega)(5\,\text{A}) - (50 \times 10^{-3}\,\text{V})}$$

$$= \boxed{0.01002\,\Omega}$$

(b) $I_s = \dfrac{V}{R_s} = \dfrac{50 \times 10^{-3}\,\text{V}}{0.01002} = \boxed{4.99\,\text{A}}$

(c) $V = \left(\dfrac{R_s R_V}{R_s + R_V}\right) I$

$$I = \left(\frac{R_s + R_V}{R_s R_V}\right) V$$

$$\frac{R_s + R_V}{R_s R_V} = \frac{0.01002\,\Omega + 4.17\,\Omega}{(0.01002\,\Omega)(4.17\,\Omega)} = \boxed{100\,\Omega^{-1}}$$

CIRCUITS
DC Elec

25. $$Q(t) = CV_C(t) = CV\left(1 - e^{-\frac{t}{RC}}\right)$$

$$E_C(t) = \frac{1}{2}CV_C^2(t) = \frac{1}{2}CV^2\left(1 - e^{-\frac{t}{RC}}\right)^2$$

$$E_C(0.05) = \frac{1}{2}(100 \times 10^{-6}\text{ F})(100\text{ V})^2$$

$$\times \left(1 - e^{-\frac{(0.05\text{ s})}{(10^3\ \Omega)(100 \times 10^{-6}\text{ F})}}\right)^2 = \boxed{0.0774\text{ J}}$$

26. (a) Since the current in the inductor cannot change instantaneously,

$$I(0) = \boxed{0}$$

(b) $$I(t) = \frac{V}{R}\left(1 - e^{-\frac{Rt}{L}}\right)$$

$$I(2) = \left(\frac{100\text{ V}}{80\ \Omega}\right)\left(1 - e^{\frac{(80\ \Omega)(2\text{ s})}{(7\text{ H})}}\right)$$

$$= \boxed{1.25\text{ A (down)}}$$

27. $$I(t) = \left(\frac{V}{R}\right)\exp\left[\frac{R - \sqrt{R^2 + 4\frac{L}{C}}}{2L}t\right]$$

$$I(1) = \left(\frac{100\text{ V}}{15\ \Omega}\right)$$

$$\times \exp\left[\frac{15\ \Omega - \sqrt{(15\ \Omega)^2 + (4)\left(\frac{4\text{ H}}{200 \times 10^{-6}\text{ F}}\right)}}{(R)(4\text{ H})}\right](1\text{ s})$$

$$= \boxed{18\text{ A (down)}}$$

28. Before the switch is moved to position B,

$$I(t) = \frac{V_{100}}{R}\left(1 - e^{-\frac{Rt}{L}}\right)$$

$$I(5 \times 10^{-4}) = \left(\frac{100\text{ V}}{80\ \Omega}\right)\left(1 - e^{-\frac{(80\ \Omega)(5 \times 10^{-4}\text{s})}{0.15\text{ H}}}\right)$$

$$= \boxed{0.2926\text{ A (clockwise)}}$$

Move the switch to position B. Solve for currents I_1 and I_2, whose sum is I_N, the current flowing in the circuit.

$$I_1(t) = I_0 e^{-\frac{Rt}{L}} \text{ (clockwise)}$$

$$I_2(t) = \frac{V_{50}}{R}\left(1 - e^{-\frac{Rt}{L}}\right) \quad \text{(clockwise)}$$

$$I_N(t) = I_1(t) + I_2(t)$$

$$= I_0 e^{-\frac{Rt}{L}} + \frac{V_{50}}{R}\left(1 - e^{-\frac{Rt}{L}}\right)$$

$$= \frac{50\text{ V}}{80\ \Omega} + \left(0.2926\text{ A} - \frac{50\text{ V}}{80\ \Omega}\right)e^{-\frac{(80\ \Omega)t}{0.15\text{ H}}}$$

$$= \boxed{0.625 - 0.332e^{-(533s^{-1})t}\text{ A (clockwise)}}$$

29. Close the switch. Solve for currents I_1 and I_2, whose sum is I_N, the current flowing in the circuit.

$$Q_1(t) = Q_0 e^{-\frac{t}{RC}}$$

$$I_1(t) = \left|\frac{dQ_1}{dt}\right| = \left(\frac{Q_0}{RC}\right)e^{-\frac{t}{RC}}\text{ (clockwise)}$$

$$Q_2(t) = CV\left(1 - e^{-\frac{t}{RC}}\right)$$

$$I_2(t) = \frac{dQ_2}{dt} = \left(\frac{V}{R}\right)e^{-\frac{t}{RC}}\text{ (clockwise)}$$

$$I_N(t) = I_1(t) + I_2(t)$$

$$= \left(\frac{Q_0}{RC}\right)e^{-\frac{t}{RC}} + \left(\frac{V}{R}\right)e^{-\frac{t}{RC}}$$

$$= \left(\frac{600 \times 10^{-6}\text{ C}}{(10^3\ \Omega)(200 \times 10^{-6}\text{ F})} + \frac{60\text{ V}}{10^3\ \Omega}\right) \times e^{-\frac{t}{(10^3\ \Omega)(200 \times 10^{-6}\text{ F})}}$$

$$= 0.063e^{-5t}\text{ A (clockwise)}$$

When $t = 0$,

$$\boxed{I = 0.063\text{ A}}$$

ALTERNATING-CURRENT CIRCUITS

1. (a) $15\cos(-180°) + j15\sin(-180°) = \boxed{-15 + j0}$

   (b) $10\cos(37°) + j10\sin(37°) = \boxed{7.99 + j6.02}$

   (c) $50\cos(120°) + j50\sin(120°) = \boxed{-25 + j43.3}$

   (d) $21\cos(-90°) + j21\sin(-90°) = \boxed{0 - j21}$

2. (a) $\sqrt{(6)^2 + (7)^2}\underline{/\tan^{-1}\left(\frac{7}{6}\right)} = \boxed{9.22\underline{/49.4°}}$

   (b) $\sqrt{(50)^2 + (-60)^2}\underline{/\tan^{-1}\left(\frac{-60}{50}\right)}$

   $= \boxed{78.10\underline{/-50.19°}}$

   (c) $\sqrt{(-75)^2 + (45)^2}\underline{/\tan^{-1}\left(\frac{45}{-75}\right)}$

   $= \boxed{87.46\underline{/149.04°}}$

   (d) $\sqrt{(90)^2 + (-180)^2}\underline{/\tan^{-1}\left(\frac{-180}{90}\right)}$

   $= \boxed{201.25\underline{/-63.43°}}$

3. (a) $\frac{1}{6}\sqrt{(7)^2 + (5)^2}\underline{/\tan^{-1}\left(\frac{5}{7}\right)} = \boxed{1.4\underline{/35.54°}}$

   (b) $\dfrac{\sqrt{(13)^2 + (17)^2}\underline{/\tan^{-1}\left(\frac{17}{13}\right)}}{\sqrt{(15)^2 + (-10)^2}\underline{/\tan^{-1}\left(\frac{-10}{15}\right)}}$

   $= \dfrac{21.4\underline{/52.59°}}{18.0\underline{/-33.69°}} = \boxed{1.19\underline{/86.28°}}$

   (c) $\boxed{0.588\underline{/34°}}$

4. (a) $V_{\text{ave}} = \frac{1}{T}\int_0^T v(t)dt$

   $= \left(\frac{1}{4\text{ ms}}\right)\int_0^{2\text{ ms}} 5\,dt + \left(\frac{1}{4\text{ ms}}\right)\int_{2\text{ ms}}^{4\text{ ms}} 3\,dt$

   $= \boxed{4\text{ V}}$

   $V_{\text{eff}} = \sqrt{\frac{1}{T}\int_0^T v^2(t)dt}$

   $= \sqrt{\left(\frac{1}{4\text{ ms}}\right)\left[\int_0^{2\text{ ms}}(5)^2 dt + \int_{2\text{ ms}}^{4\text{ ms}}(3)^2 dt\right]}$

   $= \boxed{4.12\text{ V}}$

$\text{CF} = \dfrac{V_m}{V_{\text{eff}}} = \dfrac{5\text{ V}}{4.12\text{ V}} = \boxed{1.21}$

$\text{FF} = \dfrac{V_{\text{eff}}}{V_{\text{ave}}} = \dfrac{4.12\text{ V}}{4\text{ V}} = \boxed{1.03}$

(b) $V_{\text{ave}} = \frac{1}{T}\int_0^T v(t)dt = \dfrac{1}{3\text{ ms}}\int_0^{3\text{ ms}}(50\text{ ms}^{-1})t\,dt$

   $= \boxed{75\text{ V}}$

$V_{\text{eff}} = \sqrt{\frac{1}{T}\int_0^T v^2(t)\,dt}$

   $= \sqrt{\left(\frac{1}{3\text{ ms}}\right)\int_0^{3\text{ ms}}[(50\text{ ms}^{-1})t]^2\,dt} = \boxed{86.6\text{ V}}$

$\text{CF} = \dfrac{V_m}{V_{\text{eff}}} = \dfrac{150\text{ V}}{86.6\text{ V}} = \boxed{1.73}$

$\text{FF} = \dfrac{V_{\text{eff}}}{V_{\text{ave}}} = \dfrac{86.6\text{ V}}{75\text{ V}} = \boxed{1.15}$

(c) $V_{\text{ave}} = \frac{1}{T}\int_0^T v(t)\,dt = \left(\dfrac{1}{20\text{ ms}}\right)\int_0^{5\text{ ms}} 80\,dt$

   $= \boxed{20\text{ V}}$

$V_{\text{eff}} = \sqrt{\frac{1}{T}\int_0^T v^2(t)\,dt} = \sqrt{\left(\dfrac{1}{20\text{ ms}}\right)\int_0^{5\text{ ms}}(80)^2 dt}$

   $= \boxed{40\text{ V}}$

$\text{CF} = \dfrac{V_m}{V_{\text{eff}}} = \dfrac{80\text{ V}}{40\text{ V}} = \boxed{2}$

$\text{FF} = \dfrac{V_{\text{eff}}}{V_{\text{ave}}} = \dfrac{40\text{ V}}{20\text{ V}} = \boxed{2}$

(d) $V_{\text{ave}} = \frac{1}{T}\int_0^T v(t)\,dt$

   $= \left(\dfrac{1}{10\text{ ms}}\right)\int_0^{7\text{ ms}}\left[\left(-\frac{30}{7}\text{ms}^{-1}\right)t + 30\right]dt$

   $= \boxed{10.5\text{ V}}$

$V_{\text{eff}} = \sqrt{\frac{1}{T}\int_0^T v^2(t)\,dt}$

   $= \sqrt{\left(\dfrac{1}{10\text{ ms}}\right)\int_0^{7\text{ ms}}\left[\left(-\frac{30}{7}\text{ms}^{-1}\right)t + 30\right]^2 dt}$

   $= \boxed{14.5\text{ V}}$

PROFESSIONAL PUBLICATIONS, INC. ● Belmont, CA

$$CF = \frac{V_m}{V_{eff}} = \frac{30 \text{ V}}{14.5 \text{ V}} = \boxed{2.07}$$

$$FF = \frac{V_{eff}}{V_{ave}} = \frac{14.5 \text{ V}}{10.5 \text{ V}} = \boxed{1.38}$$

5. (a) $Z = \sqrt{(\Sigma R)^2 + (\Sigma X_L - \Sigma X_C)^2}$

$$= \sqrt{R^2 + \left(\omega L - \frac{1}{\omega C}\right)^2}$$

$$= \sqrt{(80\,\Omega)^2 + \left[(20 \text{ s}^{-1})(5.0 \times 10^{-3} \text{ H}) - \frac{1}{(20 \text{ s}^{-1})(500 \times 10^{-6} \text{ F})}\right]^2}$$

$$= 128\,\Omega$$

$$\phi_{\mathbf{Z}} = \tan^{-1}\left(\frac{\Sigma X_L - \Sigma X_C}{\Sigma R}\right) = \tan^{-1}\left(\frac{\omega L - \frac{1}{\omega C}}{R}\right)$$

$$= \tan^{-1}\left(\frac{(20 \text{ s}^{-1})(5.0 \times 10^{-3} \text{ H}) - \frac{1}{(20 \text{ s}^{-1})(500 \times 10^{-6} \text{ F})}}{80\,\Omega}\right)$$

$$= -51.31°$$

$$\mathbf{Z} = \boxed{128\underline{/-51.31°}\ \Omega}$$

$$\mathbf{Y} = \frac{1}{\mathbf{Z}} = \frac{1}{128\underline{/-51.31°}\ \Omega} = \boxed{7.81 \times 10^{-3}\underline{/51.31°}\ \text{S}}$$

(b) $\mathbf{Y} = \frac{1}{\mathbf{Z}} = \frac{1}{\mathbf{Z}_C} + \frac{1}{\mathbf{Z}_L} + \frac{1}{\mathbf{Z}_R}$

$$= \frac{1}{8\underline{/-90°}\ \Omega} + \frac{1}{3\underline{/90°}\ \Omega} + \frac{1}{5\underline{/0°}\ \Omega}$$

$$= \tfrac{1}{8}\underline{/90°}\ \text{S} + \tfrac{1}{3}\underline{/-90°}\ \text{S} + \tfrac{1}{5}\underline{/0°}\ \text{S}$$

$$= \tfrac{1}{5}\text{S} + j\tfrac{1}{8}\text{S} - j\tfrac{1}{3}\text{S} = 0.289\underline{/-46.17°}\ \text{S}$$

$$\mathbf{Z} = \frac{1}{\mathbf{Y}} = \frac{1}{0.289\underline{/-46.17°}\ \text{S}}$$

$$= \boxed{3.46\underline{/46.17°}\ \Omega}$$

6. $X_L = \omega L = (2\pi)(1200 \text{ Hz})(350 \times 10^{-6} \text{ H})$

$$= 2.64\,\Omega$$

$$\mathbf{I} = \frac{\mathbf{V}}{\mathbf{Z}} = \frac{10\underline{/0°}\ \text{V}}{2.64\underline{/90°}\ \Omega} = \boxed{3.79\underline{/-90°}\ \text{A}}$$

7. $V = \sqrt{2}\,V_{rms}$

$$\mathbf{Z}_l = 4\,\Omega + j5\,\Omega \qquad \mathbf{Z}_r = 5\,\Omega - j3\,\Omega$$
$$= 6.40\ \underline{/51.34°}\ \Omega \qquad = 5.83\ \underline{/-30.96°}\ \Omega$$

120 $\sqrt{2}\ \underline{/0°}$ V
60 Hz

(b) $\mathbf{I}_l = \frac{\mathbf{V}}{\mathbf{Z}_l} = \frac{\sqrt{2} \times 120\underline{/0°}\ \text{V}}{6.40\underline{/51.34}\ \Omega} = \boxed{26.5\underline{/-51.34°}\ \text{A}}$

$\mathbf{I}_r = \frac{\mathbf{V}}{\mathbf{Z}_r} = \frac{\sqrt{2} \times 120\underline{/0°}\ \text{V}}{5.83\underline{/-30.96°}\ \Omega} = \boxed{29.1\underline{/30.96°}\ \text{A}}$

(a) $\mathbf{I}_T = \mathbf{I}_l + \mathbf{I}_r$

$$= 26.5\underline{/-51.34°}\ \text{A} + 29.1\underline{/30.96°}\ \text{A}$$

$$= 41.9\underline{/-7.82°}\ \text{A}$$

$$\mathbf{I}_{T,rms} = \frac{1}{\sqrt{2}}\mathbf{I}_T = \left(\frac{1}{\sqrt{2}}\right)(41.9\underline{/-7.82°}\ \text{A})$$

$$= \boxed{29.6\underline{/-7.82°}\ \text{A}}$$

(c) $P = \tfrac{1}{2}\sum I_R V_R = \tfrac{1}{2}I_l(I_l R_{\mathbf{Z}_l}) + \tfrac{1}{2}I_r(I_r R_{\mathbf{Z}_r})$

$$= \tfrac{1}{2}(26.5 \text{ A})^2(4\,\Omega) + \tfrac{1}{2}(29.1 \text{ A})^2(5\,\Omega)$$

$$= \boxed{3522 \text{ W}}$$

8.

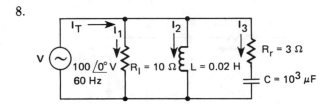

The above circuit can be simplified to

$$Z_3 = \sqrt{(\Sigma R)^2 + (\Sigma X_L - \Sigma X_C)^2}$$

$$= \sqrt{R_r^2 + \left(-\frac{1}{\omega C}\right)^2}$$

$$= \sqrt{(3\,\Omega)^2 + \left(-\frac{1}{(2\pi)(60 \text{ s}^{-1})(10^3 \times 10^{-6} \text{ F})}\right)^2}$$

$$= 4.00\,\Omega$$

$$\phi_{\mathbf{Z}} = \tan^{-1}\left(\frac{\Sigma X_L - \Sigma X_C}{\Sigma R}\right) = \tan^{-1}\left(\frac{-\frac{1}{\omega L}}{R}\right)$$

$$= \tan^{-1}\left(\frac{-\frac{1}{(2\pi)(60 \text{ s}^{-1})(10^3 \times 10^{-6} \text{ F})}}{3\,\Omega}\right)$$

$$= -41.48°$$

$$\mathbf{Z}_s = 4.00\underline{/-41.48°}\ \Omega$$

(a) $I_1 = \dfrac{V}{Z_1} = \dfrac{100\underline{/0°}\ V}{10\underline{/0°}\ \Omega} = \boxed{10\underline{/0°}\ A}$

$I_2 = \dfrac{V}{Z_2} = \dfrac{100\underline{/0°}\ V}{(2\pi)(60\ s^{-1})(0.02\ H)\underline{/90°}}$

$= \boxed{13.26\underline{/-90°}\ A}$

$I_3 = \dfrac{V}{Z_3} = \dfrac{100\underline{/0°}\ V}{4.00\underline{/-41.48°}\ \Omega} = \boxed{25\underline{/41.48°}\ A}$

(b) $I_T = I_1 + I_2 + I_3$

$= 10\underline{/0°}\ A + 13.26\underline{/-90°}\ A + 25\underline{/41.48°}\ A$

$= \boxed{28.9\underline{/6.55°}\ A}$

(c) $PF = \cos(6.55°) = \boxed{0.993\ (99.3\%)}$

9.

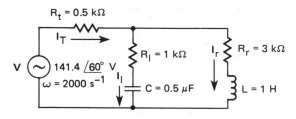

The circuit can be simplified to

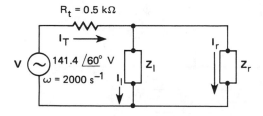

$Z_l = \sqrt{(\Sigma R)^2 + (\Sigma X_L - \Sigma X_C)^2}$

$= \sqrt{R_l^2 + \left(-\dfrac{1}{\omega C}\right)^2}$

$= \sqrt{(10^3\ \Omega)^2 + \left[-\dfrac{1}{(2000\ s^{-1})(0.5 \times 10^{-6}\ F)}\right]^2}$

$= 1414\ \Omega$

$\phi_{Z_l} = \tan^{-1}\left(\dfrac{\Sigma X_L - \Sigma X_C}{\Sigma R}\right) = \tan^{-1}\left(\dfrac{-\dfrac{1}{\omega C}}{R_l}\right)$

$= \tan^{-1}\left(\dfrac{-\dfrac{1}{(2000\ s^{-1})(0.5 \times 10^{-6}\ F)}}{10^3\ \Omega}\right) = -45°$

$Z_l = 1414\underline{/-45°}\ \Omega$

$Z_r = \sqrt{(\Sigma R)^2 + (\Sigma X_L - \Sigma X_C)^2} = \sqrt{R_r^2 + (\omega L)^2}$

$= \sqrt{(3 \times 10^3\ \Omega)^2 + [(2000\ s^{-1})(1\ H)]^2} = 3606\ \Omega$

$\phi_{Z_r} = \tan^{-1}\left(\dfrac{\Sigma X_L - \Sigma X_C}{\Sigma R}\right) = \tan^{-1}\left(\dfrac{\omega L}{R_r}\right)$

$= \tan^{-1}\left(\dfrac{(2000\ s^{-1})(1\ H)}{3 \times 10^3\ \Omega}\right) = 33.69°$

$Z_r = 3606\underline{/33.69°}\ \Omega$

The circuit can be simplified to

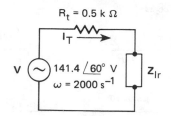

$\dfrac{1}{Z_{lr}} = \dfrac{1}{Z_l} + \dfrac{1}{Z_r}$

$= \dfrac{1}{1414\underline{/-45°}\ \Omega} + \dfrac{1}{3606\underline{/33.69°}\ \Omega}$

$= 7.072 \times 10^{-4}\underline{/45°}\ S + 2.773 \times 10^{-4}\underline{/-33.69°}\ S$

$= 8.086 \times 10^{-4}\underline{/25.35°}\ S$

$Z_{lr} = \dfrac{1}{8.086 \times 10^{-4}\underline{/25.35°}\ S} = 1237\underline{/-25.35°}\ \Omega$

The previous circuit can be simplified to

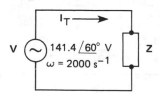

$Z = Z_{R_t} + Z_{lr} = 500\underline{/0°}\ \Omega + 1237\underline{/-25.35°}\ \Omega$

$= 1702\underline{/-18.12°}\ \Omega$

$I_t = \dfrac{V}{Z} = \dfrac{141.1\underline{/60°}\ V}{1702\underline{/-18.12°}\ \Omega} = 0.0831\underline{/78.12}\ A$

$V_C = Z_C I_l = Z_C\left(\dfrac{Z_{lr}}{Z_l} I_T\right)$

$= \left[\dfrac{1}{(2000\ s^{-1})(0.5 \times 10^{-6}\ F)}\right](\underline{/-90°})$

$\times \left(\dfrac{1237\underline{/-25.35°}\ \Omega}{1414\underline{/-45°}\ \Omega}\right)(0.0831\underline{/78.12}\ A)$

$= \boxed{72.7\underline{/7.77°}\ V}$

10.

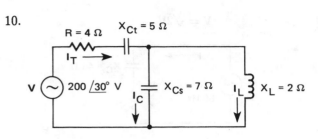

The circuit can be simplified to

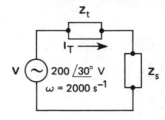

$$\mathbf{Z}_t = \mathbf{Z}_R + \mathbf{Z}_{X_{Ct}} = 4\underline{/0°}\ \Omega + 5\underline{/-90°}$$
$$= 6.4\underline{/-51.34°}\ \Omega$$

$$\frac{1}{\mathbf{Z}_s} = \frac{1}{\mathbf{Z}_{X_{Cs}}} + \frac{1}{\mathbf{Z}_{X_L}} = \frac{1}{7\underline{/-90°}\ \Omega} + \frac{1}{2\underline{/90°}\ \Omega}$$
$$= \tfrac{1}{7}\underline{/90°}\ \text{S} + \tfrac{1}{2}\underline{/-90°}\ \text{S} = 0.357\underline{/-90°}\ \text{S}$$

$$\mathbf{Z}_s = \frac{1}{0.357\underline{/-90°}\ \text{S}} = 2.8\underline{/-90°}\ \Omega$$

The previous circuit can be simplified to

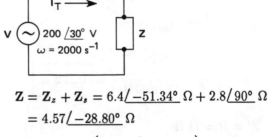

$$\mathbf{Z} = \mathbf{Z}_z + \mathbf{Z}_s = 6.4\underline{/-51.34°}\ \Omega + 2.8\underline{/90°}\ \Omega$$
$$= 4.57\underline{/-28.80°}\ \Omega$$

$$\mathbf{V}_L = \frac{\mathbf{Z}_s}{\mathbf{Z}}\mathbf{V} = \left(\frac{2.8\underline{/90°}\ \Omega}{4.57\underline{/-28.80°}\ \Omega}\right)(200\underline{/30°}\ \text{V})$$

$$= \boxed{122.5\underline{/148.80°}\ \text{V}}$$

11. Since the current cannot change instantaneously,

$$\boxed{I(0) = 0}$$

12.

```
    R_t = 6 Ω        R_L
V (~) 120 V              L
    60 Hz
I_T = 12 A
```

(a)
$$P = I_T^2(R_t + R_L)$$
$$R_L = \frac{P}{I_t^2} - R_t = \frac{1152\ \text{W}}{(12\ \text{A})^2} - 6\ \Omega = 2\ \Omega$$

(b)
$$V = I_T Z = I_T\sqrt{(\Sigma R)^2 + (\Sigma X)^2}$$
$$= I_T\sqrt{(R_t + R_L)^2 + (2\pi f L)^2}$$

$$L = \frac{\sqrt{\left(\dfrac{V}{I_T}\right)^2 - (R_t + R_L)^2}}{2\pi f}$$

$$= \frac{\sqrt{\left(\dfrac{120}{12}\right) - (6+2)^2}}{(2\pi)(60)} = \boxed{0.0159\ \text{H}}$$

13.

```
        /|
       / |
      /  | Q
     /   |
    /φ___|
     P
```

$$(\text{PF})_1 = 60\%\ \text{lagging}$$
$$P_1 = 400\ \text{kW}$$
$$\phi_1 = \cos^{-1}(0.6) = 53.13°$$
$$Q_1 = P_1\tan\phi_1 = (400\ \text{kW})(\tan(53.13°)) = 533\ \text{kW}$$

$$(\text{PF})_2 = 80\%\ \text{lagging}$$
$$P_2 = 400\ \text{kW} + 150\ \text{kW} = 550\ \text{kW}$$
$$\phi_2 = \cos^{-1}(0.8) = 36.87°$$
$$Q_2 = P_2\tan\phi_2 = (550\ \text{kW})(\tan(36.87°)) = 412\ \text{kW}$$

$$P_m = 150\ \text{kW}$$
$$Q_m = Q_1 - Q_2 = 533\ \text{kW} - 412\ \text{kW} = 121\ \text{kW}$$
$$\phi_m = \tan^{-1}\left(\frac{Q_m}{P_m}\right) = \tan^{-1}\left(\frac{121\ \text{kW}}{150\ \text{kW}}\right) = 38.89°$$
$$(\text{PF})_m = \cos\phi_m = \cos(38.89°)$$
$$= \boxed{0.778\ (77.8\%)\ \text{leading}}$$

14. $$P_m = \frac{n}{\eta}P_n = \left(\frac{3}{0.92}\right)(2000\ \text{hp})\left(0.7457\ \frac{\text{kW}}{\text{hp}}\right)$$
$$= 4863\ \text{kW}$$

$(PF)_m = 80\%$ leading

$P_m = 4863$ kW

$\phi_m = \cos^{-1}(0.8) = 36.87°$

$Q_m = P_m \tan \phi_m = (4863 \text{ kW})(\tan(36.87°))$

$\qquad = 3647$ kVAR

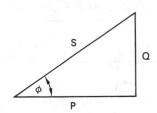

$(PF)_e = 82\%$ lagging

$S_e = 10^4$ kVA

$P_e = S_e \cos \phi_e = (10^4 \text{ kVA})(0.82) = 8200$ kW

$\phi_e = \cos^{-1}(0.82) = 34.92°$

$Q_e = P_e \tan \phi_e = (8200 \text{ kVA})(\tan(34.92°))$

$\qquad = 5725$ kVAR

(a) $\quad P_n = P_m + P_e = 4863 \text{ kW} + 8200 \text{ kW}$

$\qquad = \boxed{13,063 \text{ kW}}$

$\phi_n = \tan^{-1}\left(\dfrac{2078 \text{ kW}}{13,063 \text{ kW}}\right) = 9.04°$

$Q_n = Q_e - Q_m = 5725 \text{ kVAR} - 3647 \text{ kVAR}$

$\qquad = 2078$ kVAR

(b) $\quad (PF)_n = \cos \phi_n = \cos(9.04°)$

$\qquad = \boxed{0.988 \ (98.8\%) \text{ lagging}}$

15.

$Z = \sqrt{R^2 + (\omega L)^2}$

$R = \sqrt{\dfrac{V^4}{P^2} - (\omega L)^2}$

$\quad = \sqrt{\dfrac{(100 \text{ V})^4}{(50 \text{ W})^2} - [(2\pi)(4000 \text{ Hz})(0.001 \text{ H})]^2}$

$\quad = \boxed{198 \ \Omega}$

16. $\qquad\qquad\qquad V = \sqrt{2}V_{rms}$

(a) $\quad Z = \sqrt{(\Sigma R)^2 + (\Sigma X_L - \Sigma X_C)^2}$

$\qquad = \sqrt{R^2 + (-X_C)^2}$

$\qquad = \sqrt{(11 \ \Omega)^2 + (-46 \ \Omega)^2} = 47.3 \ \Omega$

$\phi_{\mathbf{Z}} = \tan^{-1}\left(\dfrac{\Sigma X_L - \Sigma X_C}{\Sigma R}\right) = \tan^{-1}\left(\dfrac{-X_C}{R}\right)$

$\qquad = \tan^{-1}\left(\dfrac{-46 \ \Omega}{11 \ \Omega}\right) = -76.55°$

$\mathbf{Z} = \boxed{47.3\underline{/-76.55°} \ \Omega}$

(b) $\quad \mathbf{I} = \dfrac{\mathbf{V}}{\mathbf{Z}} = \dfrac{\sqrt{2} \times 120\underline{/0°} \text{ V}}{47.3\underline{/-76.55°} \ \Omega} = \boxed{3.58\underline{/76.55°} \text{ A}}$

(c)

$L = \dfrac{X_C}{2\pi f} = \dfrac{46 \ \Omega}{(2\pi)(60 \text{ Hz})} = \boxed{0.122 \text{ H}}$

(d) $\quad Z' = R = 11 \ \Omega$

$\phi_{\mathbf{Z}'} = \tan^{-1}\left(\dfrac{\Sigma X_L - \Sigma X_C}{\Sigma R}\right)$

$\qquad = \tan^{-1}\left(\dfrac{0}{11 \ \Omega}\right) = 0° \quad$ since $X_L = X_C$

$\qquad\qquad\qquad\qquad\qquad$ at resonance

$\mathbf{Z}' = 11\underline{/0°} \ \Omega$

$\mathbf{I}_{max} = \dfrac{\mathbf{V}}{\mathbf{Z}'} = \dfrac{120\sqrt{2}\underline{/0°} \text{ V}}{11\underline{/0°} \ \Omega} = 15.4\underline{/0°} \text{ A}$

$\mathbf{V}_C + \mathbf{V}_L = \mathbf{Z}_C \mathbf{I}_{max} + \mathbf{Z}_L \mathbf{I}_{max} = (\mathbf{Z}_C + \mathbf{Z}_L)\mathbf{I}_{max}$

$\mathbf{Z}_C = X_C\underline{/\phi_C} = 46\underline{/-90°} \ \Omega$

$\mathbf{Z}_L = X_L\underline{/\phi_L} = 46\underline{/90°} \ \Omega$

$\mathbf{V}_C + \mathbf{V}_L = (46\underline{/-90°} \ \Omega + 46\underline{/90°})(15.4\underline{/0°} \text{ A})$

$\qquad = \boxed{0 \text{ V}}$

17.

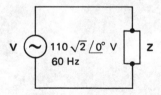

(a) $C = \dfrac{1}{\omega X_C} = \dfrac{1}{(2\pi)(60 \text{ Hz})(20 \ \Omega)}$

$\qquad = \boxed{1.326 \times 10^{-4} \text{ F}}$

$Z = \sqrt{(\Sigma R)^2 + (\Sigma X_L - \Sigma X_C)^2}$

$\quad = \sqrt{R^2 + (X_L - X_C)^2} = R \qquad \text{since } X_C = X_L$

$\phi_{\mathbf{Z}} = \tan^{-1}\left(\dfrac{\Sigma X_L - \Sigma X_C}{\Sigma R}\right) = \tan^{-1}\left(\dfrac{0}{R}\right) = 0°$

$\mathbf{Z} = R\underline{/0°}$

$\mathbf{I} = \dfrac{\mathbf{V}}{\mathbf{Z}} = \dfrac{220\underline{/0°} \text{ V}}{R\underline{/0°}} = \boxed{\dfrac{220}{R}\underline{/0°} \text{ V}}$

Ideally, $R = 0$, making I infinite. Every real circuit, however, has some resistance.

18. $\qquad\qquad V = \sqrt{2}\, V_{\text{rms}}$

The circuit can be simplified to

(a) $Z = \sqrt{(\Sigma R)^2 + (\Sigma X_L - \Sigma X_C)^2}$

$\quad = \sqrt{R_{\mathbf{Z}}^2 + \left(X_{\mathbf{Z}_C} - \left(\dfrac{1}{\omega C}\right)\right)^2}$

$C = \dfrac{1}{\omega X_{\mathbf{Z}_C}} = \dfrac{1}{(2\pi)(60 \text{ Hz})(4 \ \Omega)}$

$\quad = \boxed{6.63 \times 10^{-4} \text{ F}}$

(b) $Q = \dfrac{1}{\omega RC} = \dfrac{1}{(2\pi)(60 \text{ Hz})(4 \ \Omega)(6.63 \times 10^{-4} \text{ F})}$

$\quad = \boxed{1.00}$

(c) $\qquad\qquad Q = \dfrac{f}{f_2 - f_1}$

$f_2 - f_1 = \dfrac{f}{Q} = \dfrac{60 \text{ Hz}}{1.00} = \boxed{60 \text{ Hz}}$

19.

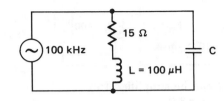

$\omega \approx \dfrac{1}{\sqrt{LC}}$

$C \approx \dfrac{1}{\omega^2 L} = \dfrac{1}{[(2\pi)(100 \times 10^3 \text{ Hz})]^2 (100 \times 10^{-6} \text{ H})}$

$\quad = \boxed{\begin{array}{l} 25.3 \times 10^{-9} \text{ F} \ (25.3 \text{ nF}) \\ (23.9 \text{ nF exactly}) \end{array}}$

20. $\quad f = \left(\dfrac{1}{2\pi}\right)\left(\dfrac{1}{\sqrt{LC}}\right)$

$\qquad = \left(\dfrac{1}{2\pi}\right)\left(\dfrac{1}{\sqrt{(1 \times 10^{-3} \text{ H})(2 \times 10^{-9} \text{ F})}}\right)$

$\qquad = 112{,}540 \text{ Hz}$

$Q = R\sqrt{\dfrac{C}{L}} = (10 \ \Omega)\left(\sqrt{\dfrac{2 \times 10^{-9} \text{ F}}{1 \times 10^{-3} \text{ H}}}\right)$

$\qquad = 0.01414$

Also,

$\qquad\qquad Q = \dfrac{\omega_0}{\omega_2 - \omega_1} = \dfrac{f_0}{f_2 - f_1}$

ω_1 and ω_2 are the 70% points.

$f_2 - f_1 = Q f_0 = (0.01414)(112{,}540 \text{ Hz}) = 1591 \text{ Hz}$

$f_1 f_2 = f_0 \mp \dfrac{f_2 - f_1}{2}$

$\qquad = 112{,}540 \text{ Hz} \mp \dfrac{1591 \text{ Hz}}{2}$

$\qquad = \boxed{111{,}745 \text{ Hz}, \ 113{,}336 \text{ Hz}}$

21. (a) $V_s = \left(\dfrac{N_s}{N_p}\right) V_p = \left(\dfrac{40}{200}\right)(550 \text{ V}) = \boxed{110 \text{ V}}$

(b) $I_p = \dfrac{V_s^2}{V_p Z_s} = \dfrac{(110 \text{ V})^2}{(550 \text{ V})(4.2 \ \Omega)} = \boxed{5.24 \text{ A}}$

(c) $I_s = \dfrac{V_s}{Z_s} = \dfrac{110 \text{ V}}{4.2 \ \Omega} = \boxed{26.2 \text{ A}}$

22. $\dfrac{N_p}{N_s} = \sqrt{\dfrac{Z_p}{Z_s}} = \sqrt{\dfrac{200\ \Omega}{\dfrac{8\ \Omega}{3}}} = \boxed{8.66}$

23.

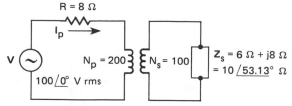

The circuit can be simplified to

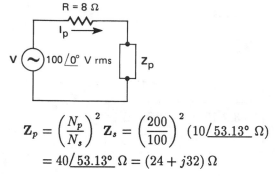

$$\mathbf{Z}_p = \left(\dfrac{N_p}{N_s}\right)^2 \mathbf{Z}_s = \left(\dfrac{200}{100}\right)^2 (10\underline{/53.13°}\ \Omega)$$
$$= 40\underline{/53.13°}\ \Omega = (24 + j32)\ \Omega$$

The circuit can be further simplified to

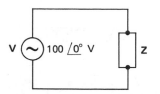

$$Z = \sqrt{(\Sigma R)^2 + (\Sigma X_L - \Sigma X_C)^2}$$
$$= \sqrt{(R + R_{\mathbf{Z}m})^2 + (X_{\mathbf{Z}m})^2}$$
$$= \sqrt{(8\ \Omega + 24\ \Omega)^2 + (32\ \Omega)^2} = 45.25\ \Omega$$

$$\phi_{\mathbf{Z}} = \tan^{-1}\left(\dfrac{\Sigma X_L - \Sigma X_C}{\Sigma R}\right) = \tan^{-1}\left(\dfrac{X_{\mathbf{Z}m}}{R + R_{\mathbf{Z}m}}\right)$$
$$= \tan^{-1}\left(\dfrac{32\ \Omega}{8\ \Omega + 24\ \Omega}\right) = 45°$$

$$\mathbf{Z} = 45.25\underline{/45°}\ \Omega$$

(a) $\mathbf{I}_p = \dfrac{\mathbf{V}}{\mathbf{Z}} = \dfrac{100\underline{/0°}\ \text{V}}{45.25\underline{/45°}\ \Omega}$

$$= \boxed{2.21\underline{/-45°}\ \text{A rms}}$$

(b) $P_s = I_s^2 R_s = I_s^2 Z_s \cos\phi = \left(\dfrac{N_p I_p}{N_s}\right)^2 Z_s \cos\phi$

$$= \left(\dfrac{(200)(2.21\ \text{A})}{100}\right)^2 (10\ \Omega)(\cos(53.13°))$$

$$= \boxed{117.2\ \text{W}}$$

THREE-PHASE ELECTRICITY

1. $R = \dfrac{V_l^2}{P} = \dfrac{(208\text{ V})^2}{3 \times 10^3\text{ W}} = \boxed{14.42\ \Omega}$

2. $I_l = \left(\dfrac{1}{\sqrt{3}}\right)\left(\dfrac{P_\Delta}{V_l \cos\phi}\right)$

$\quad = \left(\dfrac{1}{\sqrt{3}}\right)\left[\dfrac{5000 \times 10^3\text{ W}}{(4160\text{ V})(0.84)}\right] = \boxed{826\text{ A}}$

3. (a) $P = \dfrac{V_l^2}{R} = \dfrac{(120\text{ V}^2)^2}{10\ \Omega} = \boxed{1440\text{ W}}$

   (b) $S_\Delta = (3)\left(\dfrac{V_l^2}{R}\right) = (3)\left[\dfrac{(120\text{ V})^2}{10\ \Omega}\right]$

$\quad\quad = \boxed{4320\text{ W}}$

4. (a) $\mathbf{I}_p = \dfrac{\mathbf{V}_l}{\mathbf{Z}} = \dfrac{208\underline{/0°}\text{ V}}{20\underline{/25°}\ \Omega} = \boxed{10.4\underline{/-25°}\text{ A}}$

   (b) $\mathbf{I}_l = \sqrt{3}\underline{/-30°}\ \mathbf{I}_p$

$\quad\quad = \left(\sqrt{3}\underline{/-30°}\right)(10.4\underline{/-25°}\text{ A})$

$\quad\quad = \boxed{18\underline{/-55°}\text{ A}}$

   (c) $\mathbf{V}_p = \mathbf{V}_l = \boxed{208\underline{/0°}\text{ V}}$

   (d) $P_p = V_p I_p(\cos\phi)$

$\quad\quad = (208\text{ V})(10.4\text{ A})(\cos 25°) = \boxed{1960.5\text{ W}}$

   (e) $P_\Delta = 3P_p = (3)(1960.5\text{ W}) = \boxed{5882\text{ W}}$

5. (a) $V_p = \dfrac{110\text{ V}}{\sqrt{3}} = \boxed{63.51\text{ V}}$

   (b) $Z = \sqrt{(3\ \Omega)^2 + (4\ \Omega)^2} = 5$

$\quad\quad \phi \arctan\left(\dfrac{3\ \Omega}{4\ \Omega}\right) = 53.15°$

$\quad\quad \mathbf{I}_l = \dfrac{\mathbf{V}_p}{\mathbf{Z}} = \dfrac{63.51\underline{/0°}\text{ V}}{5\underline{/53.13°}\ \Omega} = \boxed{12.70\underline{/-53.13°}\text{ A}}$

   (c) $\mathbf{I}_p = \mathbf{I}_l = \boxed{12.70\underline{/-53.13°}\text{ A}}$

   (d) $P_t = \sqrt{3}\,VI\cos\phi$

$\quad\quad = \left(\sqrt{3}\right)(110\text{ V})(12.7\text{ A})(\cos 53.13°)$

$\quad\quad = \boxed{1451\text{ W}}$

CIRCUITS
3-Phase

ROTATING ELECTRICAL MACHINES

1. (a) $V_m = (NAB) \left(2\pi \left(\dfrac{n}{60 \frac{\text{sec}}{\text{min}}} \right) \left(\dfrac{p}{2} \right) \right)$

$= (12)(20)(0.10\,\text{m})(0.167\,\text{m}) \left(1.0\,\dfrac{\text{Wb}}{\text{m}^2} \right)$

$\times (2\pi) \left(\dfrac{1800\,\text{rpm}}{60 \frac{\text{sec}}{\text{min}}} \right) \left(\dfrac{4}{2} \right)$

$= \boxed{1511\,\text{V}}$

(b) $P = \dfrac{(1\,\text{kW}) \left(\dfrac{\text{hp}}{0.7457\,\text{kW}} \right)}{0.9} = \boxed{1.49\,\text{hp}}$

2. (a) $n = (2) \left(60\,\dfrac{\text{sec}}{\text{min}} \right) \left(\dfrac{f}{p} \right)$

$= (2) \left(60\,\dfrac{\text{sec}}{\text{min}} \right) \left(\dfrac{60\,\text{Hz}}{24} \right) = \boxed{300\,\text{rpm}}$

(b) $BA = \dfrac{(\sqrt{2})\,V_{\text{eff}}}{N\omega} = \dfrac{(\sqrt{2})\,(2200\,\text{V})}{(20)(2\pi)(60\,\text{Hz})}$

$= \boxed{0.413\,\text{Wb}}$

3. $I = \dfrac{P}{\eta V \cos\phi}$

$= \dfrac{(20\,\text{hp}) \left(\dfrac{0.7457 \times 10^3\,\text{W}}{\text{hp}} \right)}{(0.86)(440\,\text{V})(0.76)} = \boxed{51.86\,\text{A}}$

4. (a) $n_r = \left(\dfrac{(2) \left(60\,\dfrac{\text{sec}}{\text{min}} \right)}{p} \right) (1-s)$

$= \left(\dfrac{(2) \left(60\,\dfrac{\text{sec}}{\text{min}} \right) (60\,\text{Hz})}{4} \right) (1-0.03)$

$= \boxed{1746\,\text{rpm}}$

(b) $T = \dfrac{P}{\omega} = \dfrac{(200\,\text{hp}) \left(\dfrac{550\,\dfrac{\text{ft-lbf}}{\text{sec}}}{\text{hp}} \right)}{2\pi \left(1746\,\dfrac{\text{rev}}{\text{min}} \right) \left(\dfrac{\text{min}}{60\,\text{sec}} \right)}$

$= \boxed{602\,\text{ft-lbf}}$

(c) $I_l = \left(\dfrac{1}{\sqrt{3}} \right) \left(\dfrac{P}{\eta V_l \cos\phi} \right)$

$= \left(\dfrac{1}{\sqrt{3}} \right) \left(\dfrac{(200\,\text{hp}) \left(\dfrac{0.7457 \times 10^3\,\text{W}}{\text{hp}} \right)}{(0.85)(440\,\text{V})(0.91)} \right)$

$= \boxed{253\,\text{A}}$

5.

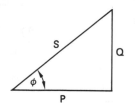

The original power angle is

$\phi_i = \arccos(0.82) = 34.92°$

$P_1 = 550\,\text{kW}$

$Q_i = P_1 \tan\phi_i = (550\,\text{kW})(\tan 34.92°) = 384.0\,\text{kVAR}$

The new conditions are

$$P_2 = (250\,\text{hp}) \left(0.7457\,\dfrac{\text{kW}}{\text{hp}} \right) = 186.4\,\text{kW}$$

$$\phi_f = \arccos(0.95) = 18.19°$$

Since both motors perform real work,

$$P_f = P_1 + P_2 = 550\,\text{kW} + 186.4\,\text{kW} = 736.4\,\text{kW}$$

The new reactive power is

$$Q_f = P_f \tan\phi_f = (736.4\,\text{kW})(\tan 18.19°)$$
$$= 242.0\,\text{kVAR}$$

The change in reactive power is

$$\Delta Q = 384.0\,\text{kVAR} - 242.0\,\text{kVAR} = 142\,\text{kVAR}$$

Synchronous motors used for power factor correction are rated by apparent power.

$S = \sqrt{(\Delta P)^2 + (\Delta Q)^2}$

$= \sqrt{(186.4\,\text{kW})^2 + (142\,\text{kVAR})^2} = \boxed{234.3\,\text{kVA}}$

6. $f = \dfrac{pn_s}{(20)\left(60 \frac{\text{sec}}{\text{min}}\right)} = \dfrac{pn}{(2)\left(60 \frac{\text{sec}}{\text{min}}\right)(1-s)}$

The slip and number of poles are unknown. Assume $s = 0$ and $p = 4$.

$$f = \dfrac{(4)\left(960 \frac{\text{rev}}{\text{min}}\right)}{(2)\left(60 \frac{\text{sec}}{\text{min}}\right)} = 32 \text{ Hz}$$

This is not close to anything in commercial use. Try $p = 6$.

$$f = \dfrac{(6)\left(960 \frac{\text{rev}}{\text{min}}\right)}{(2)\left(60 \frac{\text{sec}}{\text{min}}\right)} = 48 \text{ Hz}$$

With a 4% slip, $f = 50$ Hz.

$\boxed{50 \text{ Hz (European)}}$

7. $V = \left(\dfrac{z}{a}\right)p\phi\Omega$

For simplex lap winding, $a = p$.

$$V = \left(\dfrac{200}{4}\right)(4)(3.0 \times 10^{-4} \text{ Wb})\left(\dfrac{3600 \text{ rpm}}{60 \frac{\text{sec}}{\text{min}}}\right)$$

$= \boxed{3.6 \text{ V}}$

8. (a) $I_{f0} = \dfrac{V_{l0}}{R_f} = \dfrac{121 \text{ V}}{134 \, \Omega} = \boxed{0.903 \text{ A}}$

(b) Voltage is proportional to rotational speed.

$$n_l = \left(\dfrac{V_{ll} + R_a\left(I_{ll} + \frac{V_{ll}}{R_f}\right)}{V_{l0} + R_a\frac{V_{l0}}{R_f}}\right)n_0$$

$$= \left(\dfrac{110 \text{ V} + (0.31 \, \Omega)\left(30 \text{ A} + \frac{110 \text{ V}}{134 \, \Omega}\right)}{121 \text{ V} + (0.31 \, \Omega)\left(\frac{121 \text{ V}}{134 \, \Omega}\right)}\right)(1775 \text{ rpm})$$

$= \boxed{1750 \text{ rpm}}$

(c) $P = \dfrac{V_{ll}^2}{R_f} + \left(I_{ll} + \dfrac{V_{ll}}{R_f}\right)^2 R_a$

$$= \dfrac{(110 \text{ V})^2}{134 \, \Omega} + \left(30 \text{ A} + \dfrac{110 \text{ V}}{134 \, \Omega}\right)^2 (0.31 \, \Omega)$$

$= \boxed{385 \text{ W}}$

9. (a) $E = V_{ll} - I_{al}R_a = 110 \text{ V} - (90 \text{ A})(0.05 \, \Omega)$

$= \boxed{105.5 \text{ V}}$

(b) $I_{ll} = \dfrac{V_{ll}}{R_f} + I_{al} = \dfrac{110 \text{ V}}{60 \, \Omega} + 90 \text{ A}$

$= \boxed{91.8 \text{ A}}$

10. $\eta = \dfrac{V_{ll}I_{ll} - \left(\frac{V_{ll}^2}{V_{l0}}\right)I_{f0} - \left(I_{ll} - \frac{V_{ll}}{V_{l0}}I_{f0}\right)^2\left(R_{a0} + \frac{V_{b0}}{I_{a0}}\right)}{V_{ll}I_{ll}}$

$$= \dfrac{(240 \text{ V})(67 \text{ A}) - \left(\frac{(240 \text{ V})^2}{(240 \text{ V})}\right)(3.16 \text{ A})}{(240 \text{ V})(67 \text{ A})}$$

$$- \dfrac{\left(67 \text{ A} - \left(\frac{240 \text{ V}}{240 \text{ V}}\right)(3.16 \text{ A})\right)^2\left(0.207 \, \Omega + \frac{2 \text{ V}}{3.35 \text{ A}}\right)}{(240 \text{ V})(67 \text{ A})}$$

$= \boxed{0.75 \ (75\%)}$

$P = \eta V_{ll}I_{ll} = \dfrac{(0.75)(240 \text{ V})(67 \text{ A})}{745.7 \frac{\text{W}}{\text{hp}}} = \boxed{16.2 \text{ hp}}$

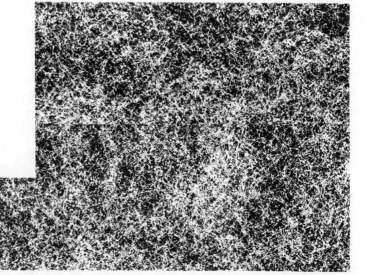

ELECTRONICS

1. (a) $I_s = \dfrac{I}{e^{\frac{qV}{\eta\kappa T}} - 1}$

For a germanium diode, $\eta \approx 1$. $T = 20 + 273 = 293$K.

$$I_s = \dfrac{70\,\mu\text{A}}{\exp\left(\dfrac{(1.6\times 10^{-19}\text{C})(-1.5\text{ V})}{(1)\left(1.38\times 10^{-23}\frac{\text{J}}{\text{K}}\right)(293\text{K})}\right) - 1}$$

$$\approx \dfrac{70\,\mu\text{A}}{-1} = \boxed{-70\,\mu\text{A}}$$

The reverse saturation current is constant for reverse voltages up to the breakdown voltage.

(b) $I \approx I_s \left[e^{\frac{40V}{\eta}} - 1 \right]$

$$= (70 \times 10^{-6}\text{ A}) \left[e^{\frac{\left(40\frac{1}{V}\right)(0.2\text{ V})}{1}} - 1 \right]$$

$$= \boxed{0.209\text{ A}}$$

(c) $\dfrac{I_{s,40°C}}{I_{s,20°C}} = 2^{\left(\frac{40°C - 20°C}{10}\right)} = (2)^2 = 4$

$$I_{s,40°C} = 4 I_{s,20°C} = (4)(70\,\mu\text{A}) = 280\,\mu\text{A}$$

$$I = I_s \left[e^{\frac{qV}{\eta\kappa T}} - 1 \right]$$

$$= (280 \times 10^{-6}\text{A}) \left[e^{\left(\frac{(1.6\times 10^{-19}\text{C})(0.2\text{ V})}{(1)\left(1.38\times 10^{-23}\frac{\text{J}}{\text{K}}\right)(313\text{K})}\right)} - 1 \right]$$

$$= \boxed{0.4617\text{ A}}$$

2. Since $20 > 5\sqrt{2}$, the applied voltage is never negative. Therefore, the diode does not rectify the applied voltage and serves no purpose. Model the circuit as an ideal diode and a $100\,\Omega$ resistance. (The forward resistance could also be used.)

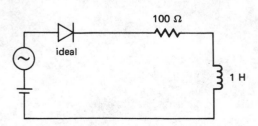

$XL = \omega L = \left(60\,\dfrac{\text{rad}}{\text{s}}\right)(1\text{ H}) = 60\,\Omega$

$Z = \sqrt{R^2 + X_L^2} = \sqrt{(100\,\Omega)^2 + (60\,\Omega)^2} = 116.6\,\Omega$

$\phi = \arctan\left(\dfrac{X_L}{R}\right) = \arctan\left(\dfrac{60\,\Omega}{100\,\Omega}\right) = 30.96°$

$\mathbf{I} = \dfrac{\mathbf{V}}{\mathbf{Z}} = \dfrac{20 + 5\sqrt{2}\underline{/0°}}{116.6\underline{/30.96°}}$

The d-c voltage across the inductor is zero. Therefore, the 20 V bias is not used.

$$\mathbf{V}_L = \mathbf{I}\mathbf{Z}_L = \left(\dfrac{5\sqrt{2}\underline{/0°}}{116.6\underline{/30.96°}}\right)(60\underline{/90°})$$

$$= \boxed{3.639\underline{/59.04°}\text{ V}}$$

3. The saturation current doubles for every 10°C.

(a) $\dfrac{I_{s2}}{I_{s1}} = 2^{\left(\frac{\Delta T}{10°C}\right)} = 2^{\frac{100°C - 25°C}{10°C}} = (2)^{7.5} = 181$

$$I_{s2} = (181)(100\,\mu\text{A}) = \boxed{18{,}100\,\mu\text{A (18.1 mA)}}$$

(b) At 0°C,

$$I_{s2} = 2^{\left(\frac{0°C - 25°C}{10°C}\right)} I_{s1} = (0.177)(100\,\mu\text{A}) = \boxed{17.7\,\mu\text{A}}$$

(c) $I_{-0.5} \approx I_s \left[e^{40V} - 1 \right]$

$$= (100\,\mu\text{A}) \left[e^{\left(40\frac{1}{V}\right)(-0.5\text{ V})} - 1 \right]$$

$$= -100\,\mu\text{A}$$

4. Clipping of the voltage wave will occur.

(a) $I = \dfrac{100\sin\omega t\text{ V}}{5000\,\Omega + 5000\,\Omega} = \boxed{10^{-2}\sin\omega t\text{ A}}$

(b)

5. Writing Kirchhoff's voltage law around the base loop,

$$V_{BB} - V_{BE} - (I_C + I_B)R_E = 0$$

$$\frac{I_C}{I_B} = \frac{\alpha}{1 - \alpha} = \frac{0.99}{1 - 0.99} = 99$$

$$4\text{ V} - V_{BE} - (I_C + \frac{I_C}{99})1000\ \Omega = 0$$

$$4\text{ V} - 0.6\text{ V} - (1010\ \Omega)I_C = 0$$

$$\boxed{I_C = 3.366 \times 10^{-3}\text{ A} \ (3.366\text{ mA})}$$

6. R_1 and R_2 form a voltage divider.

$$V_B = \left(\frac{60\text{ k}\Omega}{30\text{ k}\Omega + 60\text{ k}\Omega}\right)(24\text{ V}) = 16\text{ V}$$

Writing Kirchhoff's voltage law around the base circuit,

$$V_B - V_{BE} - (I_B + I_C)R_E = 0$$

$$\frac{I_C}{I_B} = \beta = \frac{\alpha}{1 - \alpha} = \frac{0.95}{1 - 0.95} = 19$$

$$16\text{ V} - 0.6\text{ V} - \left(\frac{I_C}{19} + I_C\right)(50,000\ \Omega) = 0$$

$$\boxed{I_C = 0.0002926\text{ A}}$$

7. (a) A common base equivalent T circuit is

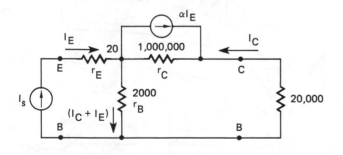

$$\alpha = \frac{\beta}{1 + \beta} = \frac{I_C}{I_E} = \frac{99}{100} = 0.99$$

$$V_L = I_C R_L = \alpha I_E R_L$$

$$= (0.99)(10 \times 10^{-6}\text{A})(2 \times 10^4\ \Omega)$$

$$= \boxed{0.198\text{ V}}$$

(b) Writing Kirchhoff's voltage law around the collector-base loop,

$$0 = (I_C + \alpha I_E)(10^6\ \Omega) + (I_C + I_E)(2000\ \Omega)$$
$$+ I_C(20,000\ \Omega)$$

$$I_E = 1.03 I_C$$

Also, note that $V_{cb} = I_C r_L = 20,000 I_C$

$$V_{eb} = I_E(20\ \Omega) + (I_C + I_E)(2000\ \Omega)$$
$$= (1.03)(20\ \Omega)I_C + (2000\ \Omega)I_C + (2000\ \Omega)(1.03)I_C$$
$$= (4080.6\ \Omega)I_C$$

$$I_C = \frac{V_{eb}}{4080.6\ \Omega} = \frac{V_{cb}}{20,000\ \Omega}$$

$$A = \frac{V_{cb}}{V_{eb}} = \frac{20,000\text{ V}}{4080.6\text{ V}} = \boxed{4.9}$$

8. A common emitter equivalent circuit is

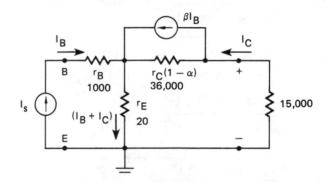

$$r_C(1 - \alpha) = (1,200,000\ \Omega)(1 - 0.97) = 36,000\ \Omega$$

$$r_B = 1000\ \Omega$$

$$r_E = 20\ \Omega$$

$$\beta = \frac{\alpha}{1 - \alpha} = \frac{0.97}{1 - 0.97} = 32.33$$

$$V_{ce} = 0.1\text{ V} = I_C(15,000\ \Omega)$$

Writing Kirchhoff's voltage law around the collector-emitter loop,

$$15,000 I_C + (I_C - \beta I_B)r_C(1 - \alpha) + (I_B + I_C)r_E = 0$$
$$15,000 I_C + (I_C - 32.33 I_B)(36,000) + (I_B + I_C)(1000) = 0$$
$$I_C(15,000 + 36,000 + 1000) - I_B(1,163,880 - 1000) = 0$$

(b) $$\frac{I_C}{I_B} = \boxed{22.36}$$

(a) $$I_B = \frac{I_C}{22.36} = \frac{\left(\dfrac{0.1}{15,000}\right)}{22.36} = \boxed{2.98 \times 10^{-7}\text{ A}}$$

9.

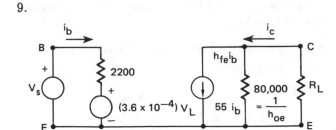

(a)
$$-i_c R_L = V_L$$
$$i_b = 0.5 \, \mu A$$

$$(55 i_b)\left(\frac{80{,}000 \, R_L}{R_L + 80{,}000}\right) = V_L = 0.1 \, V$$

$$80{,}000 \, R_L = (R_L + 80{,}000)\left[\frac{0.1}{(55)(0.5 \times 10^{-6} \, A)}\right]$$

$$80{,}000 \, R_L = 3636.36 \, R_L + (290.91 \times 10^6)$$

$$\boxed{R_L = 3809 \, \Omega}$$

(b) $V_s = i_b(2200 \, \Omega) + (3.6 \times 10^{-4})V_L$
$$= (0.5 \times 10^{-6} \, A)(2200 \, \Omega) + (3.6 \times 10^{-4})(0.1 \, V)$$
$$= (1.1 \times 10^{-3}) + (36 \times 10^{-6})$$
$$= \boxed{1.136 \times 10^{-3} \, V \ (1.136 \, mV)}$$

(c) $A = \dfrac{V_L}{V_s} = \boxed{\dfrac{0.1 \, V}{1.136 \times 10^{-3} \, V} = 88.03}$

10. $P_1(V_{DD} = 20 \, V, \ I_D = 0)$
$P_2(V_{DD} = 0, \ I_D = 10 \, mA)$

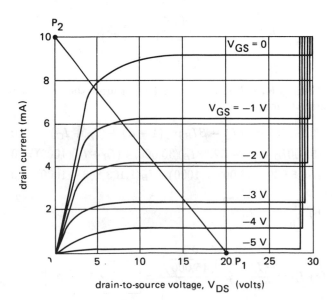

drain-to-source voltage, V_{DS} (volts)

11. $P_1(V_{DS} = 20 \, V, \ 0)$

$$P_2\left(0, I_D = \frac{20 \, V}{1500 \, \Omega} = 0.01333 \, A \ (13.33 \, mA)\right)$$

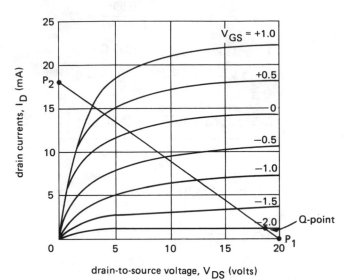

drain-to-source voltage, V_{DS} (volts)

12. (Note that the minimum V_{GS} is off the chart, and only larger values are presented. By default, these larger values must be used.)

For linear amplification, choose a Q-point on the $V_{GS} = -2.0$ V line. Higher characteristic curves are not as flat, hence, are less desirable. Power dissipation is another consideration. It should be evaluated at the extreme points of the signal swing, not just the Q-point.

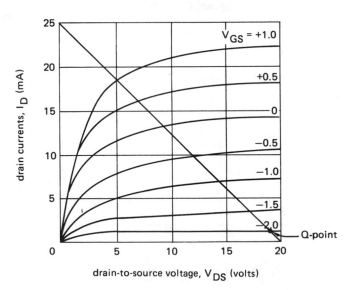

drain-to-source voltage, V_{DS} (volts)

13. (a) $A = \dfrac{-R_f}{R_{in}} = \dfrac{-10^6 \, \Omega}{50 \, \Omega} = \boxed{-20{,}000}$

(b) $v_{out} = A v_{in} = (-20{,}000)(0.1 \, V) = \boxed{-2000}$

14. This is a summing circuit.

$$v_{out} = -\left(\frac{Z_f}{Z_1}\right)v_1 - \left(\frac{Z_f}{Z_2}\right)v_2$$

$$v_{out} = -\left(\frac{1}{sC_1R_1}\right)v_1 - \left(\frac{1}{sC_2R_2}\right)v_2$$

$$v_{out} = \frac{-1}{(2\times10^{-6}\text{ F})(0.75\times10^6\text{ }\Omega)}\int 10\sin200t\,dt$$

$$-\frac{1}{(2\times10^{-6}\text{ F})(0.5\times10^6\text{ }\Omega)}\int 15\sin200t\,dt$$

$$= \left(\frac{2}{3}\right)\left(\frac{1}{200}\right)(10)\cos200t$$

$$+ \left(\frac{1}{10}\right)\left(\frac{15}{200}\right)\cos200t + C$$

$$= \boxed{\left(\frac{1}{30}\right)\cos200t + \left(\frac{3}{400}\right)\cos200t + C\text{ V}}$$

15. The circuit is a differentiator.

$$v_{out} = -RC\frac{dv_{in}}{dt}$$

$$= -(0.8\times10^6\text{ }\Omega)(2\times10^{-6}\text{ F})\left[\frac{d(100\sin30t)}{dt}\right]$$

$$= \boxed{-4800\cos30t\text{ V}}$$

16. This is a summing circuit.

$$v_{out} = \left(\frac{-6\text{ k}\Omega}{4.5\text{ k}\Omega}\right)v_1 - \left(\frac{6\text{ k}\Omega}{5\text{ k}\Omega}\right)v_2 - \left(\frac{6\text{ k}\Omega}{7.5\text{ k}\Omega}\right)v_3$$

$$= (-12\text{ V})\left(\frac{6\text{ k}\Omega}{4.5\text{ k}\Omega} + \frac{6\text{ k}\Omega}{5\text{ k}\Omega} + \frac{6\text{ k}\Omega}{7.5\text{ k}\Omega}\right) = \boxed{-40\text{ V}}$$

17. This is a summing circuit.

$$v_{out} = -\left[\left(\frac{R_f}{1\text{ }\Omega}\right)(1\text{ V}) + \left(\frac{R_f}{1\text{ }\Omega}\right)(6\text{ V}) + \left(\frac{R_f}{1\text{ }\Omega}\right)(3\text{ V})\right]$$

$$= -8\text{ V}$$

$$R_f = \boxed{0.8\text{ }\Omega}$$

$$v_{out} = -\left[\left(\frac{R_f}{1\text{ }\Omega}\right)(10\text{ V}) + \left(\frac{R_f}{1\text{ }\Omega}\right)(10\text{ V}) + \left(\frac{R_f}{1\text{ }\Omega}\right)(10\text{ V})\right]$$

$$= -15\text{ V}$$

$$R_f = \boxed{0.5\text{ }\Omega}$$

18. $$v_{out} = -\frac{1}{RC}\int v_{in}\,dt$$

$$= -\left[\frac{1}{(500,000\text{ }\Omega)(2\times10^{-3}\text{ F})}\right]\int 8\cos10t\,dt$$

$$= (-8\times10^{-4})\sin10t + d_o$$

At $T = 0$, $d_o = 12$ in.

$$v_{out} = \boxed{(-8\times10^{-4})\sin10t + 12\text{ in}}$$

19. $$\frac{v_{out}}{v_{in}} = -\frac{Z_f}{Z_{in}} = \frac{-500,000\text{ }\Omega}{\left[\dfrac{(10^6)\left(\dfrac{1}{s(2\times10^{-6})}\right)}{10^6 + \dfrac{1}{s(2\times10^{-6})}}\right]}$$

$$\frac{v_{out}}{v_{in}} = -\frac{\dfrac{1}{2}}{\left(\dfrac{1}{2s+1}\right)} = -\left(s + \frac{1}{2}\right)$$

$$v_{out} = -\left(s + \frac{1}{2}\right)v_{in}$$

$$v_{out} = \boxed{-\frac{dv_{in}}{dt} - \frac{v_{in}}{2}}$$

20. $$\frac{v_{out}}{v_{in}} = -\frac{Z_f}{Z_{in}}$$

$$= -\frac{\left(\dfrac{(600,000)\left[\dfrac{1}{s(2\times10^{-6})}\right]}{600,000 + \left[\dfrac{1}{s(2\times10^{-6})}\right]}\right)}{2,000,000}$$

$$\frac{v_{out}}{v_{in}} = -\frac{0.3}{1.2s+1}$$

$$v_{out} = (1.2s + 1) = -0.3v_{in}$$

$$(1.2)\left(\frac{dv_{out}}{dt}\right) + v_{out} = -0.3v_{in}$$

$$\boxed{(1.2)\left(\frac{dv_{out}}{dt}\right) + v_{out} + 0.3v_{in} = 0}$$

LIGHT, ILLUMINATION, AND OPTICS

1.
$$\lambda = \lambda' \left[\frac{1 - \frac{v}{c}}{\sqrt{1 - \left(\frac{v}{c}\right)^2}} \right]$$

Solving for v/c,

$$\frac{v}{c} = \frac{1 \pm \left(\frac{\lambda}{\lambda'}\right)^2}{1 + \left(\frac{\lambda}{\lambda'}\right)^2} = \frac{1 \pm \left(\frac{4350}{6580}\right)^2}{1 + \left(\frac{4350}{6580}\right)^2}$$

$$= 0.3917$$

$$v = (0.3917)c = (0.3917)\left(3 \times 10^8 \frac{m}{s}\right)$$

$$= \boxed{1.1751 \times 10^8 \text{ m/s}}$$

2.
$$E_t = E_s$$

$$\frac{\phi_t}{A_t} = \frac{\phi_s}{A_s}$$

Because both lamps are omnidirectional sources,

$$\frac{\phi_t}{4\pi r_t^2} = \frac{\phi_s}{4\pi r_s^2}$$

$$\phi_t = \phi_s \left(\frac{r_t}{r_s}\right)^2$$

$$I_t = I_s \left(\frac{r_t}{r_s}\right)^2 = \left(20 \frac{lm}{sr}\right)\left(\frac{2 \text{ m} - 0.8 \text{ m}}{0.8 \text{ m}}\right)^2$$

$$= \boxed{45 \text{ lm/sr}}$$

3. The luminous efficiency is 16 lm/W.

$$\frac{\Phi}{Q} = \eta_l$$

$$\Phi = Q\eta_l = (100 \text{ W})\left(16 \frac{lm}{W}\right) = 1600 \text{ lm}$$

$$E = \frac{\Phi}{A} = \frac{\Phi}{4\pi r^2} = \frac{1600 \text{ lm}}{(4\pi)(5 \text{ m})^2}$$

$$= \boxed{5.093 \text{ lm/m}^2}$$

4.
$$\text{irradiance} = \frac{Q}{4\pi r^2} = \frac{400 \text{ W}}{(4\pi)(20 \text{ m})^2}$$

$$= \boxed{0.0796 \text{ W/m}^2}$$

5.

$$\sin \theta = \frac{y}{L} \approx \frac{y}{x} = \frac{m\lambda}{d}$$

$$y \approx \frac{Lm\lambda}{d} \approx \frac{(3 \text{ m})(1)(5 \times 10^{-7} \text{ m})}{\left(\frac{1}{2000} \frac{1}{cm}\right)\left(\frac{m}{100 \text{ cm}}\right)}$$

$$\approx \boxed{0.3 \text{ m}}$$

6.
$$\sin \theta = \frac{m\lambda}{d}$$

$$\theta = \arcsin\left(\frac{m\lambda}{d}\right)$$

$$= \arcsin\left(\frac{(1)(5890 \times 10^{-10} \text{ m})}{\left(\frac{1}{1000} \frac{1}{cm}\right)\left(\frac{m}{100 \text{ cm}}\right)}\right)$$

$$= \boxed{3.377°}$$

7.
$$n = \frac{c_{vacuum}}{c_{medium}} = \frac{186,300 \frac{miles}{sec}}{124,000 \frac{miles}{sec}} = \boxed{1.5024}$$

8.
$$n_a \sin \theta_a = n_g \sin \theta_g$$

$$\theta_g = \arcsin\left(\frac{n_a}{n_g}\sin\theta_a\right) = \arcsin\left(\frac{1}{1.70}\sin 90°\right)$$

$$= \boxed{36.03°}$$

9.
$$n = \frac{\sin\left[\frac{1}{2}(\alpha + \theta)\right]}{\sin\left(\frac{1}{2}\alpha\right)} = \frac{\sin\left[\frac{1}{2}(60° + 45°)\right]}{\sin\left[\frac{1}{2}(60°)\right]}$$

$$= \boxed{1.587}$$

10.
$$\frac{1}{f} = \frac{1}{i} + \frac{1}{o}$$

$$i = \frac{1}{\dfrac{1}{f} - \dfrac{1}{o}}$$

(a)
$$i = \frac{1}{\dfrac{1}{5\text{ cm}} - \dfrac{1}{3\text{ cm}}} = \boxed{-7.5\text{ cm}}$$

(b)
$$i = \frac{1}{\dfrac{1}{5\text{ cm}} - \dfrac{1}{7\text{ cm}}} = \boxed{17.5\text{ cm}}$$

(c)
$$i = \frac{1}{\dfrac{1}{5\text{ cm}} - \dfrac{1}{12\text{ cm}}} = \boxed{8.57\text{ cm}}$$

11.
$$\frac{1}{f} = \frac{1}{i} + \frac{1}{o}$$

$$i = \frac{1}{\dfrac{1}{f} - \dfrac{1}{o}} = \frac{1}{\dfrac{1}{1\text{ m}} - \dfrac{1}{1.5\text{ m}}}$$

$$= 3\text{ m (i.e., the screen distance)}$$

The frosted glass screen has been placed at the image plane of the lens system.

$$m = -\frac{i}{o} = -\frac{3}{1.5} = \boxed{-2}$$

WAVES AND SOUND

1. loudness $= 10 \log \left(\dfrac{I}{I_o} \right)$

$$= 10 \log \left(\dfrac{5 \times 10^{-10} \, \dfrac{W}{m^2}}{1 \times 10^{-12} \, \dfrac{W}{m^2}} \right) = \boxed{26.99 \text{ dB}}$$

2. (a) $a = \sqrt{\dfrac{E_{\text{aluminum}} g_c}{\rho_{\text{aluminum}}}}$

$$= \sqrt{\dfrac{\left(10 \times 10^6 \, \dfrac{lbf}{in^2} \right) \left(\dfrac{12 \text{ in}}{ft} \right)^2 \left(32.2 \, \dfrac{lbm\text{-}ft}{lbf\text{-}sec^2} \right)}{173 \, \dfrac{lbm}{ft^3}}}$$

$$= \boxed{16{,}371 \text{ ft/sec}}$$

(b) $a = \sqrt{k g_c R T}$

$$= \sqrt{(1.4)\left(32.2 \, \dfrac{lbm\text{-}ft}{lbf\text{-}sec^2} \right)\left(53.3 \, \dfrac{ft\text{-}lbf}{lbm\text{-}^\circ R} \right)(120+460)^\circ R}$$

$$= \boxed{1181 \text{ ft/sec}}$$

(c) $a = \sqrt{k g_c R T}$

$$= \sqrt{(1.4)\left(32.2 \, \dfrac{lbm\text{-}ft}{lbf\text{-}sec^2} \right)\left(53.3 \, \dfrac{ft\text{-}lbf}{lbm\text{-}^\circ R} \right)(100+460)^\circ R}$$

$$= \boxed{1160 \text{ ft/sec}}$$

(d) $T = \left[\left(\dfrac{9}{5} \right)(50^\circ C) + 32 \right] {}^\circ F = 122^\circ F$

$$a = \sqrt{\dfrac{E}{\rho}} = \sqrt{\dfrac{\left(333.2 \times 10^3 \, \dfrac{lbf}{in^2} \right) \left(\dfrac{12 \text{ in}}{ft} \right)^2}{1.917 \, \dfrac{slug}{ft^3}}}$$

$$= \boxed{5003 \text{ ft/sec}}$$

(e) $T = \left[\left(\dfrac{9}{5} \right)(60^\circ C) + 32 \right] {}^\circ F = 140^\circ F$

$$a = \sqrt{\dfrac{E}{\rho}} = \sqrt{\dfrac{\left(330 \times 10^3 \, \dfrac{lbf}{in^2} \right) \left(\dfrac{12 \text{ in}}{ft} \right)^2}{1.908 \, \dfrac{slug}{ft^3}}}$$

$$= \boxed{4991 \text{ ft/sec}}$$

3. $a = \sqrt{k g_c R T}$

$$= \sqrt{(1.4)\left(32.2 \, \dfrac{lbm\text{-}ft}{lbf\text{-}sec^2} \right)\left(53.3 \, \dfrac{ft\text{-}lbf}{lbm\text{-}^\circ R} \right)(70+460)^\circ R}$$

$$= 1128.5 \text{ ft/sec}$$

$$\lambda = \dfrac{a}{f} = \dfrac{1128.5 \, \dfrac{ft}{sec}}{20 \, \dfrac{1}{sec}} = \boxed{56.42 \text{ ft}}$$

4. Since each octave involves a 2:1 frequency ratio,

$$f_{\text{high C}} = 4 f_{\text{middle C}} = (4)(260 \text{ Hz}) = \boxed{1040 \text{ Hz}}$$

5. For musical instruments, harmonics are an octave apart.

$$f_2 = 2f_1 = (2)(528 \text{ Hz}) = \boxed{1056 \text{ Hz}}$$

$$f_3 = 2f_2 = (2)(1056 \text{ Hz}) = \boxed{2112 \text{ Hz}}$$

(This assumes the piccolo is a cylinder open at both ends; $\lambda = 2L/n$.)

6. Assume the string is tuned to the fundamental frequency. Let n be the harmonic number.

$$f = \dfrac{n v_{\text{transverse}}}{2L}$$

$$v_{\text{transverse}} = \dfrac{2Lf}{n} = \sqrt{\dfrac{T g_c}{m}}$$

$$T = \dfrac{m}{g_c} \left(\dfrac{2Lf}{n} \right)^2$$

$$= \left(\dfrac{0.07 \text{ lbm}}{(1 \text{ ft})\left(32.2 \, \dfrac{lbm\text{-}ft}{lbf\text{-}sec^2} \right)} \right) \left(\dfrac{(2)(1 \text{ ft})\left(262 \, \dfrac{1}{sec} \right)}{1} \right)^2$$

$$= \boxed{596.9 \text{ lbf}}$$

7. Let n be the harmonic number. Let T be the tension in string.

$$v_{\text{transverse}} = \frac{2Lf}{n} = \sqrt{\frac{Tg_c}{m}}$$

$$T = \left(\frac{m}{g_c}\right)\left(\frac{2Lf}{n}\right)^2 = \frac{(0.038\,\text{lbm})\left(\dfrac{(2)(4\,\text{ft})\left(150\,\dfrac{1}{\text{sec}}\right)}{1}\right)^2}{(4\,\text{ft})\left(32.2\,\dfrac{\text{lbm-ft}}{\text{lbf-sec}^2}\right)}$$

$$= \boxed{424.8\,\text{lbf}}$$

8. Assume the fundamental tone, $n = 1$, is produced.

$$f = \frac{nv_{\text{transverse}}}{2L}$$

$$v_{\text{transverse}} = \frac{2Lf}{n} = \frac{(2)(3\,\text{ft})\left(198\,\dfrac{1}{\text{sec}}\right)}{1}$$

$$= \boxed{1188\,\text{ft/sec}}$$

9. $f_{\text{beat}} = f_{\text{fork}} - f_{\text{string}}$

$$f_{\text{string}} = f_{\text{fork}} - f_{\text{beat}} = 512\,\text{Hz} - 2\,\text{Hz} = 510\,\text{Hz}$$

To eliminate beats, the string must be tightened until its frequency matches the frequency of the tuning fork.

$$f'_{\text{string}} = 512\,\text{Hz}$$

$$v_t = \sqrt{\frac{T}{m}}$$

$$T = v_t{}^2 m$$

$$f = \frac{nv_t}{2L}$$

$$v_t = \frac{2Lf}{n}$$

$$T = \frac{4L^2 f^2 m}{n^2}$$

$$\frac{\Delta T}{T} = \frac{(f'_{\text{string}})^2 - (f_{\text{string}})^2}{(f_{\text{string}})^2} = \frac{(512\,\text{Hz})^2 - (510\,\text{Hz})^2}{(510\,\text{Hz})^2}$$

$$= \boxed{0.00786}$$

10. The speakers presumably emit sound waves in phase with each other. Since these waves must cancel at the point of silence, they must be $n\lambda/2$ out of phase.

$$\frac{d_{\text{left}}}{\lambda} = \frac{d_{\text{right}}}{\lambda} + \frac{2n + 1}{2} \quad [n \text{ is any positive integer}]$$

$$\frac{d_{\text{left}}}{\dfrac{v}{f}} = \frac{d_{\text{right}}}{\dfrac{v}{f}} + \frac{2n + 1}{2}$$

$$f = \frac{(2n + 1)v}{(2)(d_{\text{left}} - d_{\text{right}})} = \frac{(2n + 1)\left(1100\,\dfrac{\text{ft}}{\text{sec}}\right)}{(2)(10\,\text{ft} - 8\,\text{ft})}$$

$$f = \boxed{(275)(2n + 1)\,\text{Hz}}$$

11. Assume a closed tube resonating at its fundamental frequency.

$$\lambda = \frac{a}{f} = \frac{4L}{2n - 1}$$

$$a = \frac{4Lf}{2n - 1} = \frac{(4)(4\,\text{ft})\left(256\,\dfrac{1}{\text{sec}}\right)}{(2)(1) - 1} = \boxed{4096\,\text{ft/sec}}$$

PHYSICS
Sound

MODELING OF ENGINEERING SYSTEMS

1.

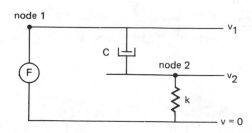

At node 1: $F = C(v_1 - v_2) = C(x_1' - x_2')$

At node 2: $F = k(x_2 - 0) = kx_2$

2.

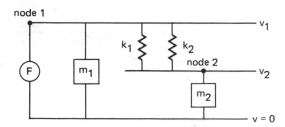

At node 1: $F = m_1 a_1 + (k_1 + k_2)(x_1 - x_2)$

$$F = \boxed{m_1 x_1'' + (k_1 + k_2)(x_1 - x_2)}$$

At node 2: $0 = m_2 a_2 + (k_1 + k_2)(x_2 - x_1)$

$$0 = \boxed{m_2 x_2'' + (k_1 + k_2)(x_2 - x_1)}$$

3. This is a rotational system.

T = applied torque = FL

θ = rotated angle = $\dfrac{x_1}{L} = \dfrac{x_2}{l}$

α = angular acceleration = $(kx_2)l$

$\quad = (kl\sin\theta)l \approx kl^2\theta$ for small values of θ

I = moment of inertia of beam about the end for a

$\quad$ slender rod $= \dfrac{1}{3}mL^2$

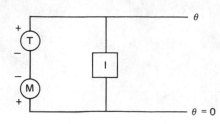

$$T - M = I\alpha$$

$$\boxed{T = I\alpha + M = I\ddot{\theta} + kl^2\theta}$$

In terms of the variables listed in the problem,

$$FL = \frac{1}{3}mL^2\left(\frac{\ddot{x}_1}{L}\right) + klx_2$$

$$\boxed{F = \frac{1}{3}m\ddot{x}_1 + k\frac{l}{L}x_2}$$

4.

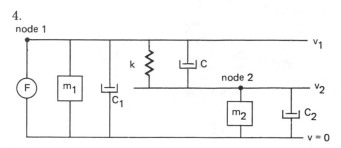

At node 1: $F = m_1 a_1 + C_1(v_1 - 0) + C(v_1 - v_2)$
$\qquad\qquad + k(x_1 - x_2)$

$$F = \boxed{m_1 x_1'' + C_1 x_1' + C(x_1' - x_2') + k(x_1 - x_2)}$$

At node 2: $0 = m_2 a_2 + C_2(v_2 - 0) + C(v_2 - v_1)$
$\qquad\qquad + k(x_2 - x_1)$

$$0 = \boxed{m_2 x_2'' + C_2 x_2' + C(x_2' - x_1') + k(x_2 - x_1)}$$

5.

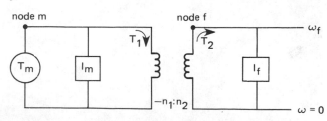

At node m: $T_m = I_m \alpha_m + T_1$

$\qquad\qquad\quad T_m = I_m \theta_m'' + T_1$

At node f: $T_2 = I_f \alpha_f = I_f \theta_f''$

At the transformer:

$$T_1 = \left(-\frac{n_1}{n_2}\right)T_2 = -\left(\frac{n_1}{n_2}\right)I_f \ddot{\theta}_f$$

SYSTEMS
Modeling

A combined equation of motion in terms of the variables listed in the problem statement is

$$T_m = I_m \ddot{\theta}_m - \left(\frac{n_1}{n_2}\right) I_f \ddot{\theta}_f$$

6.

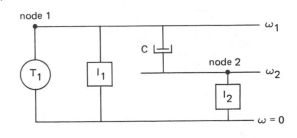

At node 1: $T_1 = I_1 \alpha_1 + C(\omega_1 - \omega_2)$

$\quad\quad\quad\quad\quad T_1 = I_1 \theta_1'' + C(\theta_1' - \theta_2')$

At node 2: $0 = I_2 \alpha_2 + C(\omega_2 - \omega_1)$

$\quad\quad\quad\quad\quad 0 = I_2 \theta_2'' + C(\theta_2' - \theta_1')$

7. $\sum i_{\text{out}} = 0$ at the node connecting the resistor and inductor.

$$0 = \frac{V_2 - V_1}{R} + \frac{1}{L} \int V_2 \, dt$$

$$0 = \frac{V_2' - V_1'}{R} + \frac{V_2}{L}$$

8. At the transformer,

$$\frac{I_1}{I_2} = \frac{1}{n}$$

At the right loop,

$$-LI_2' - V_1 \left(\frac{1}{n}\right) = 0$$

$$-LnI_1' - \frac{V_1}{n} = 0$$

$$V_1 + Ln^2 I_1' = 0$$
$$I_2 = nI_1$$

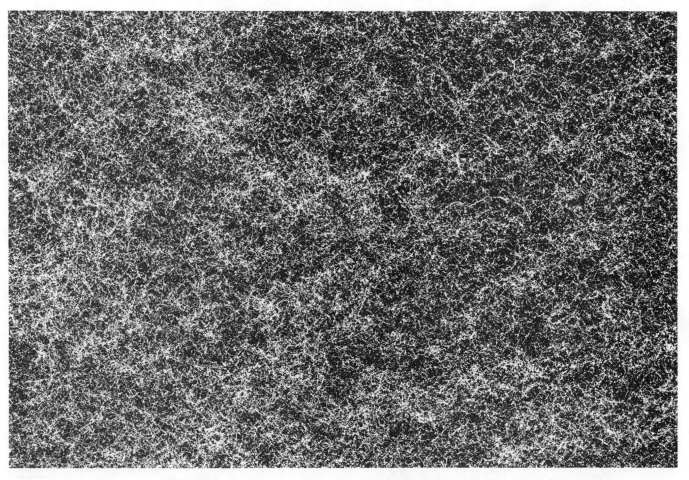

ANALYSIS OF ENGINEERING SYSTEMS

1. At $t = 0.5$ sec,

$$\omega t = \left(10 \, \frac{\text{rad}}{\text{sec}}\right)(0.5 \text{ sec}) = 5 \text{ rad}$$

For a normalized curve of natural response corresponding to $\zeta = 0.3$,

$$\frac{x(t)}{\omega} \approx -0.22 \text{ sec/rad} \quad (\text{at } t = 0.5 \text{ sec})$$

$$x(t = 0.5 \text{ sec}) = \left(-0.22 \, \frac{\text{sec}}{\text{rad}}\right)\left(10 \, \frac{\text{rad}}{\text{sec}}\right) = \boxed{-2.2}$$

2. (a)

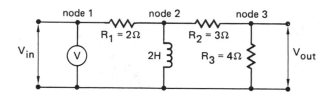

At node 2,

$$0 = \frac{V_2 - V_1}{R_1} + \frac{V_2 - V_3}{R_2} + \frac{1}{L}\int V_2\, dt$$

$$= \frac{V_2 - V_1}{2} + \frac{V_2 - V_3}{3} + \frac{1}{2}\int V_2\, dt$$

$$0 = \frac{V_2 - V_1}{2} + \frac{V_2 - V_3}{3} + \frac{V_2}{2s} \qquad [\text{Eq. 1}]$$

At node 3,

$$0 = \frac{V_3 - V_2}{R_2} + \frac{V_3}{R_3}$$

$$= \frac{V_3 - V_2}{3} + \frac{V_3}{4} \qquad [\text{Eq. 2}]$$

From Eq. 1,

$$V_2 = \frac{3sV_1 + 2sV_3}{5s + 3}$$

From Eq. 2,

$$V_2 = \frac{7V_3}{4}$$

$$\frac{3sV_1 + 2sV_3}{5s + 3} = \frac{7V_3}{4}$$

$$T(s) = \frac{V_{\text{out}}}{V_{\text{in}}} = \frac{V_3}{V_1} = \frac{12s}{27s + 21} = \boxed{\frac{4s}{9s + 7}}$$

The black box representation of this system is

$$V_{\text{in}} \longrightarrow \boxed{\frac{4s}{9s + 7}} \longrightarrow V_{\text{out}}$$

(b)
$$0 = \frac{V_{\text{out}} - V_{\text{in}}}{R} + C\frac{dV_{\text{out}}}{dt}$$

$$0 = \frac{V_{\text{out}} - V_{\text{in}}}{R} + CsV_{\text{out}}$$

$$0 = V_{\text{out}} - V_{\text{in}} + CRsV_{\text{out}}$$

$$T(s) = \frac{V_{\text{out}}}{V_{\text{in}}} = \frac{1}{1 + CRs}$$

$$= \frac{1}{1 + (3 \times 10^{-6}\,\text{F})\left(\frac{1}{2} \times 10^6\,\Omega\right)s} = \boxed{\frac{2}{3s + 2}}$$

$$V_{\text{in}} \longrightarrow \boxed{\frac{2}{3s + 2}} \longrightarrow V_{\text{out}}$$

3.

$$V_{\text{in}} \longrightarrow \bigcirc \longrightarrow \boxed{-100} \longrightarrow \boxed{-80} \longrightarrow \boxed{-100} \longrightarrow V_{\text{out}}$$

The overall gain without feedback is

$$K = (-100)(-80)(-100) = \boxed{-800{,}000}$$

$$\frac{V_{\text{out}} - hV_{\text{out}}}{995\,\Omega} = \frac{hV_{\text{out}}}{5\,\Omega}$$

$$h = \frac{5\,\Omega}{995\,\Omega + 5\,\Omega} = 0.005$$

$K < 0$, but the feedback adds, so the feedback is negative.

$$K_{\text{loop}} = \frac{K}{1 - Kh} = \frac{-800{,}000}{1 - (-800{,}000)(0.005)} = \boxed{-200}$$

4.
$$v_{out} = G_{loop} v_i$$

$$\frac{\Delta v_{out}}{v_{out}} = \frac{\Delta G_{loop}}{G_{loop}}$$

(Assume the input signal has no uncertainty.)

$$\frac{\Delta G_{loop}}{G_{loop}} \frac{G}{\Delta G} = \frac{1}{1+G}$$

$$\frac{\Delta v_{out}}{v_{out}} = \frac{\Delta G_{loop}}{G_{loop}} = \left(\frac{\Delta G}{G}\right)\left(\frac{1}{1+G}\right)$$

$$\frac{\Delta G}{G} = 0.1$$

$$\frac{\Delta v_{out}}{v_{out}} = (0.1)\left(\frac{1}{1+1000}\right) = 9.99 \times 10^{-5}$$

$$\approx \boxed{0.01\%}$$

5. (a)

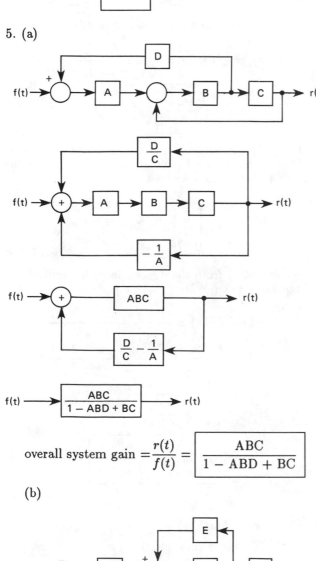

$$\text{overall system gain} = \frac{r(t)}{f(t)} = \boxed{\frac{ABC}{1 - ABD + BC}}$$

(b)

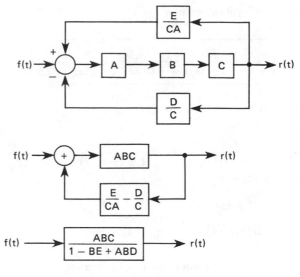

$$\text{overall system gain} = \frac{r(t)}{f(t)} = \boxed{\frac{ABC}{1 - BE + ABD}}$$

6.

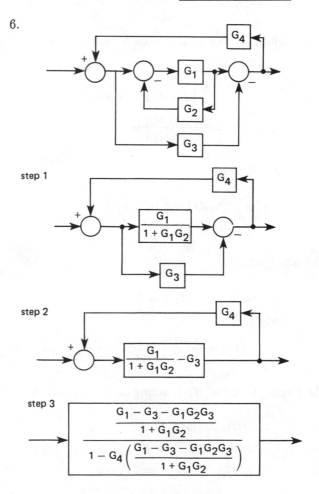

step 1

step 2

step 3

$$\frac{\dfrac{G_1 - G_3 - G_1 G_2 G_3}{1 + G_1 G_2}}{1 - G_4\left(\dfrac{G_1 - G_3 - G_1 G_2 G_3}{1 + G_1 G_2}\right)}$$

$$\boxed{\frac{G_1 - G_3 - G_1 G_2 G_3}{1 + G_1 G_2 - G_1 G_4 + G_3 G_4 + G_1 G_2 G_3 G_4}}$$

The overall system gain is

$$\boxed{\frac{G_1 - G_3 - G_1 G_2 G_3}{1 + G_1 G_2 - G_1 G_4 + G_3 G_4 + G_1 G_2 G_3 G_4}}$$

From step 2,

$$G(s) = \frac{G_1}{1 + G_1 G_2} - G_3 = \frac{-5}{1 + (-5)(2)} - 4 = -\frac{31}{9}$$
$$H(s) = G_4 = 3$$

$$G(s)H(s) = \left(-\frac{31}{9}\right)(3) = -\frac{31}{3} < 0$$

The system has the negative feedback; therefore,

$$\text{sensitivity } S = \frac{1}{1 + G(s)H(s)} = \frac{1}{1 + \left(-\frac{31}{3}\right)} = -\frac{3}{28}$$

7. By definition, the system transfer function is

$$T(s) = \mathcal{L}\left[\frac{r_{\text{sys}}(t)}{f_{\text{sys}}(t)}\right] = \frac{R_{\text{sys}}(s)}{F_{\text{sys}}(s)}$$

$$= \frac{\text{output signal}}{\text{input signal}}$$

The output signal is

$$r_{\text{sys}}(t) = \delta'(t) + \delta(t)$$
$$R_{\text{sys}}(s) = s + 1$$

The input signal is

$$f_{\text{sys}}(t) = r''(t) + 3r'(t) + r(t)$$
$$F_{\text{sys}}(s) = s^2 + 3s + 1$$
$$T(s) = \frac{R_{\text{sys}}(s)}{F_{\text{sys}}(s)} = \frac{s + 1}{s^2 + 3s + 1}$$

The forcing function, $f(t)$, acting on this system is

$$f(t) = \delta(t)$$
$$F(s) = \mathcal{L}[f(t)] = \mathcal{L}[\delta(t)] = 1$$
$$R(s) = T(s)F(s) = (T(s))(1)$$

$$= T(s) = \boxed{\frac{s + 1}{s^2 + 3s + 1}}$$

8. Using the final value theorem, obtain the steady-state step response by substituting 0 for s in the transfer function.

If $b_p \neq 0$,

$$\boxed{R(s) = T(0) = \frac{b_p}{a_n}}$$

If $b_p = 0$, the numerator is zero.

$$\boxed{R(s) = 0}$$

9. (a) Substitute $j\omega$ for s in the transfer function to obtain the steady-state response for a sinusoidal input.

$$R(s) = T(j\omega) = \frac{1 + j\omega}{1 + 2j\omega - 2\omega^2} = \frac{1 + j\omega}{(1 - 2\omega^2) + 2j\omega}$$

The amplitude is the absolute value of $R(s)$.

$$|R(s)| = \sqrt{T^2(j\omega)} = \frac{\sqrt{1 + \omega^2}}{\sqrt{(1 - 2\omega^2)^2 + 4\omega^2}}$$

$$= \boxed{\sqrt{\frac{1 + \omega^2}{1 + 4\omega^4}}}$$

(b) The phase angle of the steady-state response is $\text{Arg}[R(s)]$. Find this by substituting $T(j\omega)$ into the form $(a + jb)/c$ whose Arg is

$$\arctan\left(\frac{\dfrac{b}{c}}{\dfrac{a}{c}}\right) = \arctan\left(\frac{b}{a}\right)$$

First eliminate j from the denominator by multiplying by its complex conjugate $[(1 + 2\omega^2) - 2j\omega]$.

$$R(s) = T(j\omega) = \frac{(1 + j\omega)(1 - 2\omega^2 - 2j\omega)}{1 + 4\omega^4}$$

$$= \frac{1 - 2\omega^2 + 2\omega^2 + j(-\omega - 2\omega^3)}{1 + 4\omega^4}$$

$$= \frac{1 + j(-\omega - 2\omega^3)}{1 + 4\omega^4}$$

$$\text{Arg}[R(s)] = \text{Arg}[T(j\omega)]$$

$$= \boxed{\arctan\left[\frac{-(\omega + 2\omega^3)}{1}\right]}$$

10. (a) $R(s) = T(s)F(s)$

$$f(t) = \mu_0 \quad \text{(unit step)}$$
$$F(s) = \mathcal{L}[f(t)] = \frac{1}{s}$$
$$R(s) = T(s)\frac{1}{s} = \frac{1}{s(s + a)(s + b)}$$

PROFESSIONAL PUBLICATIONS, INC. ● Belmont, CA

From the transform table,

$$r(t) = \mathcal{L}^{-1}[R(s)] = \frac{1}{ab} + \frac{be^{-at} - ae^{-bt}}{ab(a-b)}$$

$$= r_1 + r_2(t)$$

The steady-state response is $\lim_{t \to \infty} r(t)$. Since $\lim_{t \to \infty} r_2(t) = 0$, r_1 is the steady-state response.

$$r_1 = \boxed{\frac{1}{ab}}$$

(b) The steady-state response in the frequency domain for a step input of magnitude h is $R(s) = hT(0)$. Substituting $h = 1$ and $s = 0$ in $T(s)$,

$$R(s) = (h)[T(0)] = \frac{1}{(0+a)(0+b)} = \boxed{\frac{1}{ab}}$$

11.

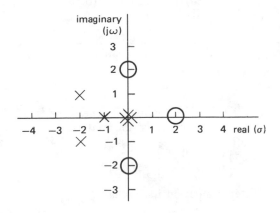

$$\frac{(s^2 + 4)(s - 2)}{s^2(s^2 + 4s + 5)(s + 1)} = \frac{(s - 2j)(s + 2j)(s - 2)}{s^2(s + 1)(s + 2 - j)(s + 2 + j)}$$

For zeros, set the numerator equal to zero. For poles, set the denominator equal to zero.

12. The amplitude and phase of the steady-state response are given by $T(j\omega_o)$, where $T(s)$ is the transfer function. Since $j\omega_o$ is a zero of $T(s)$,

$$R(s) = T(j\omega_o) = 0$$

Therefore, the steady-state response is zero. The system entirely blocks the angular frequency, ω_o.

13. $T(s) = \dfrac{3s + 18}{s^2 + 12s + 3200} = \dfrac{as + b}{s^2 + (\text{BW})s + \omega_n^{\,2}}$

(a) bandwidth = BW = $\boxed{12 \text{ rad/sec}}$

(b) peak frequency = $\omega_n = \sqrt{3200}$

$$= \boxed{56.57 \text{ rad/sec}}$$

(c) half-power points = $\omega_n \pm \dfrac{\text{BW}}{2}$

$$= 56.57 \,\frac{\text{rad}}{\text{sec}} \pm \frac{12 \,\frac{\text{rad}}{\text{sec}}}{2}$$

$$= \boxed{\begin{array}{l} 62.57 \text{ rad/sec and} \\ 50.57 \text{ rad/sec} \end{array}}$$

(d) quality factor = $Q = \dfrac{\omega_n}{\text{BW}} = \dfrac{56.57 \,\frac{\text{rad}}{\text{sec}}}{12 \,\frac{\text{rad}}{\text{sec}}}$

$$= \boxed{4.71}$$

14. The system has a pole in the right-hand half of the s-plane ($s = 1$).

$$s^2 + s - 2 = (s - 1)(s + 2)$$

$$\boxed{\text{The system is unstable.}}$$

GENERAL SYSTEMS MODELING

1. (a) The network corresponding to the project is

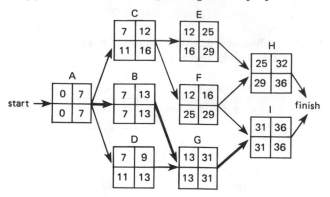

(b) The critical path is **start-A-B-G-I-finish**.

(c) The latest starts are the same as the earliest starts for these nodes.

(d) The earliest finish = the latest finish = $\boxed{36}$

2. The decision tree diagram is

month:

| | 1 | 2 | 3 | expected value of each path, $(p_1)(p_2)(p_3)$ |

$(0.6)(0.75)(0.75) = 0.3375$

$(0.6)(0.25)(0.45) = 0.0675$

$(0.4)(0.45)(0.75) = 0.1350$

$(0.4)(0.55)(0.45) = 0.0990$

The black nodes represent the expected proportion of people who use brand A. At the end of the third month, the expected market share of brand A is

$$\sum p\{A\} = 0.3375 + 0.0675 + 0.1350 + 0.0990$$

$$= \boxed{0.639 \ (63.9\%)}$$

3. (a) The expected profit is

$$\sum p\{x_i\}(x_i) = (0.25)(15) + (0.3)(1) + (0.45)(-6)$$

$$= \boxed{1.35}$$

(b) The loss matrix is

	θ_1	θ_2	θ_3
a_1 (invest)	-15	-1	6
a_2 (do not invest)	0	0	0

The maximum losses are

action	max loss
a_1	6
a_2	0

The minimax criterion would direct the company not to invest since that action minimizes the maximum loss.

If the Bayes principle is used, the expected losses are

$$a_1 : \sum p\{\theta_i\} L(a_1, \theta_i)$$

$$= (0.25)(-15) + (0.3)(-1) + (0.45)(6) = -1.35$$

$$a_2 : 0$$

The Bayes principle for minimizing the expected loss would direct the company to invest.

4. There are two switches made in the second and third months. The one-step transition matrix is

$$\mathbf{P} = \begin{bmatrix} 0.75 & 0.25 \\ 0.45 & 0.55 \end{bmatrix}$$

The two-step transition matrix is

$$\mathbf{P}^2 = \mathbf{P} \times \mathbf{P} = \begin{bmatrix} 0.75 & 0.25 \\ 0.45 & 0.55 \end{bmatrix} \begin{bmatrix} 0.75 & 0.25 \\ 0.45 & 0.55 \end{bmatrix}$$

$$= \begin{bmatrix} 0.675 & 0.325 \\ 0.585 & 0.415 \end{bmatrix}$$

$$p_A^2 = p_A^0 p_{AA}^2 + p_B^0 p_{BA}^2$$

$$= (0.6)(0.675) + (0.4)(0.585) = \boxed{0.639}$$

5. Assume this is an $M/M/1$ system.

$$\lambda = 4 \frac{\text{jobs}}{\text{hr}}$$

$$\mu = \left(\frac{1 \text{ job}}{6 \text{ min}} \right) \left(60 \frac{\text{min}}{\text{hr}} \right) = 10 \frac{\text{jobs}}{\text{hr}}$$

$$\rho = \frac{\lambda}{\mu} = \frac{4}{10} = 0.4$$

(a) $p(0) = 1 - \rho = \boxed{0.6 \ (60\% \text{ idle})}$

(b) If the secretary is working on one job and there are three jobs waiting to be started, there are four in the system.

$$p(4) = p(0)(\rho^4) = (0.6)(0.4)^4 = \boxed{0.0154}$$

(c) The average number of jobs waiting to be started is

$$L_q = \frac{\rho\lambda}{\mu - \lambda} = \frac{(0.4)(4)}{10 - 4} = \boxed{0.267 \text{ jobs}}$$

(d) $p\{t > h\} = e^{-\mu h}$

$$p\{t > 0.5\} = e^{-(10)(0.5)} = e^{-5} = \boxed{0.0067}$$

6. Let x_A and x_B be the numbers of barrels of formulas A and B, respectively. Formulate this as a linear programming problem.

$$\text{Max } Z = 7x_A + 2x_B$$
$$\text{such that} \quad x_A + x_B \geq 100$$
$$x_B \leq 55$$
$$2x_A + x_B \leq 180$$
$$x_A \geq 0$$
$$x_B \geq 0$$

This is a linear, two-dimensional system and can be solved graphically.

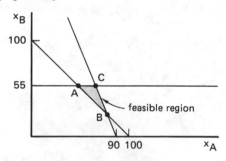

Since the feasible region is bounded, there are only three possible solutions: A, B, and C.

A is the intersection of $x_B = 55$ and $x_A + x_B = 100$.

$$A = (45, 55)$$
$$Z_A = (7)(45) + (2)(55) = 425$$

B is the intersection of $x_A + x_B = 100$ and $2x_A + x_B = 180$.

$$B = (80, 20)$$
$$Z_B = (7)(80) + (2)(20) = 600$$

C is the intersection of $x_B = 55$ and $2x_A + x_B = 180$

$$C = (62.5, 55)$$
$$Z_C = (7)(62.5) + (2)(55) = 547.5$$

$$\boxed{\begin{aligned} Z_{max} &= 600 \\ x_A &= 80 \text{ barrels} \\ x_B &= 20 \text{ barrels} \end{aligned}}$$

7.

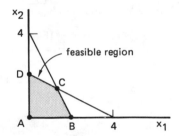

This is a linear, two-dimensional system, and the feasible region is bounded. Therefore, the possible solutions are $A = (0,0)$, $B = (2,0)$, $D = (0,2)$, and C, which is the intersection of $2x_1 + x_2 = 4$ and $x_1 + 2x_2 = 4$: $C = (4/3, 4/3)$.

$$Z_A = (12)(0) + (18)(0) = 0$$
$$Z_B = (12)(2) + (18)(0) = 24$$
$$Z_C = (12)\left(\frac{4}{3}\right) + (18)\left(\frac{4}{3}\right) = 40$$
$$Z_D = (12)(0) + (18)(2) = 36$$

$$\boxed{Z_{max} = 40 \text{ at } (x_1, x_2) = \left(\frac{4}{3}, \frac{4}{3}\right)}$$

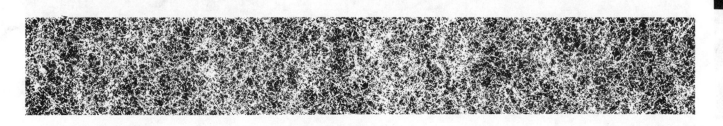

COMPUTER HARDWARE

1. (a) areal density = (no. bits per track)

   × (no. tracks per radial inch)

$$= \left(10{,}000 \frac{\text{bits}}{\text{track}}\right)\left(\frac{1000 \text{ tracks}}{\frac{D}{2}}\right)$$

$$= \left(10{,}000 \frac{\text{bits}}{\text{track}}\right)\left(\frac{2000 \text{ tracks}}{6 \text{ in}}\right)$$

$$= \boxed{3.33 \times 10^6 \text{ bits/radial in}}$$

(b) access time = average latency + average seek time

The average latency can be approximated as one-half time for the full revolution.

$$\left(\frac{1}{2} \text{ rev}\right)\left(\frac{1}{2000} \frac{\text{min}}{\text{rev}}\right)\left(60 \frac{\text{sec}}{\text{min}}\right)\left(1000 \frac{\text{ms}}{\text{sec}}\right) = 15 \text{ ms}$$

average access time = 15 ms + 20 ms

$$= \boxed{35 \text{ ms}}$$

2. Since eight tracks are used for data recording, 1 byte of data is stored across the width of the tape. Therefore, 1600 bytes of data are stored per inch of length. The length required for 1 megabyte is

$$\frac{1{,}048{,}576 \text{ bytes}}{\left(1600 \frac{\text{bytes}}{\text{in}}\right)\left(12 \frac{\text{in}}{\text{ft}}\right)} = \boxed{54.6 \text{ ft}}$$

This neglects any interblock gaps.

3. The time required is

$$t = \frac{(20 \text{ Mb})\left(1{,}048{,}576 \frac{\text{bytes}}{\text{Mb}}\right)\left(10 \frac{\text{bits}}{\text{byte}}\right)}{\left(2400 \frac{\text{bits}}{\text{sec}}\right)\left(60 \frac{\text{sec}}{\text{min}}\right)\left(60 \frac{\text{min}}{\text{hr}}\right)}$$

$$= \boxed{24.3 \text{ hrs}}$$

DATA STRUCTURES AND PROGRAM DESIGN

1.

```
          ┌─────────┐
          │  start  │
          └────┬────┘
               │
         ┌─────┴──────┐
         │ input A, B, C │
         └─────┬──────┘
               │
         ◇ A > B and A > C ◇──y──┐ B = B + C ──→ place1 ←──┐
               │ n                                          │
         ◇ B > A and B > C ◇──y──┐ A = A + C ──→ place2     │
               │ n                                          │
         ◇ B > A and B < C ◇──y──┐ B = B − A ───────────────┘
               │ n
          ┌────┴────┐
          │   end   │
          └─────────┘
```

2.

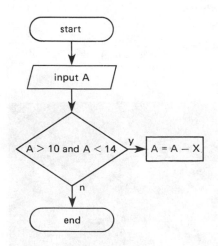

3.

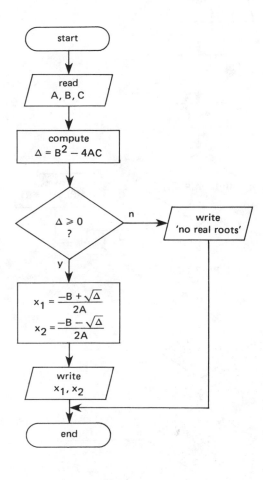

PROFESSIONAL PUBLICATIONS, INC. ● Belmont, CA

4.

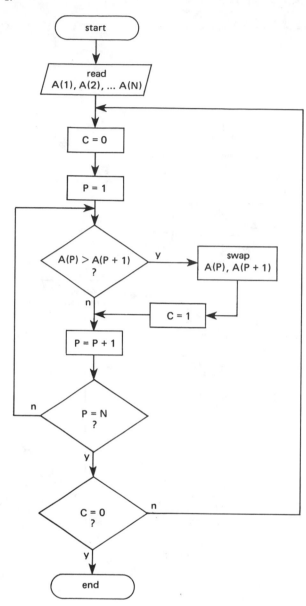

Assume that the sought-for element is the first element in the list. This will produce a worst-case search. Let $a(n)$ be the number of comparisons needed to find the element in the list. Since the binary search procedure discards (disregards) half of the list after each comparison,

$$a(n) = 1 + a\left(\frac{n}{2}\right) \qquad \text{[Eq. 1]}$$

But, $n = 2^p$.

$$a(n) = a(2^p) = 1 + a\left(\frac{2^p}{2}\right) = 1 + a(2^{p-1}) \qquad \text{[Eq. 2]}$$

Equation 1 can be applied to the second term of Eq. 2.

$$a(2^{p-1}) = 1 + a\left(\frac{2^{p-1}}{2}\right) = 1 + a(2^{p-2}) \qquad \text{[Eq. 3]}$$

Combining Eqs. 2 and 3,

$$a(n) = a(2^p) = 1 + 1 + a(2^{p-2})$$

Equation 1 can continue to be applied to the last term until the list contains only two elements, after which the search becomes trivial. This results in a string of p 1's.

$$a(n) = 1 + 1 + 1 + \cdots + 1 = p$$

(An argument for $a(n) = p + 1$ can be made depending on the how the search logic handles a miss with 2 elements in the list. If it merely selects the remaining element without comparing it, then $a(n) = p$. If a singular element is investigated, then $a(n) = p + 1$.)

Now,

$$n = 2^p$$
$$\log_2(n) = \log_2(2^p) = p$$

But $p = a(n) = \log_2(n)$.

Using logarithm identities,

$$p = \log_2(n) = \frac{\log_{10}(n)}{\log_{10}(2)}$$

5. To simplify, assume that the number of elements in the list is some integer power of 2.

$$n = 2^p$$

FORTRAN

1. The algebraic translation is

$$I = JK + L - M$$

$$= (3)(4) + 6 - 9 = \boxed{9}$$

2. The algebraic translations are

(a) X is true if $J > L$ or $K \geq (M - J)$. Otherwise, X is false.

Since J is greater than L, $\boxed{\text{X is true.}}$

(b) X is true if $J > L$ or $K < (M - J)$. Otherwise, X is false.

Since J is greater than L, $\boxed{\text{X is true.}}$

3. The algebraic translation of the FORTRAN formula for Q is

$$Q = \frac{R}{S^T} - T = \frac{18}{(6)^2} - 2 = -1.5$$

Since Q is negative, go to statement 10.

    10   Q=10.

Go to statement 40.

    40   END

Since the program ends at statement 40 with Q still equal to 10, $\boxed{Q = 10.}$

4.
```
      PROD=1.0
      DO 15 I=1, 10
      READ(5,10) B
10    FORMAT(F7.4)
      PROD=PROD*B
      IF (PROD .GT. 5.0) GO TO 20
15    CONTINUE
20    (any subsequent statement)
```

5.
```
      SUM=0.0
      DO 15 I=1,10
      READ(5,10) B
10    FORMAT(F7.4)
      C=AMOD(B/7.0)
      SUM=SUM+C
15    CONTINUE
```

6.
```
      A = 4104.
      B = 6802.
      C = 5798.
      WRITE (6, 10) A, B, C
10    FORMAT ('  ',3 (E7.2/))
```

7.
```
2.300 E2 (starting in column 12)
```

8.
```
      REAL FUNCTION STHETA (OMEGA,T)
      STHETA = SQRT (OMEGA*T)
      RETURN
      END
```

9.
```
      REAL FUNCTION FEET(RINCHS)
      FEET = RINCHS/12.0
      RETURN
      END
```

10.
```
      SUBROUTINE ADD (B , SUM)
      IF (B.LT.10.0) GO TO 5
      SUM = SUM+B
5     RETURN
      END
```

11.
```
COMMON A,B
```

PROFESSIONAL PUBLICATIONS, INC. ● Belmont, CA

12.

```
      INTEGER A,B,C
      READ(5,10) A,B,C
10    FORMAT(3I4)
      IF(A.LT.B.AND.A.LT.C) I=A
      IF(B.LT.C.AND.B.LT.A) I=B
      IF(C.LT.A.AND.C.LT.B) I=C
      WRITE(6,20) I
20    FORMAT(' ','THE SMALLEST INTEGER IS',I4)
      STOP
      END
```

13.

```
      REAL K
      DATA K,X/1.3,14.5
      F=K*X+0.05
      F=0.1*FLOAT(INT(F*10.0))
      WRITE(6,200) K,X,F
200   FORMAT(' ','K=',F6.3,' ','X=',F6.3,' ','F=',F6.3)
      STOP
999   END
```

An equivalent program can be written by substituting
the following code for the data statement.

```
      READ(5,100) K,X
100   FORMAT(2F4.1)
```

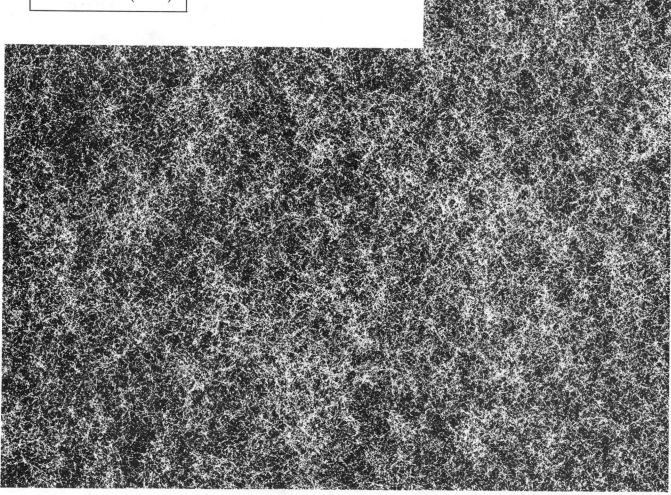

ANALOG COMPUTERS

1. First, differentiate both sides.

$$V_0' = -V_1 - 4V_2$$

Next, rearrange.

$$-\tfrac{1}{2}V_0' = V_1 + 4V_2 + \tfrac{1}{2}V_0'$$

Draw the circuit.

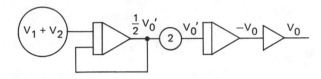

2. *step 1*:

$$x''' + 2x'' + 6x' + \tfrac{2}{7}x = 7$$

step 2:

$$-x''' = 2x'' + 6x' + \tfrac{2}{7}x - 7$$

step 3:

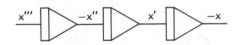

step 4:

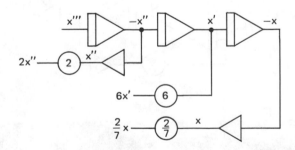

steps 5 and 6:

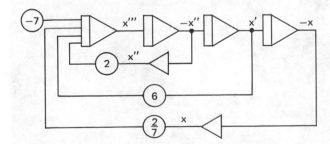

3.

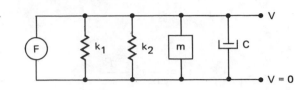

$$F = ma + Cv + (k_1 + k_2)x$$
$$mx'' + Cx' + (k_1 + k_2)x - F = 0$$

step 1:

$$x'' + \frac{C}{m}x' + \left(\frac{k_1 + k_2}{m}\right)x - \frac{F}{m} = 0$$

step 2:

$$-x'' = \frac{C}{m}x' + \left(\frac{k_1 + k_2}{m}\right)x - \frac{F}{m}$$

step 3:

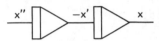

step 4:

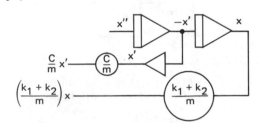

steps 5 and 6:

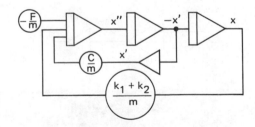

4.

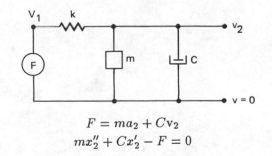

$$F = ma_2 + Cv_2$$
$$mx_2'' + Cx_2' - F = 0$$

PROFESSIONAL PUBLICATIONS, INC. ● Belmont, CA

step 1:

$$x_2'' + \frac{C}{m}x_2' - \frac{F}{m} = 0$$

step 2:

$$-x_2'' = \frac{C}{m}x_2' - \frac{F}{m}$$

step 3:

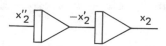

step 4:

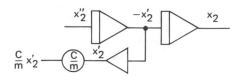

steps 5 and 6:

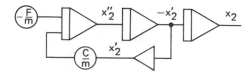

Notice that $F = k(x_1 - x_2)$ is a second equation that applies to this system. If x_2 is available, x_1 can be found.

$$x_1 = \frac{F}{k} + x_2$$

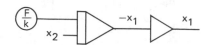

5.

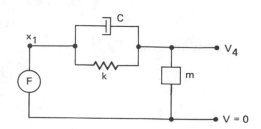

$$F = ma_4$$
$$mx_4'' - F = 0$$

step 1:

$$x_4'' - \frac{F}{m} = 0$$

step 2:

$$x_4'' = \frac{F}{m}$$

step 3:

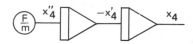

Notice that $F = C(v_1 - v_4) + k(x_1 - x_4)$ is a second equation that applies to the system. If x_4 is available, x_1 can be found.

$$F = Cx_1' - Cx_4' + kx_1 - kx_4$$
$$x_1' = -x_4' + \frac{k}{C}x_1 - \frac{k}{C}x_4 - \frac{F}{C}$$

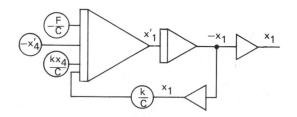

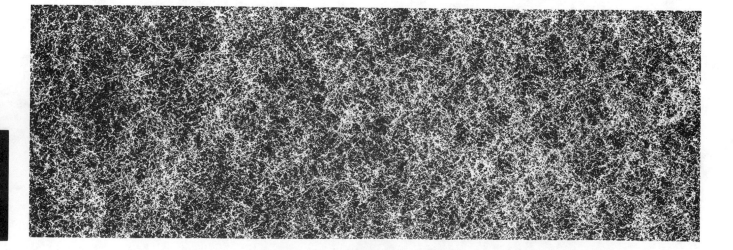

ATOMIC THEORY AND NUCLEAR ENGINEERING

1. (a) $E_{quantum} = \dfrac{hc}{\lambda}$

$= \left[\dfrac{(6.625 \times 10^{-34} \text{ J·s}) \left(3.00 \times 10^8 \frac{m}{s} \right)}{6.563 \times 10^{-10} \text{ m}} \right]$

$\times \left(\dfrac{eV}{1.6021 \times 10^{-19} \text{ J}} \right)$

$= \boxed{1.89 \text{ eV}}$

(b) $\boxed{\text{A photon is identical to a quantum.}}$

2. $E_{radiation} = \dfrac{hc}{\lambda}$

$= \left[\dfrac{(6.625 \times 10^{-34} \text{ J·s}) \left(3.00 \times 10^8 \frac{m}{s} \right)}{1.216 \times 10^{-10} \text{ m}} \right]$

$\times \left(\dfrac{eV}{1.6021 \times 10^{-19} \text{ J}} \right)$

$= \boxed{10.20 \text{ eV}}$

3. $E_{photon} = \dfrac{hc}{\lambda}$

$= \left[\dfrac{(6.625 \times 10^{-34} \text{ J·s}) \left(3.00 \times 10^8 \frac{m}{s} \right)}{0.22 \text{ m}} \right]$

$\times \left(\dfrac{eV}{1.6021 \times 10^{-19} \text{ J}} \right)$

$= \boxed{5.64 \times 10^{-6} \text{ eV}}$

4. $\Delta E = \Delta mc^2 = (m_{initial} - m_{final})c^2$

$= [(4)(1.008145 \text{ amu}) - 4.00387 \text{ amu}]$

$\times \left(\dfrac{1.660 \times 10^{-27} \text{ kg}}{\text{amu}} \right) \left(3.00 \times 10^8 \frac{m}{s} \right)^2$

$\times \left(\dfrac{eV}{1.6021 \times 10^{-19} \text{ J}} \right) \left(\dfrac{MeV}{10^6 \text{ eV}} \right)$

$= \boxed{26.77 \text{ MeV}}$

5. $m_{final} - m_{initial} = (16.9991 \text{ amu} + 1.00797 \text{ amu})$

$- (4.0026 \text{ amu} + 14.0067 \text{ amu})$

$= -0.00223 \text{ amu}$

$\Delta E = (m_{final} - m_{initial})c^2$

$= (-0.00223 \text{ amu}) \left(\dfrac{1.660 \times 10^{-27} \text{ kg}}{\text{amu}} \right)$

$\times \left(3.00 \times 10^8 \frac{m}{s} \right)^2 \left(\dfrac{eV}{1.6021 \times 10^{-19} \text{ J}} \right) \left(\dfrac{MeV}{10^6 \text{ eV}} \right)$

$= \boxed{-2.08 \text{ MeV} \quad \text{(an energy release)}}$

6. $T_{\frac{1}{2}} = \dfrac{0.6931}{\lambda}$

$\lambda = \dfrac{0.6931}{T_{\frac{1}{2}}} = \dfrac{0.6931}{15 \text{ hr}} = 0.04621 \dfrac{1}{\text{hr}}$

$t_2 - t_1 = \dfrac{-\ln \left(\frac{m_2}{m_1} \right)}{\lambda} = \dfrac{-\ln \left(\frac{9 \text{ g}}{48 \text{ g}} \right)}{0.04621 \frac{1}{\text{hr}}}$

$= \boxed{36.23 \text{ hr}}$

7. In $m_t = m_0 e^{-\lambda t}$, the term $e^{-\lambda t}$ may be thought of as the probability of any given nucleus not decaying by time t. Therefore, the probability that a nucleus decays in time t is $p(t)$.

$p(t) = 1 - e^{-\lambda t}$

$\lambda = \dfrac{0.6931}{T_{\frac{1}{2}}} = \dfrac{0.6931}{4 \text{ months}} = 0.1733 \dfrac{1}{\text{month}}$

(a) $p(4 \text{ months}) = 1 - e^{-(0.1733 \text{ month}^{-1})(4 \text{ months})}$

$= \boxed{0.50}$

(b) $p(8 \text{ months}) - p(4 \text{ months})$

$= 1 - e^{-(0.1733 \text{ month}^{-1})(8 \text{ months})}$

$- (1 - e^{-(0.1733 \text{ month}^{-1})(4 \text{ months})})$

$= \boxed{0.25}$

(c) $p(12 \text{ months}) = 1 - e^{-(0.1733 \text{ month}^{-1})(12 \text{ months})}$

$= \boxed{0.875}$

8. $\dfrac{m_t}{m_0} = e^{\left(-\frac{0.6931\ \text{yr}}{\lambda}\right)t} = e^{-\left(\frac{0.6931\ \text{yr}}{1620\ \text{yr}}\right)(4000\ \text{yr})}$

$\qquad = 0.1806$

The remaining fraction is

$$1 - 0.1806 = \boxed{0.819}$$

9. Since the activity halves each minute, the half-life, $T_{\frac{1}{2}}$, is $\boxed{1\ \text{min}}$

10. A 90 eV electron has kinetic energy of 90 eV.

$$E_k = m_0(k-1)c^2$$

$k = \dfrac{E_k}{m_0 c^2} + 1$

$\quad = \dfrac{(90\ \text{eV})\left(1.6021 \times 10^{-19}\ \frac{\text{J}}{\text{eV}}\right)}{(9.1091 \times 10^{-31}\ \text{kg})\left(3 \times 10^8\ \frac{\text{m}}{\text{s}}\right)^2} + 1$

$\quad = 1.0001759$

$v = c\sqrt{1 - \left(\dfrac{1}{k}\right)^2}$

$\quad = \left(3 \times 10^8\ \dfrac{\text{m}}{\text{s}}\right)\sqrt{1 - \left(\dfrac{1}{1.0001759}\right)^2}$

$\quad = 5.63 \times 10^6\ \text{m/s}$

$v_w = \dfrac{c^2}{v} = \dfrac{\left(3.00 \times 10^8\ \frac{\text{m}}{\text{s}}\right)^2}{5.63 \times 10^6\ \frac{\text{m}}{\text{s}}} = 1.60 \times 10^{10}\ \text{m/s}$

$p = \dfrac{E_t}{v_w} = \dfrac{k m_0 c^2}{v_w}$

$\quad = \dfrac{(1.0001759)(9.1091 \times 10^{-31}\ \text{kg})\left(3.00 \times 10^8\ \frac{\text{m}}{\text{s}}\right)^2}{1.60 \times 10^{10}\ \frac{\text{m}}{\text{s}}}$

$\quad = 5.125 \times 10^{-24}\ \text{N·s}$

$\lambda = \dfrac{h}{p} = \dfrac{6.625 \times 10^{-34}\ \text{J·s}}{5.125 \times 10^{-24}\ \text{N·s}} = \boxed{1.29 \times 10^{-10}\ \text{m}}$

11. $k = \dfrac{1}{\sqrt{1 - \left(\frac{v}{c}\right)^2}}$

$\quad = \dfrac{1}{\sqrt{1 - \left(\frac{2 \times 10^7\ \frac{\text{m}}{\text{s}}}{3 \times 10^8\ \frac{\text{m}}{\text{s}}}\right)^2}} = 1.00223$

$v_w = \dfrac{c^2}{v} = \dfrac{\left(3.00 \times 10^8\ \frac{\text{m}}{\text{s}}\right)^2}{2 \times 10^7\ \frac{\text{m}}{\text{s}}} = \boxed{4.5 \times 10^9\ \text{m/s}}$

$p = \dfrac{E_t}{v_w} = \dfrac{k m_0 c^2}{v_w}$

$\quad = \dfrac{(1.00223)(9.1091 \times 10^{-31}\ \text{kg})\left(3 \times 10^8\ \frac{\text{m}}{\text{s}}\right)^2}{4.5 \times 10^9\ \frac{\text{m}}{\text{s}}}$

$\quad = 1.8259 \times 10^{-23}\ \text{N·s}$

$\lambda = \dfrac{h}{p} = \dfrac{6.625 \times 10^{-34}\ \text{J·s}}{1.8259 \times 10^{-23}\ \text{N·s}} = \boxed{3.628 \times 10^{-11}\ \text{m}}$

12. $\qquad \epsilon_x \epsilon_p \approx h$

$\qquad \epsilon_p \approx \dfrac{h}{\epsilon_x} = \dfrac{6.625 \times 10^{-34}\ \text{J·s}}{0.0005\ \text{m}}$

$\qquad = \boxed{1.325 \times 10^{-30}\ \text{kg·m/s}}$

13. $\qquad \epsilon_x \epsilon_p \approx h$

Since the mass is constant,

$\qquad \epsilon_x m_e \epsilon_v \approx h$

$\qquad \epsilon_x \approx \dfrac{h}{m_e \epsilon_v} \approx \dfrac{6.625 \times 10^{-34}\ \text{J·s}}{(9.11 \times 10^{-31}\ \text{kg})\left(0.1\ \frac{\text{m}}{\text{s}}\right)}$

$\qquad \approx \boxed{7.272 \times 10^{-3}\ \text{m}}$

14. $\boxed{\text{A neutron}}$

15. $E_{\text{pion}} = \dfrac{E_0}{3} = \dfrac{2E_{\text{proton}}}{3}$

$\qquad = \dfrac{(2)(1.00727663\ \text{amu})\left(931.1\ \frac{\text{MeV}}{\text{amu}}\right)}{3}$

$\qquad = \boxed{625.3\ \text{MeV}}$

16. (a) $k = \dfrac{1}{\sqrt{1 - \left(\frac{v}{c}\right)^2}} = \dfrac{1}{\sqrt{1 - (0.4)^2}}$

$\qquad = 1.091089$

$$E_t = E_0 + E_k = kE_0$$

$$= (1.091089)(1.00727663 \text{ amu}) \left(931.1 \frac{\text{MeV}}{\text{amu}}\right)$$

$$= \boxed{1023.31 \text{ MeV}}$$

(b) $k = \dfrac{1}{\sqrt{1 - (0.8)^2}} = 1.6667$

$$E_t = (1.6667)(0.000548597 \text{ amu})\left(931.1 \frac{\text{MeV}}{\text{amu}}\right)$$

$$= \boxed{0.8513 \text{ MeV}}$$

(c) A photon has no rest mass.

$$E = \frac{hc}{\lambda} = \frac{1.24 \times 10^{-6} \text{ eV·m}}{5000 \times 10^{-10} \text{ m}}$$

$$= \boxed{2.48 \text{ eV}}$$

17. $v_{\text{rel}} = \left| \dfrac{v_1 - v_2}{1 - \dfrac{v_1 v_2}{c^2}} \right| = \left| \dfrac{0.8c - (-0.7c)}{1 - \dfrac{(0.8c)(-0.7c)}{c^2}} \right|$

$$= \boxed{0.961538c}$$

18. After the acceleration,

$$E_v = km_0 c^2 = kE_0$$

$$k = \frac{E_v}{E_0} = \frac{985 \text{ MeV}}{938 \text{ MeV}} = 1.0501066$$

$$\Delta m = m_v - m_0 = km_0 - m_0 = (k-1)m_0$$

$$= (1.0501066 - 1)(1.6725 \times 10^{-27} \text{ kg})$$

$$= \boxed{8.340 \times 10^{-29} \text{ kg}}$$

19. A pion in motion has the same average life as a pion at rest if both lives are measured in the reference frame of the pion.

$$T_{\text{avg,v}} = T_{\text{avg,0}} = 1.9 \times 10^{-16} \text{ s}$$

From a stationary observer's viewpoint,

$$T_{\text{avg,0}} = kT_{\text{avg,v}} = \left(\frac{E_v}{E_0}\right) T_{\text{avg,v}}$$

$$= \left(\frac{175 \text{ MeV}}{140 \text{ MeV}}\right)(1.9 \times 10^{-16} \text{ s})$$

$$= \boxed{2.375 \times 10^{-16} \text{ s}}$$

20. $E_0 = m_0 c^2$

$$= (1.6725 \times 10^{-27} \text{ kg})\left(3.00 \times 10^8 \frac{\text{m}}{\text{s}}\right)^2$$

$$= 1.5053 \times 10^{-10} \text{ J}$$

$$k = \frac{E_t}{E_0} = \frac{E_0 + E_k}{E_0}$$

$$= \frac{1.5053 \times 10^{-10} \text{ J} + (30 \times 10^9 \text{ eV})\left(1.6021 \times 10^{-19} \frac{\text{J}}{\text{eV}}\right)}{1.5053 \times 10^{-10} \text{ J}}$$

$$= 32.9292$$

$$m_v = km_0 = (32.9292)(1.6725 \times 10^{-27} \text{ kg})$$

$$= 5.5074 \times 10^{-26} \text{ kg}$$

$$v = c\sqrt{1 - \left(\frac{1}{k}\right)^2}$$

$$= \left(3.00 \times 10^8 \frac{\text{m}}{\text{s}}\right) \sqrt{1 - \left(\frac{1}{32.9292}\right)^2}$$

$$= 2.9986 \times 10^8 \text{ m/s}$$

This is essentially the speed of light.

$$r = \frac{mv}{qB}$$

$$= \frac{(5.5074 \times 10^{-26} \text{ kg})\left(2.9986 \times 10^8 \frac{\text{m}}{\text{s}}\right)}{(1.60210 \times 10^{-19} \text{ C})(8000 \text{ gauss})\left(\dfrac{1 \text{ T}}{10\,000 \text{ gauss}}\right)}$$

$$= 128.85 \text{ m}$$

$$d = 2r = (2)(128.85 \text{ m}) = \boxed{257.7 \text{ m}}$$

21. (a) $k = \dfrac{1}{\sqrt{1 - \left(\dfrac{v}{c}\right)^2}} = \dfrac{1}{\sqrt{1 - (0.6)^2}} = 1.25$

$$m_v = km_0 = (1.25)(1.6725 \times 10^{-27} \text{ kg})$$

$$= 2.0906 \times 10^{-27} \text{ kg}$$

$$f = \frac{Bq}{2\pi m_v} = \frac{(1.8 \text{ T})(1.60210 \times 10^{-19} c)}{(2\pi)(2.0906 \times 10^{-27} \text{ kg})}$$

$$= \boxed{2.195 \times 10^7 \text{ Hz}}$$

(b) $\Delta m = m_v - m_0 = (k-1)m_0$

$$= (1.25 - 1)(1.6725 \times 10^{-27} \text{ kg})$$

$$= \boxed{4.1813 \times 10^{-28} \text{ kg} \quad (25\%)}$$

ENGINEERING LAW

1. The three different forms of company ownership are the (a) sole proprietorship, (b) partnership, and (c) corporation.

A *sole proprietor* is his/her own boss. This satisfies the proprietor's ego and facilitates quick decisions, but unless the proprietor is trained in business, the company will usually operate without the benefit of expert or mitigating advice. The sole proprietor also personally assumes all the debts and liabilities of the company. A sole proprietorship is terminated upon the death of the proprietor.

A *partnership* increases the capitalization and the knowledge base beyond that of a proprietorship, but offers little else in the way of improvement. In fact, the partnership creates an additional disadvantage of one partner's possible irresponsible actions creating debts and liabilities for the remaining partners.

A *corporation* has sizable capitalization (provided by the stockholders) and a vast knowledge base (provided by the board of directors). It keeps the company and owner liability separate. It also survives the death of any employee, officer, or director. Its major disadvantage is the administrative work required to establish and maintain the corporate structure.

2. To be legal, a contract must contain an *offer*, some form of *consideration* (which does not have to be equitable), and an *acceptance* by both parties. To be enforceable, the contract must be voluntarily entered into, both parties must be competent and of legal age, and the contract cannot be for illegal activities.

3. A written contract will identify both parties, state the purpose of the contract and the obligations of the parties, give specific details of the obligations (including relevant dates and deadlines), specify the consideration, state the boilerplate clauses to clarify the contract terms, and leave places for signatures.

4. A consultant will either charge a fixed fee, a variable fee, or some combination of the two. A one-time fixed fee is known as a *lump-sum fee*. In a *cost plus fixed fee* contract, the consultant will also pass on certain costs to the client. Some charges to the client may depend on other factors, such as the salary of the consultant's staff, the number of days the consultant works, or the eventual cost or value of an item being designed by the consultant.

5. A *retainer* is a (usually) nonreturnable advance paid by the client to the consultant. While the retainer may be intended to cover the consultant's initial expenses until the first big billing is sent out, there does not need to be any rational basis for the retainer. Often, a small retainer is used by the consultant to qualify the client (i.e., to make sure the client is not just shopping around and getting free initial consultations) and as a security deposit (to make sure the client does not change consultants after work begins).

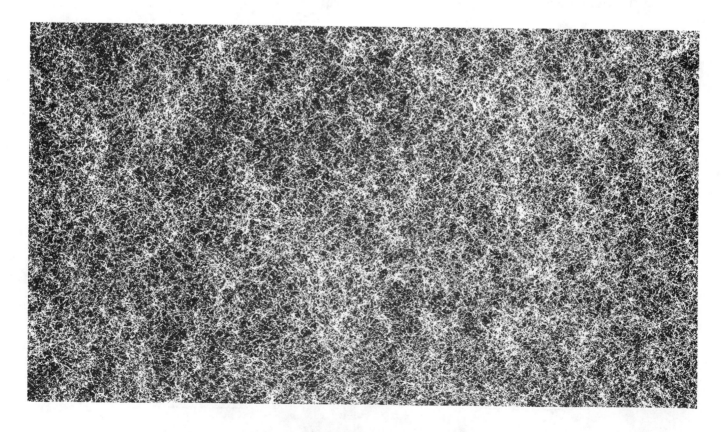

ENGINEERING ETHICS

Introduction to the Answers

Case studies in law and ethics can be interpreted in many ways. The problems presented are simple thumbnail outlines. In most real cases, there will be more facts to influence a determination than are presented in the case scenarios. In some cases, a state may have specific laws affecting the determination; in other cases, prior case law will have been established.

The determination of whether an action is legal can be made in two ways. The obvious interpretation of an illegal action is one that violates a specific law or statute. An action can also be *found to be illegal* if it is judged in court to be a breach of a written, verbal, or implied contract. Both of these approaches are used in the following solutions.

These answers have been developed to teach legal and ethical principals. While being realistic, they are not necessarily based on actual incidents or prior case law.

1. (a) Stamping plans for someone else is illegal. The registration laws of all states permit a registered engineer to stamp/sign/seal only plans that were prepared by him personally or were prepared under his direction, supervision, or control. This is sometimes called being in *responsible charge*. The stamping/signing/sealing, for a fee or gratis, of plans produced by another person, whether that person is registered or not and whether that person is an engineer or not, is illegal.

(b) The act is unethical. An illegal act, being a concealed act, is intrinsically unethical. In addition, stamping/signing/sealing plans that have not been checked violates the rule contained in all ethical codes that requires an engineer to protect the public.

2. Unless the engineer and contractor worked together such that the engineer had personal knowledge that the information was correct, accepting the contractor's information is illegal. Not only would using unverified data violate the state's registration law (for the same reason that stamping/signing/sealing unverified plans in problem 1 was illegal), but the engineer's contract clause dealing with assignment of work to others would probably be violated.

The act is unethical. An illegal act, being a concealed act, is intrinsically unethical. In addition, using unverified data violates the rule contained in all ethical codes that requires an engineer to protect the client.

3. It is illegal to alter a report to bring it "more into line" with what the client wants unless the alterations represent actual, verified changed conditions. Even when the alterations are warranted, however, use of the unverified remainder of the report is a violation of the state registration law requiring an engineer only to stamp/sign/seal plans developed by or under him. Furthermore, this would be a case of fraudulent misrepresentation unless the originating engineer's name was removed from the report.

Unless the engineer who wrote the original report has given permission for the modification, altering the report would be unethical.

4. Assignment of engineering work is legal (1) if the engineer's contract permitted assignment, (2) all prerequisites (i.e., notifying the client) were met, and (3) the work was performed under the direction of another licensed engineer.

Assignment of work is ethical (1) if it is not illegal, (2) if it is done with the awareness of the client, and (3) if the assignor has determined that the assignee is competent in the area of the assignment.

5. (a) The registration laws of many states require a hearing to be held when a licensee is found guilty of unrelated, but nevertheless unforgivable, felonies (e.g., moral turpitude). The specific action (e.g., suspension, revocation of license, public censure, etc.) taken depends on the customs of the state's registration board.

(b) By convention, it is not the responsibility of technical and professional organizations to monitor or judge the personal actions of their members. Such organizations do not have the authority to discipline members (other than to revoke membership), nor are they immune from possible retaliatory libel/slander lawsuits if they publicly censure a member.

6. The action is legal because, by verifying all the assumptions and checking all the calculations, the engineer effectively does the work himself. Very few engineering procedures are truly original; the fact that someone else's effort guided the analysis does not make the action illegal.

The action is probably ethical, particularly if the client and the predecessor are aware of what has happened (although it is not necessary for the predecessor to be told). It is unclear to what extent (if at all) the predecessor should be credited. There could be other extenuating circumstances that would make referring to the original work unethical.

7. Gifts, per se, are not illegal. Unless accepting the phones violates some public policy or other law, or is in some way an illegal bribe to induce the engineer to favor the contractor, it is probably legal to accept the phones.

Ethical acceptance of the phones requires (among other considerations) that (1) the phones be required for the job, (2) the phones be used for business only, (3) the phones are returned to the contractor at the end of the job, and (4) the contractor's and engineer's clients know and approve of the transaction.

ETHICS

8. There are two issues here: (1) the assignment and (2) the incorporation of work done by another. To avoid a breach, the contracts of both the design and checking engineers must permit the assignments. To avoid a violation of the state registration law requiring engineers to be in responsible charge of the work they stamp/sign/seal, both the design and checking engineers must verify the validity of the changes.

To be ethical, the actions must be legal and all parties (including the design engineer's client) must be aware that the assignments have occurred and that the changes have been made.

9. The name is probably legal. If the name was accepted by the state's corporation registrar, it is a legally formatted name. However, some states have engineering registration laws that restrict what an engineering corporation may be named. For example, all individuals listed in the name (e.g., "Cooper, Williams, and Somerset — Consulting Engineers") may need to be registered. Whether having "Associates" in the name is legal depends on the state.

Using the name is unethical. It misleads the public and represents unfair competition with other engineers running one-person offices.

10. Unless the engineeer's accusation is known to be false or exaggerated, or the engineer has signed an agreement (e.g., confidentiality, non-disclosure, etc.) with his employer forbidding the disclosure, the letter to the newspaper is probably not illegal.

The action is probably unethical. (If the letter to the newspaper is unsigned it is a concealed action and is definitely unethical.) While whistle-blowing to protect the public is implicitly an ethical procedure, unless the engineer is reasonably certain that manufacture of the toxic product represents a hazard to the public, he has a responsibility to the employer. Even then, the engineer should exhaust all possible remedies to render the manufacture nonhazardous before blowing the whistle. Of course, the engineer may quit working for the chemical plant and be as critical as the law allows without violating engineer-employer ethical considerations.

11. Unless the engineer's payment was explicitly linked in the contract to the amount of time spent on the job, taking the full fee would not be illegal or a breach of the contract.

An engineer has an obligation to be fair in estimates of cost, particularly when the engineer knows no one else is providing a competitive bid. Taking the full fee would be ethical if the original estimate was arrived at logically and was not meant to deceive or take advantage of the client. An engineer is permitted to take advantage of economies of scale, state-of-the-art techniques, and break-through methods. (Similarly, when a job costs more than the estimate, the engineer may be ethically bound to stick with the original estimate.)

12. In the absence of a nondisclosure or noncompetition agreement or similar contract clause, working for the competitor is probably legal.

Working for both clients is unethical. Even if both clients know and approve, it is difficult for the engineer not to "cross-pollinate" his work and improve one client's position with knowledge and insights gained at the expense of the other client. Furthermore, the mere appearance of a conflict of interest of this type is a violation of most ethical codes.

13. In the absence of a sealed-bid provision mandated by a public agency and requiring all bids to be opened at once (and the award going to the lowest bidder), the action is probably legal.

It is unethical for an engineer to undercut the price of another engineer. Not only does this violate a standard of behavior expected of professionals, it unfairly benefits one engineer because a similar chance is not given to the other engineer. Even if both engineers are bidding openly against each other (in an auction format), the client must understand that a lower price means reduced service. Each reduction in price is an incentive to the engineer to reduce the quality or quantity of service.

14. It is generally legal for an engineer to advertise his services. Unless the state has relevant laws, the engineer probably did not engage in illegal actions.

Most ethical codes prohibit unprofessional advertising. The unfortunate location due to a random drawing might be excusable, but the engineer should probably refuse to participate. In any case, the balloons are a form of unprofessional advertising, and as such, are unethical.

15. (a) As stated in the scenario statement, the client's actions are legal for that country. The fact that the actions might be illegal in another country is irrelevant. Whether or not the strike is legal depends on the industry and the laws of the land. Some or all occupations (e.g., police and medical personnel) may be forbidden to strike. Assuming the engineer's contract does not prohibit his own participation, the engineer should determine the legality of the strike before making a decision to participate.

If the client's actions are inhuman, the engineer has an ethical obligation to withdraw from the project. Not doing so associates the profession of engineering with human misery.

(b) The engineer has a contract to complete the project for the client. (It is assumed that the contract between the engineer and client was negotiated in good faith, that the engineer had no knowledge of the work conditions prior to signing, and that the client did not falsely induce the engineer to sign.) Regardless of the reason for withdrawing, the engineer is breaching his contract. In the absence of proof of illegal actions by the client, withdrawal by the engineer requires a return of all fees received. Even if no fees have been received, withdrawal exposes the engineer to other delay-related claims by the client.

16. A contract for an illegal action cannot be enforced. Therefore, any confidentiality or nondisclosure agreement that the engineer has signed is unenforceable if the production process is illegal, uses illegal chemicals, or violates laws protecting the environment. If the production process is not illegal, it is not legal for the engineer to expose the client.

Society and the public are at the top of the hierarchy of an engineer's responsibilities. Obligations to the public take precedence over the client. If the production process is illegal, it would be ethical to expose the client.

17. It is probably legal for the engineer to use the facilities, particularly if the employer is aware of the use. (The question of whether the engineer is trespassing or violating a company policy cannot be answered without additional information.)

Moonlighting, in general, is not ethical. Most ethical codes prohibit running an engineering consulting business while receiving a salary from another employer. The rationale is that the moonlighting engineer is able to offer services at a much lower price, placing other consulting engineers at a competitive disadvantage. The use of someone else's equipment only compounds the problem. Since the engineer does not have to pay for using the equipment, he does not have to charge his clients for it. This places him at an unfair competitive advantage compared to other consultants who have invested heavily in equipment.

ETHICS

QUICK — *I need additional study materials!*

Please send me the review materials I have checked. I understand any item may be returned for a full refund within 30 days. I have provided my bank card number as method of payment, and I authorize you to charge your current prices and shipping/handling charge against my account. (Don't forget a solutions manual for your reference manual.)

Solutions Manuals:

For the E-I-T Exam:
- ☐ Engineer-In-Training Reference Manual ☐
  - ☐ Engineer-In-Training Sample Examinations
  - ☐ Engineering Fundamentals Quick Reference Cards
  - ☐ E-I-T Mini-Exams
  - ☐ 1001 Solved Engineering Fundamentals Problems

For the P.E. Exams:
- ☐ Civil Engineering Reference Manual ☐
  - ☐ Civil Engineering Sample Examination
  - ☐ Civil Engineering Quick Reference Cards
  - ☐ Seismic Design of Building Structures
  - ☐ Timber Design for the Civil P.E. Exam
- ☐ Mechanical Engineering Reference Manual ☐
  - ☐ Mechanical Engineering Sample Examination
  - ☐ Mechanical Engineering Quick Reference Cards
- ☐ Electrical Engineering Reference Manual ☐
  - ☐ Electrical Engineering Sample Examination
  - ☐ Electrical Engineering Quick Reference Cards
- ☐ Chemical Engineering Reference Manual ☐
  - ☐ Chemical Engineering Practice Exam Set

Recommended for all Exams:
- ☐ Expanded Interest Tables
- ☐ Engineering Law, Design Liability, and Professional Ethics
- ☐ Engineering Unit Conversions

For fastest service call
415•593•9119

SHIP TO:

NAME _____ COMPANY _____

STREET _____ APT _____

CITY _____ STATE _____ ZIP _____

DAYTIME PHONE NUMBER _____

CHARGE TO *(REQUIRED FOR IMMEDIATE PROCESSING)*:

VISA/MC/AMEX NUMBER _____ EXP. DATE _____

NAME ON CARD _____

SIGNATURE _____

❖ *Send more information* ❖

Please send me descriptions and prices of all available E-I-T and P.E. review books. I understand there will be no obligation on my part.

NAME _____

ADDRESS _____

CITY _____

STATE _____ **ZIP** _____

A friend of mine is taking the exam, too. Send additional literature to:

NAME _____

ADDRESS _____

CITY _____

STATE _____ **ZIP** _____

I disagree...

I think there is an error on page _____. Here is the way I think it should be:

Title of this book: _____

Edition: _____ Printing: _____

Contributed by (optional): _____ ☐ Please tell me if I am correct.

NAME _____

ADDRESS _____

CITY _____

STATE _____ ZIP _____

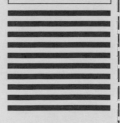

BUSINESS REPLY MAIL

FIRST CLASS MAIL   PERMIT NO. 33   BELMONT, CA

POSTAGE WILL BE PAID BY ADDRESSEE

PROFESSIONAL PUBLICATIONS INC
1250 FIFTH AVE
BELMONT CA 94002-9979

BUSINESS REPLY MAIL

FIRST CLASS MAIL   PERMIT NO. 33   BELMONT, CA

POSTAGE WILL BE PAID BY ADDRESSEE

PROFESSIONAL PUBLICATIONS INC
1250 FIFTH AVE
BELMONT CA 94002-9979

BUSINESS REPLY MAIL

FIRST CLASS MAIL   PERMIT NO. 33   BELMONT, CA

POSTAGE WILL BE PAID BY ADDRESSEE

PROFESSIONAL PUBLICATIONS INC
1250 FIFTH AVE
BELMONT CA 94002-9979